International Explorations in Outdoor and Environmental Education

Volume 10

This series focuses on contemporary trends and issues in outdoor and environmental education, two key fields that are strongly associated with education for sustainability and its associated environmental, social and economic dimensions. It also has an international focus to encourage dialogue across cultures and perspectives. The scope of the series includes formal, nonformal and informal education and the need for different approaches to educational policy and action in the twenty first century. Research is a particular focus of the volumes, reflecting a diversity of approaches to outdoor and environmental education research and their underlying epistemological and ontological positions through leading edge scholarship. The scope is also both global and local, with various volumes exploring the issues arising in different cultural, geographical and political contexts. As such, the series aims to counter the predominantly "white" Western character of current research in both fields and enable cross-cultural and transnational comparisons of educational policy, practice, project development and research. The purpose of the series is to give voice to leading researchers (and emerging leaders) in these fields from different cultural contexts to stimulate discussion and further research and scholarship to advance the fields through influencing policy and practices in educational settings. The volumes in the series are directed at active and potential researchers and policy makers in the fields. Book proposals for this series may be submitted to the Publishing Editor: Claudia Acuna E-mail: Claudia.Acuna@springer.com

More information about this series at http://www.springer.com/series/11799

Malcolm Skilbeck

Loving and Studying Nature

Celebrating the Earth through History, Culture and Education

Springer

Malcolm Skilbeck
Faculty of Arts & Education
Deakin University
Geelong, VIC, Australia

ISSN 2214-4218 ISSN 2214-4226 (electronic)
International Explorations in Outdoor and Environmental Education
ISBN 978-3-030-80750-4 ISBN 978-3-030-80751-1 (eBook)
https://doi.org/10.1007/978-3-030-80751-1

This Springer imprint is published by the registered company Springer Nature Switzerland AG
The registered company address is: Gewerbestrasse 11, 6330 Cham, Switzerland

Frontispiece: In the soils of Western Australia's impoverished sand belt, a biological "hot spot" threatened by farming and urbanization, grow some of the earth's most striking flowering and foliage plants. *Hakea Victoria*, is one of them, its foliage captured by the brilliant wildflower artist Philippa Nikulinsky. *Hakea Victoria and Western spinebill,* 2003, watercolour, pencil and gouache on Arches hot press paper, 36cm x 25cm. Courtesy Philippa Nikulinsky.

Series Editors' Foreword

Loving and Studying Nature: Celebrating the Earth Through History, Culture and Education records Malcolm Skilbeck's journey of discovery, through a number of episodes, that has taken him several years to put together. We know this because we started talking with him about his interest in researching the field naturalist club in Broken Hill more than a decade ago. At that time, we discussed the relationship between the work of the field naturalist club and the teaching of nature study in Broken Hill schools. Over the years, the project grew to an exploration of a range of aspects of human relationships with nature over time, tracing some of the ways in which Western peoples have perceived and responded to their understanding of St Augustine's two poles: the earthly (nature) and the heavenly (supernature).

Laurel Richardson (2001, p.34) notes that we are always writing about our lives in our research stories, and this volume is very much grounded in Malcolm's life and his persistent desire to understand, as the late American physicist and television personality Julius Sumner Miller would say, "Why is it so?". Malcolm's starting point for this volume is his experiences of the "ravaged land" of the mining city of Broken Hill in outback Australia when he was growing up there in the mid-twentieth century and the Barrier Field Naturalist Club's revegetation project that changed that landscape.

Like Alastair Stewart (2020), in another volume in this book series, Malcolm demonstrates that he is alert to the claim that "the Eurocentric model of understanding and relating to the land has been an ecological failure in Australia" (Stewart, 2020, p. 71). However, whereas Alastair maintains his focus on the Australian landscape and Australian First Nations' peoples' relationships with country, seeking to "(re)create education that utilises new concepts, new imaginings, for contemplating the particulars of Australian landscapes, their more-than-human inhabitants, and how the landscapes and inhabitants have changed over time and what we might do to restore them" (pp. 95–96), Malcolm takes us on a journey, in episodes, through history of human-nature relationships, particularly from a Judaeo-Christian and Western perspective, but reaching into ancient Persia, the Mogul Empire and Australian First Nations' culture, as well as Europe.

Several books in recent times have encouraged readers to think differently about human relationships with nature and developed their arguments in historical contexts. For example, Bill Gammage's *The Biggest Estate on Earth: How Aborigines Made Australia* (2011) reshapes thinking about Aboriginal land management. Simon Schama's *Landscape and Memory* (1996) aims to reveal the myths, memories and obsessions that underlie the Western world's interaction with nature. In a similar but more philosophical vein, Arran Gare's *The Philosophical Foundations of Ecological Civilization: A Manifesto for the Future* (2017) argues for "A manifesto for the future" for an ecological civilization, which builds on his previous writings about the ecological crisis and (post)modern civilization. Carolyn Merchant has written several books that, for example, look at *The Death of Nature: Women, Ecology and the Scientific Revolution* (1980); nature, gender and science in New England in *Ecological Revolutions* (1989); and associations of women with nature, starting with Eve in the Garden of Eden, in *Reinventing Eden* (1993) and *Earthcare* (1996). Each of these can be seen as having connections to this volume, but it is also very different.

Using Sara Crawley's (2002) terms, Malcolm's writing "is both personal and theoretical, interweaving lived experience with how that experience, writ large, both is and can be represented as being" (p. 72), with the chapters presenting knowledge "constructed from where the researcher/theoretician is situated" (Stanley, 2000, pp. 11–12). This book brings together discussions of human relationships with nature, and some related ideas, across the past two centuries and more. The episodes (chapters) follow Malcolm's various threads of thought, finding interconnections which others may not have seen, and some may find provocative. It is also beautifully illustrated (with much assistance from Malcolm's partner, Helen Connell, in the onerous task of seeking permissions for the multitude of images).

Ever the educator, Malcolm presents us with an ambitious project to encourage people to change their attitudes and behaviours and to value and love nature. As he writes in the first chapter:

> it is not only *ways* of knowing, understanding, feeling and expression in different symbolic forms that are of interest. The substance of shifting attitudes and their associated values also need to be teased out, however well or ill-informed they may be. Moreover, knowledge, understanding and appreciation of the natural world flow from and give impetus to many different kinds of practical action, ranging from the establishment of national parks to breeding programmes for critically endangered species. Conversely, ignorance and prejudice often combine to thwart or resist action which is well-grounded in knowledge and supported by well-reasoned preferences and values.

In Chap. 16 Malcolm discusses how nature study was introduced into Australian primary schools at the beginning of the twentieth century. The purpose was to encourage students' "understanding and appreciation of the natural world, to see the value of conserving and not carelessly exploiting the resources of the earth, to preserve areas of 'wilderness' and to enjoy the beauty of flowers, plants and animals, birds especially." For the teachers in Broken Hill in the 1930s and 1940s, the Barrier Field Naturalist Club would have been an invaluable source of information for the teaching of nature study. Sadly, in the 1960s, nature study was replaced by science in the primary school curriculum (Kass, 2018) and the nature-focused purpose was replaced by simplified Newtonian science activities.

Nature study can be seen as a precursor of environmental education, which first emerged as a field around 1970 (Gough, 1997). For example, according to the Tbilisi Declaration on Environmental Education, the goals for environmental education are (UNESCO, 1978, pp.26–27):

(a) to foster clear awareness of, and concern about, economic, social, political and ecological interdependence in urban and rural areas;
(b) to provide every person with opportunities to acquire the knowledge, values, attitudes, commitment and skills needed to protect and improve the environment;
(c) to create new patterns of behaviour of individuals, groups and society as a whole towards the environment.

We should note here that there is a point of career overlap between Malcolm and Annette in the history of environmental education in Australia. In 1980, when Malcolm was Director of the Curriculum Development Centre (CDC), it published Australia's first national statement on environmental education for schools which was written by Annette as the coordinator of environmental education at the CDC (Greenall, 1980). This statement, which was endorsed by all state and territory education authorities, promoted environmental education as "an orientation in the curriculum", which continues to the present day with sustainability as a cross-curriculum priority in the Australian Curriculum: "Sustainability [is] a priority for study that connects and relates relevant aspects of content across learning areas and subjects" (ACARA, 2021). The description of education for sustainability in the Australian Curriculum shares much in common with the above description of nature study a century ago:

> Education for sustainability develops the knowledge, skills, values and world views necessary for people to act in ways that contribute to more sustainable patterns of living. It enables individuals and communities to reflect on ways of interpreting and engaging with the world. Sustainability education is futures-oriented, focusing on protecting environments and creating a more ecologically and socially just world through informed action. Actions that support more sustainable patterns of living require consideration of environmental, social, cultural and economic systems and their interdependence. (ACARA, 2021)

Indeed, this is a very holistic vision and goal, consistent with Malcolm's argument in the latter chapters of this volume. Unfortunately, education for sustainability is not compulsory and its occurrence in the curriculum in schools is haphazard. It generally develops a shallow understanding of sustainability, with little professional support provided for teachers, and it is a threatened entity within the Australian Curriculum. A review of the Australian Curriculum in 2014 recommended that "ACARA reconceptualize the cross-curriculum priorities and instead embed teaching and learning about … sustainability explicitly, and only where educationally relevant, in the mandatory content of the curriculum" (Donnelly and Wiltshire, p. 247). This recommendation foregoes developing a coherent and progressive educational approach for developing and applying knowledge and skills in schools to foster a holistic, interdisciplinary, focused and whole-school approach to education for sustainability, which is consistent with the research literature and best practice. Fortunately, this recommendation was not acted upon, but in June 2020 the education ministers announced another review of the Australian Curriculum with the terms of reference including to:

- revisit and improve if necessary, the organising frameworks for the cross- curriculum priorities with reference to current research
- declutter the content of the Australian Curriculum by improving the relationship of the cross-curriculum priorities to learning area content, removing any repetition of content between the cross-curriculum priorities and the learning areas and replacing the current 'icon tagging' for cross-curriculum priorities on the Australian Curriculum website with a more user-orientated approach. (ACARA, 2020, p. 5)

The review is not yet completed, but it does not auger well for sustainability maintaining a place in the curriculum. Neither does the *Alice Springs (Mparntwe) Education Declaration* (Education Council, 2019), which is the agreed statement on goals for schooling in Australia. This Declaration simply sees education as preparing "young people to thrive in a time of rapid social and technological change, and complex environmental, social and economic challenges" (p. 2). The goal to accompany this statement was for "All young Australians become confident and creative individuals, successful lifelong learners, and active and informed members of the community" (p.6). This is a truncated version of the goal in the *Melbourne Declaration on Educational Goals for Young Australians* (MCEETYA, 2008) which it replaced. In the 2008 version the goal was for students to become "active and informed members of the community who … have empathy for the circumstances of others and work for the common good, in particular sustaining and improving natural and social environments" (p. 9). The *Alice Springs (Mparntwe) Education Declaration* is also notably silent on climate change and reduces consideration of sustainability in the curriculum, which was highlighted in the Melbourne Declaration, to encouraging students to "engage with complex ethical issues and concepts such as sustainability" (2019, p. 15).

While at a policy level things do not look good for sustainability in the Australian Curriculum, the sustainable schools movement has been sustained, even after the withdrawal of government funding (Larri & Colliver, 2020), and there continues to be an Australian Association for Environmental Education, as well as associations or chapters in each state and territory to support those interested in environmental education. However, without strong support in teacher education programs and/or through professional development opportunities, there is no consistent approach to implementing sustainability in schools.

In the last paragraph Malcolm asks a key question that has haunted environmental educators globally for decades: "Are we then, as a species, unwilling or unable to sustain the conditions that we know to be necessary for our continued survival and that of innumerable forms of life?" Arran Gare (2017) argues that the problem is "the rise of neoliberalism and the consequent paralysing of efforts to grapple with the ecological crisis manifest in the deeply rooted nihilism of the civilization of modernity" (p.20). As an antidote he argues,

> It is life which gives meaning to liberty and happiness. Adopting the perspective of speculative naturalism, 'life' should now be extended to the life of the 'ecological' communities, both human and non-human, within which people are participating, which they are internally related to as well as being the conditions for their existence...
>
> Reformulating culture through speculative naturalism will create new subjectivities, subjectivities committed to addressing and overcoming the threats to democracy, civilization, humanity and terrestrial life from ecological destruction, and in doing so, creating a new civilization: an ecological civilization. (Gare, 2017, p.217, pp.218-9)

Carolyn Merchant's most recent book traces the problems of prediction and control of autonomous nature from ancient times – when nature was seen as unpredictable, unruly and recalcitrant – through the post Scientific Revolution scientists who sought ways of predicting and controlling the world around them, to the twenty-first century. Here she sees humanity as being "in the throes of a paradigm shift, one that is triggered by two factors: the rise of the new sciences of chaos and complexity and by climate change as the most widespread catastrophe for the human future" (2016, p.1). She argues that, in order to live within this new chaos and complexity paradigm, a new, partnership ethic is needed: "Understanding nature as a complex system that includes humanity within it allows for the possibility that both the earth as we know it today and humanity can survive and thrive together in the coming decades" (p.153). She concludes the book with a discussion of a partnership ethic:

> By merging anthropocentric with ecocentric ethics – the ethic that includes all of nature within it – we can develop an integrated, interactive ethic based on partnership between the human and the nonhuman worlds... *A partnership ethic holds that the greatest good for the human and nonhuman communities is in their mutual living interdependence.* (p.162)

Her ethic contains five precepts for a human community in a sustainable partnership with a nonhuman community – which is in a particular place, a place in which connections to the larger world are recognized through economic and ecological exchanges:

- Equity between the human and nonhuman communities
- Moral consideration for both humans and other species
- Respect for both cultural diversity and biodiversity
- Inclusion of women, minorities, and nonhuman nature in the code of ethical accountability
- An ecologically sound management that is consistent with the continued health of both the human and the nonhuman communities

Annette (Gough, 1997) has previously argued for the relevance of Merchant's partnership ethic for future developments of environmental education using a post-structuralist pedagogy. Nowadays we would probably think in terms of an intersectional approach which would enable consideration of social and environmental justice issues through encouraging understanding the multiple positionings of race, class, gender, ethnicity, and nonhuman communities, as well as posthuman, and is consistent with understanding the total environment in a less distorted way (Gough, 2021; Lloro-Bidart & Finewood, 2018; Maina-Okori et al., 2018).

Carolyn Merchant's partnership ethic proposal, her earlier writings on women, nature, religion and science, and the writers in similar spaces open up opportunities for conversations with Malcolm Skilbeck's discussion about ways forward. That things need to change is not debatable, the educational responses around how to provide opportunities for people to become empowered for change have long been discussed. We know that the School Strike for Climate movement is gaining momentum with youth around the world, and that the public pays attention to scientists such as David Attenborough and Brian Cox. The continuing challenge is getting governments and education systems to take the environmental crisis seriously and address global problems that are part of the United Nations' Paris Agreement (2015b) and the 2030 Agenda for Sustainable Development (United Nations, 2015a). The recent *Asia and the Pacific SDG Progress Report 2021* (UN ESCAP) on progress towards the Sustainable Development Goals does not paint a positive picture. Perhaps by understanding better where we have come from in human-nature relationships through reading books such as this, people may better understand what we can do. We can hope.

Acknowledgement

We edit this series on the unceded lands of the Woi wurrung and Boon wurrung peoples of the Kulin Nations; we respectfully acknowledge their Elders, past, present and emerging and what they have taught us about this land and sea.

RMIT University
Melbourne, VIC, Australia
Annette Gough

La Trobe University
Melbourne, VIC, Australia
Noel Gough

References

Australian Curriculum and Assessment and Reporting Authority. (ACARA). (2020). *Review of the Australian Curriculum*. Retrieved from https://www.acara.edu.au/curriculum/curriculum-review/

Australian Curriculum and Assessment and Reporting Authority. (ACARA). (2021). *Australian Curriculum: Sustainability*. Retrieved from https://www.australiancurriculum.edu.au/f-10-curriculum/cross-curriculum-priorities/sustainability/

Crawley, S. L. (2002). 'They still don't understand why I hate wearing dresses!' An c, Boats and Butchness. *Cultural Studies<=> Critical Methodologies, 2*(1), 69–92.

Education Council. (2019). *Alice Springs (Mparntwe) Education Declaration*. Education Council. Retrieved from http://www.educationcouncil.edu.au/Alice-Springs%2D%2DMparntwe%2D%2DEducation-Declaration.aspx

Gammage, B. (2011). *The Biggest estate on earth: How Aborigines made Australia*. Allen & Unwin.
Gare, A. (2017). *The philosophical foundations of ecological civilization: A manifesto for the future*. Routledge.
Gough, A. (1997). *Education and the environment: Policy, trends and the problems of marginalisation*. Australian Council for Educational Research.
Gough, A. (2021, February 23). Education in the anthropocene. In *Oxford Research Encyclopedia of Education*. Oxford University Press. https://doi.org/10.1093/acrefore/9780190264093.013.1391
Greenall, A. (1980). *Environmental education for schools: Or how to catch environmental education*. Curriculum Development Centre.
Kass, Dorothy. (2018). *Educational reform and environmental concern: A history of school nature study in Australia*. Routledge.
Larri, L., & Colliver, A. (2020). Moving green to mainstream: Schools as models of sustainability for their communities – The Australian Sustainable Schools Initiative (AuSSI). In A. Gough, J.C.-K. Lee, & E.P.K. Tsang (Eds.), *Green schools globally: Stories of impact on education for sustainable development* (pp. 61-82). Springer.
Lloro-Bidart, T., & Finewood, M. H. (2018). Intersectional feminism for the environmental studies and sciences: Looking inward and outward. *Journal of Environmental Studies and Sciences, 8*, 142–151. https://doi.org/10.1007/s13412-018-0468-7
Maina-Okori, N. M., Koushik, J. R., & Wilson, A. (2018). Reimagining intersectionality in environmental and sustainability education: A critical literature review. *The Journal of Environmental Education, 49*(4), 286–296. https://doi.org/10.1080/00958964.2017.1364215
Merchant, C. (1980). *The death of nature: Women, ecology and the scientific revolution*. Harper and Row.
Merchant, C. (1989). *Ecological Revolutions: Nature, gender, and science in New England*. University of North Carolina Press.
Merchant, C. (1996). *Earthcare: Women and the environment*. Routledge.
Merchant, C. (2003). *Reinventing Eden: The fate of nature in Western culture*. Routledge.
Merchant, C. (2016). *Autonomous nature: Problems of prediction and control from ancient times to the scientific revolution*. Routledge.
Ministerial Council on Education Employment Training and Youth Affairs (MCEETYA). (2008). *Melbourne Declaration on educational goals for young Australians*. Retrieved from www.curriculum.edu.au/verve/_resources/National_Declaration_on_the_Educational_Goals_for_Young_Australians.pdf
Richardson, L. (2001). Getting personal: Writing stories. *Qualitative Studies in Education, 14*(1): 33–38.
Schama, S. (1996). *Landscape and memory*. Routledge.
Stanley, L. (2000). Children of our time: Politics, ethics and feminist research processes. In H. Hodkinson (Ed.), *Feminism and educational research methodologies* (pp. 5–35). The Manchester Metropolitan University.
Stewart, A. (2020). *Developing place-responsive pedagogy in outdoor environmental education: a rhizomatic curriculum autobiography*. Springer.
UNESCO. (1978). *Intergovernmental Conference on Environmental Education: Tbilisi (USSR), 14–26 October 1977* (Final report). UNESCO.
United Nations. (2015a). *Transforming our world: the 2030 Agenda for Sustainable Development*. Resolution adopted by the General Assembly on 25 September 2015. Retrieved from www.un.org/ga/search/view_doc.asp?symbol=A/RES/70/1&Lang=E
United Nations. (2015b). *Paris Agreement*. Retrieved from https://unfccc.int/sites/default/files/english_paris_agreement.pdf
United Nations Economic and Social Commission for Asia and the Pacific (ESCAP). (2021). *Asia and the Pacific SDG Progress Report 2021*. Retrieved from https://www.unescap.org/sites/default/d8files/knowledge-products/ESCAP_Asia_and_the_Pacific_SDG_Progress_Report_2021.pdf#page=16

Preface

My purpose in undertaking this study is to investigate and reflect on crucial ways in which nature has been apprehended, understood and valued in different cultures and over time. Its setting is the civilization of the West, but not to the exclusion of others. For millennia religious beliefs of various kinds have been a major influence on attitudes towards the natural world, for centuries dominated by a supernaturalist disdain of it. Perceptions and attitudes have changed around the world, and they continue to do so. But how, in what ways and in which directions? And what are the prospects for the kind of awakening that is needed to universalize a conservationist ethic, to sustain rather than destroy the bounty and beauty of nature?

The scope of such an investigation, to say nothing of the challenge to shift attitudes, is potentially immense. It can, however, be meaningfully approached by selecting for discussion and appraisal specific events, incidents, figures, movements and settings. These may be seen as turns not only in thought and belief, but also in action, potentially at least. Sometimes these attitudinal changes are like a vast flood on a flat plain, so slow moving as to be imperceptible except to the most acute observers. Then some impulse in the flow or a shift in terrain hastens the movement, the waters surging and driving forward. The emergence of the modern "nature ethic" has something of this character. It is a gathering force, yet to be fully released and channelled but already evident in many changes in public attitudes and expectations. Policy changes and action will follow if public understanding and feeling for the natural world increase and find expression.

Preparation of this text has extended over a long period and drawn on a wide variety of sources: scholarly, scientific, literary, artistic and others. References and notes are appended to each chapter but inevitably there are gaps and omissions. If the book contributes, in however small a way, to the gathering tide of support for conserving and protecting the natural world, it will have achieved its goal.

A note on terminology: In general I have used gender neutral terminology, wherever it seems appropriate, but the terms "man" and "mankind" cannot always sensibly be avoided. Reference in places to the terms "capitalism", "industrial capitalism" and "consumer capitalism" may seem vexatious to some readers. "Capitalism", the generic term, refers to the economic, social and political system, dominant in the

West and in most other parts of the modern world, whereby wealth is concentrated in individuals and enterprises and freely invested in production and trade for profit-making purposes and to facilitate further investment. "Industrial capitalism", the subject of a comprehensive critique by the philosophers and political economists Karl Marx and Friedrich Engels, expanded rapidly in the nineteenth century when machine-production in factories by wage-labour led to a huge increase in the volume of production, international trade and the growth of large urban centres. "Consumer capitalism" is a modern theory and an economic, social and political condition whereby consumer wants and demands are manipulated through mass marketing. Consumption, hence production, hence profits, is increased to the benefit of manufacturers and sellers of products ("consumer goods"), and success or progress is measured more by increase of wealth than its distribution.

Geelong, VIC, Australia Malcolm Skilbeck

Acknowledgements

I owe a debt to many scholars and writers as indicated in notes in each of the chapters.

Colleagues, friends and family members have been variously supportive and constructively critical, whether in advice, comment or conversation. Among them I wish to name particularly David Aspin, Rémy Boucharlat, Barbara Briggs, the late Susan Brimble, Judith Chapman, Cherry Collins, Ute Dickel, Rosalind Ellinger, Douglas Ellinger, Christine Fensham, Peter Fensham, Annette Gough, Greg Hancock, Alan Lonsdale, Richard Mabey, Craig McGregor, Philip Pilgrim, Ann Raybould, Patricia Selkirk, Richard Selleck, Brigit Skilbeck, Lucy Skilbeck and Paul Skilbeck. Above all, Helen Connell, my wife, provided invaluable support throughout the work.

Staff from the following libraries have been most helpful: Broken Hill City Library; Geelong Regional Libraries; National Library of Australia; Deakin University Library; State Archives and Records, New South Wales Government; State Library of New South Wales; State Library Victoria; University College London Institute of Education Library; University of Melbourne Library; University of Sydney Library.

During early research and later preparation for publication, I have much valued the support of Springer series co-editors Annette Gough and Noel Gough. My thanks also to Springer staff, editor, Nick Melchior, associate editor Lay Peng Ang, and project co-ordinator (books), Malini Arumugam.

Contents

Chapter 1
Introduction and Overview: Visions of Nature

It is a curious thing that among the myriad creations on this planet, while the entire plant life is deprived from taking any attitude toward nature and practically all animals can also have no "attitude" to speak of, there should be a creature called man who is both self-conscious and conscious of his surroundings and can therefore take an attitude towards it.

Lin Yutang (1938) *The Importance of Living*. p. 265.

1.1 Shifting Attitudes, Changing Values

Throughout recorded history, humans have been ready to take not only attitudes, but very often quite decisive action, targeting and manipulating plants, animals, the natural environment and fellow human beings. They have, as well, comprehensively expressed attitudes – in language, religion, art, craft, music, drama, dance, literature, poetry, customs, habits, rituals and, most recently, science. By these expressive means, consciousness itself has been and continues to be extended and enriched, while understandings have been developed and communicated. Through these different modes of expression and communication, the concept of nature has taken on a wide diversity of meanings and values.

Consciousness and self-consciousness are differentially shared by humans with their cousins among the great apes and perhaps other species as well. Awareness of the natural world, in the form of responses and interactions, is synonymous with life itself although there is a wide range of qualitative differences in experience, both personal and cultural. For some people and communities, this awareness has meant a sense of partnership or harmonious co-existence. More often, however, "man" has seen himself as a unique and dominant force by virtue of a deep sense of superiority: in the image of the crew in Tennyson's *The Lotos Eaters*, "the roof and crown of things". Humanity's rise to dominance over the earth has tragically resulted in a huge and accelerating loss of other forms of life. With some notable exceptions, it

M. Skilbeck, *Loving and Studying Nature*, International Explorations in Outdoor and Environmental Education 10, https://doi.org/10.1007/978-3-030-80751-1_1

is only relatively recently that thinkers, among them the contemporary philosophers, Peter Singer and Tom Regan, have thrown down ethical challenges to the presumption of human superiority and rightful dominance in the natural world[1].

Just when, in the pattern of evolution, humans emerged and began to express the rich – and dangerous – array of capabilities and forms of expression which distinguish them from even their closest cousins, is not known beyond vast stretches of time. It is clear, however, that tool-making, magic, speech, complex belief systems and some form of artistic expression in communal settings by the time of the Upper Palaeolithic (c.10,000 to c.35,000 years ago) were decisive in the growth of human consciousness and the formation of attitudes toward the natural world[2].

It is through various ways of expressing and communicating feelings, attitudes and beliefs that humans came to some understanding of objects and things, and to impute relationships. By looking, touching, listening, imitating, experimenting and creating, and through expressive relationships with others, "man" the social animal was enabled to participate in a culture of shared – or conflicted – meanings. Most importantly, through sensory experience, image-making and communication, distinctive and fundamental human attributes of reason and reflection on experience developed – the ability to modify and change attitudes. The long-held and tenacious belief in Western culture that "man" is a superior being, both separate from and peculiarly fitted to dominate nature to satisfy his wants and needs, is neither inevitable nor unchangeable. Nevertheless, the need to change this belief – to develop new attitudes appropriate to the survival of humanity as well as other species – has yet to be fully recognized in our social and political organization and our economic affairs.

With modernist roots in the mid to late eighteenth century Romantic figures, notably Rousseau and Wordsworth, a deep feeling for and appreciation of nature in something like its pristine state gradually gathered force during the nineteenth century. Despite many setbacks and challenges, it had, by the end of the twentieth century, secured a firm hold on public consciousness with strong if still uneven legislative backing in many countries.

Scientific methods and knowledge are perhaps the most obvious, as they are now for many people, the most powerful and convincing ways of re-shaping beliefs, attitudes and behaviour. However, direct and immediate personal experience through encounters with animals and growing and caring for plants, reading, writing, drawing and personal reflection all play a part in getting to know and understand nature. For this book, I have chosen examples separated by long periods of time, in different places and from several distinct ways of knowing about, understanding, actively relating to and expressing feelings about nature. They form the subject of separate chapters whose sequence, the first two chapters excepted, roughly approximates the chronological order of their origin, growth and impact in the life of humanity. Thus, art preceded religion, considered as a sophisticated code of beliefs and doctrines. Literary expression of a high order preceded science which, in modern times, has to a considerable extent eclipsed art, religion and literature as an assured and pervasive way of perceiving and understanding the natural world. A considerable part of the text addresses religious beliefs and attitudes – those mainly of Jews, Christians and Muslims. Their force has been enormous and continues even in modern, largely secular, life. They range from assertion of man's dominance or separateness to a readiness to accept nature on equal terms.

However, it is not only ways of knowing, understanding, feeling and expression in different symbolic forms that are of interest. The substance of shifting attitudes and their associated values also need to be teased out, however well or ill-informed they may be. Moreover, knowledge, understanding and appreciation of the natural world flow from, and give impetus to, many different kinds of practical action, ranging from the establishment of national parks to breeding programmes for critically endangered species. Conversely, ignorance and prejudice often combine to thwart or resist action which is well-grounded in knowledge and supported by well-reasoned preferences and values. This is nowhere better illustrated than in the contemporary debates over climate change, on the one hand, and the effect of human action on the overall environment, on the other. Humans are virtually unique in their ability to "take attitudes". In doing so they have a marvelous capacity to express and communicate them and to act in ways ranging from the perverse and self-destructive to the most brilliantly creative and constructive.

There are innumerable episodes to relate about the love and study of nature and its disregard and destruction, in the long and chequered history of western civilization. In 1965, at the behest of the state's Hydro-Electricity Commission, the government of Tasmania authorized the flooding of Lake Pedder in the southwest of the state, an area of outstanding natural beauty and ecological significance. The decision led to national controversy and international concern which resulted eventually in the declaration by UNESCO of the whole southwestern region of the island state as a World Heritage site. A leader in the battle to save the lake was Olegas Truchanas, a Lithuanian migrant, conservationist and nature photographer who, when opening an exhibition of paintings in 1971 on the theme of Lake Pedder, called for a new understanding of the value of the natural world. "Tasmania is not the only place in the world where long-term, careful argument has been defeated by short-term economic advantage … the time is rapidly approaching when natural environment, natural unspoiled vistas are sadly beginning to look like left-overs from a vanishing world."[3] Truchanas, who drowned when kayaking in the Gordon River which was further threatened by hydro works, urged us to change our awareness, our understanding and our behaviour. Continuing his appraisal of the significance of the Lake Pedder art, he said: "If we can revise our attitudes towards the land under our feet; if we can accept a role of steward and depart from the role of conqueror; if we can accept the view that man and nature are inseparable parts of the unified whole – then Tasmania can be a shining beacon in a dull, uniform and largely artificial world."[4] The nature photography of Ansel Adams in the USA; Olegas Truchanas; his colleague, Peter Dombrovskis, in Australia and many others, have played an important part in campaigns to conserve the natural environment (Fig. 1.1).

In this selection, practical actions and attitudes that exemplify love and reverence are considered but, also, misunderstanding of the natural world and hostility to it. The human quest for convenience and comfort has entailed vast destruction, pollution and poisoning of the natural environment. Its effects on life in all its forms we now understand and to some extent are able to ameliorate. We know that more and

Fig. 1.1 Morning Mist, Rock Island Bend, Franklin River Tasmania, 1979 by Peter Dombrovskis whose photography, together with that of his mentor and fellow Lithuanian migrant, Olegas Truchanas, has helped to develop environmental consciousness in the wider public, not only in Australia. This image of Rock Island Bend was widely and effectively used in the successful international campaign launched by later Greens Party leader Bob Brown to prevent damming the Franklin River. Now the river flows through a UNESCO World Wilderness Heritage Area of 1.58 million ha, 20% of the island state. National Library of Australia. nla.obj-147241885. Courtesy Liz Dombrovskis

more exact knowledge is needed if we are to preserve and not further destroy the environment. We are also coming to understand that systematic and not merely incidental changes in our collective behavior and expectations are required. What is perhaps less understood is that behavioral changes, to be effective and long lasting, depend on the development of more considerate and caring attitudes – a respect for all forms of life. Moreover, beliefs, especially those of ancient origin, need to be reconsidered in light of new knowledge and changing experience.

For centuries, Europe and its global outreach through voyages of discovery, trade, colonial settlement and communications has been the most powerful force for change, in ways of knowing, believing and living. Other cultures, notably Asian, have held sway in large regions and are reaching out around the world, especially through trade and technology. The spread of western culture has had negative as well as positive impacts; for example, through conquest, industrialization and the assertion of superior attitudes and values. The same may be expected of the emergence of China, Japan, South Korea and other Asian nations since they appear to be following the western models in their quest for greater economic power, political and cultural influence.

The episodes that constitute this volume, with the exception of the rise and decline of the Mogul Empire (Chap. 14) and the prehistoric cave painters of Southern Europe (Chap. 4) fall within the religious domain of Judaeo-Christianity. Consequently, attention must be given to the beliefs and values of the dominant forces in this culturally complex religion and how they affected the attitudes and behavior of people living in the expanding Christian world. An attempt to do this is made in several chapters, from which it is evident that this religion, historically of immense importance to believers and non-believers alike, has generated a wide and varied array of attitudes and behaviors. These extend from great appreciation and indeed love of nature, to disregard and negative appraisals of the natural world by contrast with the supernatural. Moreover, attitudes and beliefs have changed over time, some quite radically.

Together with Christianity, which is a material as well as a spiritual dynamic, in the form of institutions and structures, the socio-economic dynamism of capitalism, or more precisely consumer capitalism, is an immensely powerful force now, around the whole world. For the sake of the convenience and comfort of the many, we have accepted that vast riches can be concentrated in the hands of wealthy individuals and financial institutions, and traded in stock exchanges. What is more, we have learnt to live with the destruction of nature's forests and waterways, the pollution of the atmosphere and the oceans, the growth of huge urban slums and industrial wastelands and the extinction and depletion of numerous species whose habitats we have destroyed and whose lives we continue to poison. Of course, this is not the responsibility alone of the financial giants of material production, trade and distribution. We consumers have enthusiastically sought more, even to the extent of discarding last year's mobile telephone for this year's, technically near enough to being the same but in a case of the latest fashion in colour coating.

It seems endemic in the western consciousness, the constraints of Christianity's gospel notwithstanding, to want ever more of the products of the material world and greater choice among them. This expectation is driven by demographic change and the prevailing economic growth model. During the nineteenth century, as the efficiency of industrial production in factories yielded greater volumes of all kinds of goods at lower unit costs, the idea of progress, at least in material possessions, became a new, secular gospel. It was fortified by the Enlightenment confidence in the power of the human intellect, freed from the ancient bondage of Mediaeval thought and superstition, and powered by the thrilling discoveries of science. That Charles Darwin and his followers seemed to bishops, such as Samuel Wilberforce, to have brought humans down to the level of their cousins, the great apes, was not cause to abandon the idea of a continuum of improvement, spiritually as well as materially.

Certainly, scientific inquiry and experimentation have brought about the most profound changes in our understanding of the physical universe; for example, cosmology, nuclear physics, quantum mechanics and technology have yielded robots, artificial intelligence and a virtual revolution in electronic communication. But these same developments have resulted in atomic and hydrogen bombs, rocketry capable of delivering them world-wide and are a practical demonstration of our ability to destroy large cities and obliterate all the people in them. The nineteenth century idea of progress received a rude shock in the course of two world wars and global depression in the twentieth century.

In the course of these developments and the quest to feed and house rapid population growth in huge urban agglomerations, and thanks to the intensification of industrial processing of the earth's resources, the natural world has taken a tremendous battering. It is not unduly alarmist to recognize that this battering has the potential to be terminal for many species, including humanity. Equally, we are not obliged to be fatalistic, to wring our hands in despair. On the contrary, humans in past ages have been faced by the most frightening challenges: the incidence of plague across mediaeval Europe, for example, and have surmounted them. Never has our knowledge and capability to meet challenges been as great as it now is. We can, and must, rise to the challenges whose parameters and features we are able to address with a reasonable prospect of success. There are numerous examples of successful action: to curb forest destruction and to re-vegetate; to protect habitats; to aid species regeneration; to control emissions of poisonous gases; to clean up polluted waterways and to lighten the impact of the human footprint on the earth and its oceans.

This book sets forth historical examples of humans living in a sensitive, intelligent partnership with other species, of valuing of life in its many and varied forms, and of exultation in the richness and beauty of nature. Our forebears had no greater capabilities in respect of these matters than we have. In reflecting on their actions, their beliefs and values, we may be encouraged in our own lives, mentally as well as practically, to join contemporary movements, to respond to the looming threats to nature and to the earth itself, and to overcome them. Successful efforts to ward off threatened extinction of the European and American bison, such as the project at Yellowstone Park, are evidence of growing concern and practical action (Fig. 1.2).

1.2 Historical Episodes in Love and Study of Nature

The first two chapters record events in and around the inland Australian mining city of Broken Hill. The delicate plant life of a vast, semi-arid landscape was in the second half of the nineteenth century laid bare by the depredation of introduced rabbits and hard-hoofed sheep, horses and cattle. The discovery of fabulously rich deposits of silver, lead and zinc led to a global mining industry and a city whose inhabitants further destroyed vegetation for use in the mines and miners' households. That is the phase of destruction, relayed in Chap. 2, *A Ravaged Land*. It is followed, in Chap. 3, by restoration and *Recovery*. Thanks to the horticultural skills, energy, determination and understanding of natural forces on the part of a small band of field naturalists, a great conservation project was undertaken. It involved, also, leadership, and the practical and financial support by the mining companies; one of them was to yield capital which underpinned the global enterprise, BHP (Broken Hill Proprietary). The success of this regeneration project is evident in the revived

Fig. 1.2 Yellowstone, Montana, in 1872, was the first national park to be established to protect wilderness. Royal National Park just south of Sydney was the second, in 1879. These grazing bison near a hot spring in Yellowstone are rare survivors (Chap. 4). Daniel Mayer, CC BY-SA 3.0 via Wikimedia Commons

growth of indigenous vegetation in a kilometer-wide zone now encompassing Broken Hill, but also in regeneration projects undertaken by mining enterprises around the world.

Chapter 4, *Bison and Beyond – Artists encounter animals*, requires a leap of the imagination, taking us into the time of prehistoric cave dwellers in southern France and northeastern Spain. On the rocky walls and ceilings of these caves are galleries of animal paintings – bison, horse, mammoth, rhino and others, together with human figures, some of them hunting the fleeing prey. These galleries include some of the world's finest animal art. But why were they painted in dark, remote underground spaces? What is their meaning? And are there lessons we can learn, insights to be gained by their study?

Archaeologists, art theorists and other scholars have offered interpretations, ranging from a form of sympathetic magic, to strengthen the power of the hunter over his prey, to the idea of shamanistic ceremonies, but there is no written documentation or oral tradition to support these interpretations. Picasso found, in the style of animal representation, evidence of artistic mastery. Taking his cue, in a personal reading, I have treated paintings in the Altamira cave in Spain as evidence of respect for the animal and appreciation of its beauty and vitality. The ancient cave people hunted bison but did not exterminate them. That dastardly feat was accomplished in historical times, in northern Europe and the American prairies by modern humans who hunted them and destroyed their habitats. But modern humans also enjoy and appreciate animals – by painting, sculpting and modelling them to embellish galleries, public places and homes.

Chapter 5, *To be Religious*, is the first of several relatively short, closely linked chapters which explore the idea of religion and aspects of the dominant religion of western civilization, Judaeo-Christianity. This choice is dictated by the cultural and political power of Judaeo-Christianity in the course of almost two millennia, but equally by the ambivalent attitudes expressed towards the natural world. In particular, the argument that the first chapter of the first book, *Genesis,* of the *Old Testament* led to a pervasive belief in man's dominance over nature is considered. Some critics have claimed this as the ancient source of the contemporary environmental crisis. Others argue that what was meant is the current environmental creed of stewardship.

It is the variety, and diversity, of religious experience that is the focus; it is not so much the doctrines of the theologians but the typical beliefs of different practitioners. Not all religions are theistic. Some reject the idea of a supernatural being or beings, finding satisfaction for their religious feelings in a sense of belonging to the natural world, in which their idea of godhead finds expression. Etymologically, "religion" derives from a sense of belonging, but in the historical development of religious beliefs and practices, it came to mean much more, as shown in the Judaeo-Christian *Bible*, the Muslim *Q'ran* and subsequent texts.

Chapter 6, *Christianity Spreads from Ancient Roots*, sets the *Old Testament* in the context of other ancient religions, showing how many of its themes and images had antecedents and parallels in other Middle Eastern belief systems. Christianity, as it developed up until Mediaeval Europe, absorbed ideas from pagan Greek philosophy, both Platonic and Aristotelian. If a revelation from God, it was well informed by other sources. Some of the practices and beliefs absorbed from other religions left a residue of ambiguous attitudes towards nature, both reverential and sacrificial.

Christianity contrasted in many respects with its predecessors; for example, in the energetic charge to love and care for others through fellowship and charity. This did not, however, necessarily entail care and respect for all other creatures. The institutionalization of Christianity through the rise and spread of the Catholic Church led to the replacement of initial enthusiasm and preoccupation with excitable charisma, by doctrine and socio-political order. Man's role as God's regent on earth did not mean the later Protestant notion of individual, direct communion with deity but a relationship through an institution which governed his spiritual life and ordered as much as it could of his daily affairs. It was one of the greatest of the "doctors of the Church", Bishop Augustine of Hippo, who played the most influential role in laying out the claimed truths of the *Bible*, vanquishing the pagan religions of Rome and projecting the course for all believers in their passage from sinful, degraded earth to the lap of the divine being in heaven. Augustine did not hesitate to evoke fear and terror and the eternal torments of hell, inspiration for Dante's poetic "Purgatory" and many a hellfire sermon. Love of the present world, the earth and its creatures, was to be substantially eschewed by the heavenward bent.

In Chap. 7, *Souls: Pagan, Christian, Animal,* we see that souls were not of great interest to the scribes of the Old Testament; they came to the fore in the *New Testament* and in its interpreters in the Christianized west. As to what the soul is, and what its phenomenological status may be, there has been endless dispute. In religiously-minded humans, it is a metaphysical entity, not material, occupying a physical body but not like an organ, more like a suffusion of spirituality; in some religions and for some philosophers (Socrates, for example), it is immortal and capable of transmigration, leaving one body and entering another. More prosaically, "soul" is the English translation of Hebrew "Nepeth," living being.

In the course of affirmations and debates about the nature of the soul, not surprisingly questions were asked about souls in animals. Conceived as "living beings" there was no issue, but were animals possessed of the kind of metaphysical soul possessed by or attributed to humans? Do animals have capacity for goodness and potential heavenly destiny? Answers ranged from absolute denial to various intermediate states of soul. Augustine, unequivocally denying soul to animals (plants too), asserted "their life and death are a matter subordinate to our needs." Even so, he admired their qualities. Not until quite recently has the Catholic Church officially accepted that animals have souls and therefore rights, and that restraints are required to lessen human impact on the environment and animal habitats.

Imagining Nature in Myths and Creation Stories, Chap. 8, has always been a feature of Western story telling. We have a need to speculate about origins and where there is no evidence, literary or scientific, we construct tales. This art continues, now informed by scientific discoveries and theories which give to ancient myths of the cosmos a verisimilitude that is, however, denied by sceptics more concerned with verification than with the function of such stories in bonding people into communities and setting out principles of conduct.

The book, *Genesis,* of the *Old Testament* is the best known and most influential Western creation story, a repository of principles of conduct, of declared rights and opportunities for God-fearing humans. The Garden of Eden is portrayed as a divine place of happiness, harmony between humans and the natural world, restful, abundant in every respect and beautiful. The story tells us who we are, why we are here on earth, how we became sinful and the punishment we must endure. Thus creation is not an event but a process and the *Bible* is the story of that process, further extended by the Church, its theologians and bishops. In ancient times, there was a cross-national trade in creation stories, their myths, symbols and images, suggesting that in earlier times they satisfied basic human needs, touching on deep feelings.

What has been of enduring importance in the tremendous myth of a garden of Eden, is the ideal of an ordered unspoilt landscape, productive and beautiful, a source of delight which serves basic human needs. Hence in literature over many centuries, Eden has been, and remains, an image of supreme value for human life, captured by numerous poets, perhaps best of all the English Puritan, John Milton.

The Eden story has had many uses. For Muslims, it inspired in the *Q'ran* a detailed exposition of the pleasures, by no means all of them spiritual, awaiting the faithful in heaven. The tree of life, one of two in the Garden of Eden, is a fitting image – for nature lovers – of spiritual and material nourishment whereas the other, the tree of knowledge, may serve as a check on hubris as we venture ever further into the mysteries of nature. But the meaning of these ancient symbols is not fixed; it is subject to continuing interpretation. A garden, of Eden's qualities, is a rich source of knowledgeable, imaginative inquiry and action.

Chapter 9 – *Hierarchical Nature. Did God Love his Creatures Equally?* – raises the question as to whether, in Judaic religion, man was charged to master, exploit (and thereby destroy) nature. In the image of Cain and Abel, man kills man, thereby destroying some of nature and calling down the wrath of God. The story of Noah in the Ark tells of saving a small remnant of nature, selected humans and pairs of all the other creatures, while God almost despairs of the bulk of (His) creation. Humanity's survival of the devastating flood was conditional on ensuring the survival and well-being of all the other species. A fitting image for modern ecology! But why then sacrifice animals, presumably some of the Ark occupants, in thanks? Are we now past such primitive, barbaric practices as the slaughter of animals? The problem seems to be a conviction that nature is hierarchical and those at or near the top (humans) can use those below in pursuit of their beliefs.

With Chap. 10, *Walafrid Strabo, Benedictine Monk, Virgilian Poet and Mediaeval Gardener*, we trace the emergence of two intertwined strands which bring together heavenly piety and devotion to earth. Augustine and the Church had, to a substantial degree, sought to bifurcate them as polar opposites. True, people had to live as best they could on earth but their feelings and thoughts were not to dwell there beyond necessity. It was the great monastic order founded by hermit Benedict in the sixth century at Subiaco that demonstrated a union. The Benedictine monasteries were thriving earthly cultures, with rules of conduct, spiritual rituals, and tracts of land to be cultivated and nurtured through religious labour.

On a small island on the edge of Lake Constance, Abbot Walafrid Strabo in the ninth century, led a community of Benedictine monks while writing Europe's first known garden manual, in Virgilian Latin. That poet's evocation of the pleasure of dipping one's hand in the earth and enjoying the fruits of labour demonstrated satisfaction with earthly existence. The poems include images of Christian faith but are overwhelmingly a testament to a secular faith that was to grow and spread: care for the earth and enjoyment of its fruits. As well as practical advice on how to grow vegetables and flowers, Walafrid's poems celebrate the cycle of nature and the living power of the plant world in its own right. God's gift, the earth, is bounteous and beautiful. If well understood and properly cared for, it will remain so.

Chapter 11 is the story of the famous *St Francis the Poverello – Little Poor Man of Assisi – and his followers*. Once familiar to children everywhere, Francis was renowned for preaching to the birds, as one of his followers did to the fish. Pleasant tales, not so much of an ardent eccentric but of someone who seemingly had magical powers. St Francis today is revered, celebrated not only by the Catholic Church as the patron saint of animals, the environment and ecology. The Franciscans, beyond the established mediaeval monastic norms of chastity, poverty and obedience, went barefoot into the woods and countryside, preaching and communing with nature and joining the peasantry in manual labour. At least, this was the faith and practice of Francis and his immediate followers who eschewed the learning of cathedral schools and universities while living the most austere country life as followers of Christ.

What we know of Francis is a collection of stories told and embellished over many centuries: *The Little Flowers of St Francis*. The tone is one of exemplary piety, but the appeal lies as much, for many readers more, in the narrative of joyful encounters with animals. Not arcane theological doctrines but episodes of miraculous cures, ecstatic states, levitation and unlikely conversions appealed to the popular imagination. Most of all, tales of Francis making nests for doves, preaching to the birds and calming a savage wolf, speak to a profound bond between humans and all the creatures. A delight in the freshness and beauty of nature bespeaks a belief in the goodness of creation that is to be treasured. An "additional chapter" to *The Little Flower* is The Canticle of Brother Sun in which Francis praises God by praising his creation: "Be praised my Lord, with all Your creatures … for our Sister, Mother Earth".

Chapter 12, *The Protestant Reformation, Consumer Capitalism and the Religious Society of Friends*, may be regarded as opening floodgates in the sixteenth century for a veritable torrent which transformed the western world. In the collapse of the Mediaeval synthesis, both within religion and between it and the political, social and scientific worlds, forces emerged which continue to the present day to shape values, beliefs and everyday life. The Catholic Church fought back; in the vanguard, the Society of Jesus, the Jesuits, who in rebuilding Catholic doctrine in Europe, carried the faith to the New Worlds being opened by voyagers, to travelers, merchants and colonists.

A strongly contested argument, advanced by German and English scholars, led respectively by Max Weber and Richard Tawney, was that the Protestants released and fortified ancient forces of capitalist economics. Certainly, industrial capitalism followed and itself drove a revolution in human consumption and tastes but debate continues over the significance of such affinities as were suggested by Calvinist principles of prudence, disciplined work in the world, and self denial in the pursuit of economic gain. Moreover, even if some sects showed socially disturbing zeal, the leaders of the major branches of Protestantism sought new kinds of social and political order in alliance with secular rulers and states. Capitalism thrived and, aided by economic theorists from Adam Smith onwards, evolved.

"Man" was by no means dethroned; on the contrary, the new socio-economic trends reinforced the idea of human power and dominance. The Protestant and Catholic wings of Christianity alike found ways to accommodate their principles to those of capitalism and to growing secularism; for example, acceptance of charging of interest on financial transactions condemned as "usury" by Mediaeval theologians. One sect, that of The Religious Society of Friends, stands out in affirming that free of church, its priests and rituals, humans are, or may be, guided by an "Inner Light" towards the divine.

In time, what was meant by "divine" for some radical Friends no longer included a deity. For centuries, the Friends as a body had taken great interest in the natural world. Many were amateurs, then professionals and scientists; some of them were among the most active of the plant collectors and enthusiastic plantsmen of the eighteenth and nineteenth centuries. We recall that the Broken Hill field naturalists, Albert and Margaret Morris, were Quakers. "Creative responsibility" for nature is a modern Quaker testimony to care for the environment. The Society of Friends has for long been in the forefront of movements guided by both love and understanding of the earth.

Chapter 13, *The Reverend Gilbert White and the Religion of Nature* tells the story of a pioneer who is still regarded as a great leader among naturalists. It may be hazardous to impute to an eighteenth century Anglican clergyman belief in "the religion of nature", yet from a small village in the countryside of south-west England, late in the eighteenth century, he launched an enormously influential book, still in print, recording a lifetime's observations of the surrounding natural world: plants, creatures, fields, crops, pastures, skies, winds and weathers. If he did not quite worship, then he ordered them in the rituals of calendars, notebooks and letters, and mused as if led by an "inner light" of deep sympathy with the creatures and love of all creation.

White lived at a time when Renaissance humanism blossomed in poetry and art and, together with scientific inquiry, celebrated humanity and the natural world. Science and philosophy eroded, where they did not destroy, faith in many of Christianity's hallowed doctrines. These movements appear not to have ruffled the calm surface of Gilbert White's rural life any more than did the gathering political crisis of revolutions on the European continent or American independence.

The intellectual force of the "heresy" of Deism and the idea of natural religion, of God in and very largely of nature, passed White by. But there is evidence that he was influenced by "natural theologians" who from the seventeenth century found evidence in nature of design and of a designer (God). Gilbert White, the pioneer naturalist, is but one strand of the rich body of vital thought and action in the eighteenth century, but reading his classic book and reflecting on his life is a stimulus to widening the inquiry into how "nature" was understood by his contemporaries and successors.

"Nature" in the eighteenth century was in wide currency among writers as a normative concept. Life was to be lived according to "nature", not as in barefooted Francis and his followers preaching to the birds but as civility and culture according to the ideals of ancient Rome. This was nature thoroughly tamed in both its human and wilderness manifestations. Among the polite arts for the rising middle class to cultivate was gardening and poet, Alexander Pope, showed the way in his prolific verse and successful gardening on the banks of the River Thames at Twickenham.

Chapter 14, *Pathways to an Earthly Paradise – the Mogul Emperors*, might seem a digression too far from the new energies rapidly emerging as nature is studied, dissected and extolled in eighteenth century Europe. However, part of the upheaval was a strengthening belief that of Augustine's two cities, earth and heaven, it is earth that best lends itself to study and concern. Of all dynasties, it was that of the Muslim Moguls in India, Kashmir and Afghanistan that most clearly presents us with enticing images of an earthly paradise. These consisted of the magnificent marble and stone structures, grassy parklands and exquisite gardens with their ponds, fountains, pavilions, shady trees and flowering plants, veritable oases and pleasure domains in the hot sands and dust of their chosen (occupied) lands. The rich, earthly culture of the Mogul emperors had its roots in the blending of Mongolian and Persian ways of life in the reign of Timur (Tamberlaine) in fifteenth century Samarkand (Fig. 1.3).

Fig. 1.3 Timur granting audience on the occasion of his accession. *Zafarnama*, or *Book of Victory*, c. 1467 from The John Work Garrett Library of the Johns Hopkins University. CC BY-SA 2.0 via Wikimedia Commons

In the twenty-first century, the Muslim religion is frequently challenged, socially and politically in the West, as it was militarily by the Christian crusaders when, after the Prophet's death, it spread across North Africa and into southern Europe. Probably very few Christians read the *Q'ran* and the Islamic scriptures, and extremist Islamist movements have been at the centre of widely publicised violent conflict in the Middle East. Consequently, a strong negative view of the religion has developed, fostered by extremist groups in western countries and not always adequately countered by leading politicians. A study of the regime of the Mogul Emperors and Muslim clerics, and especially their tastes and cultural interests, may not seem much of a counter to the bigotry and hostility to Islam in our midst, but it does show that deep in that culture is a love of nature in all its forms, a vision of earthly existence informed but not overwhelmed by ancient religious texts. Christians and Muslims alike accepted the Judaic *Old Testament* as the foundation document and doctrinal base of their religion. Eden was cherished by both and served as an earthly image of their sought-for destiny, Paradise.

It is through the Moguls, too, that we make a connection, albeit a modest one, with the Asian world. The Moguls claimed descent, through the Turkish Timur (Tamberlaine), from Genghis Khan, the great Mongol tribal leader whose forces surged westward, ultimately overpowering and absorbing the once mighty Persian Empire. Simply put, the Mogul Empire was the product of Mongol energy and imperialism and Persian artistic culture and political organization.

The Mogul dynasty stretched from Babur, the first emperor, and his son, Humayun, in the sixteenth century to the last, Bahadur Shah, in the nineteenth century, who was overthrown and banished by the British Imperium for his alleged role in the Indian "mutiny." Third in line and generally regarded as the greatest, Akbar, consolidated the empire and cemented its culture. The Moguls were distinguished by their love of art, poetry and the beauties of nature. The Taj Mahal, built under Shah Jahan, the fifth emperor, and perhaps designed by him, is exquisite, an unqualified architectural masterpiece, constructed in dazzling white marble inlaid with semi-precious stones. It is unparalleled anywhere in the world. Like all the great Mogul buildings – mausoleums, palaces, pavilions – it is set in a garden, designed and planted on principles first established in ancient Persia, perhaps by Cyrus the Great at Pasargadae.

When the Moguls moved north from the heat of the dry Indian plains to the well-watered mountains of Kashmir, under Emperor Jahangir, son of Akbar, they established wondrous sloping water gardens on the foothills above Lake Dal. If one is prepared to risk the hazards of travel in today's fiercely disputed Kashmir, it is possible to visit one of the greatest of these gardens still in much of its original form. The dynasty declined from this cultural peak. Aurangzeb, the last of the power-wielding Moguls, displayed the fierce and cruel temperament of his ancient Mongol ancestors, was religiously intolerant, unlike his Mogul forebears, and a militarist. A dynasty that had cultivated the arts, built and treasured great gardens, enjoyed and valued the natural world and found paradise in all its heavenly Q'ranic detail on the earth, declined into a mere memory of lost grandeur. Except that as a last flourish of

Imperial rule, the British, at the beginning of the twentieth century, decided to build a great centre of administrative and political power in a new city – New Delhi. Eminent architects, Edwin Lutyens and Herbert Baker, were commissioned to design it. Charged with responsibility for the Viceroy's domain, Lutyens, who had partnered Gertrude Jekyll in designing grand gardens and mansions for England's wealthy, decided to create palace and garden in the spirit of Mogul and traditional Hindu designs, embodying many of their features. Thus, the great formal garden in which the now President's Palace sits may be regarded as last in the splendid Mogul line, as fine in its way as any of its antecedents. For some, at least, Paradise was reclaimed on earth.

In Chap. 15, *Voyagers, Explorers, Travellers and Collectors*, we return to the West but a West rapidly expanding as its tentacles stretched around the globe. The ancient geographers, in the tradition of the Greek Ptolemy, theorised a great land mass in the southern hemisphere to counterbalance Europe. It required centuries of European voyages to disprove this but in the meanwhile they found new lands. Traders, explorers, missionaries and settlers followed, leading the West to a position of dominance in world affairs. One of the great attractions was the prospect of mineral wealth. Silver and gold were only the beginning; there were also forests to axe, land to pasture and cultivate, and power to exert over rivals. Among the explorers, settlers and missionaries were naturalists interested in new forms of animal and plant life, scientists wishing to deepen understanding of the dynamics of the natural world. Missionaries, notably the Jesuits, were knowledgeable collectors and, together with young men sent out by botanical gardens and commercial plant nurseries, provided quantities of seeds and plant specimens to European institutions and enterprises. Women were a minority among the collectors, but Britain and Australia provided two botanical painters, respectively Marianne North and Ellis Rowan, whose travels in remote regions uncovered species new to western science.

The discoveries included not only coastlines, islands, southern Africa, the Americas and an island-continent, Australia; reports by navigators and accompanying naturalists opened the European mind to "primitive" people and their cultures. This was an invitation to further speculate on the Enlightenment creature, "the Noble Savage," someone not to despise as the sophisticated westerner's inferior but perhaps an indicator of another, even superior way of life. James Cook, one of the greatest of the eighteenth century navigators, showed a more sympathetic understanding of the Australian Aboriginal way of life than the bulk of settlers who followed him to colonise the land. Cook's voyage was among many that provided avenues for collectors to venture into, and to stock European gardens with the numerous species familiar today, but it also opened the way to new exploitable crops, including quinine, rubber and tea.

In Chap. 16, *Ways of Valuing and Appreciating Nature*, we enter into a domain of deep carers for the earth. Land exploration and the collection of plants and animals for human use and pleasure signified the strengthening recognition of the earth as not only home but what seemed to be an infinite resource. In the twentieth century, that "infinite" was found to be an illusion. From another perspective, however, earth could be used, pleasantly and respectfully. This chapter comprises several different kinds of understanding and use of the earth, none of them destructive and all contributing to a growing nature ethic, grounded in study, love and understanding and appreciation of nature's beauty.

Religion has featured in earlier chapters, often as a source of attitudes suggesting a righteous role for humans as lords and masters of the natural world. The activities of a group of clergymen in the course of the nineteenth and into the twentieth century present another image of attachment to and care for the earth – the English vicarage garden. These people may have cultivated a rich, supernatural domain in their devotion to heaven but it is not that for which they are remembered and even celebrated. Instead, they demonstrated the pleasures of caring for the earth in the small vicarage domains for which they were responsible. Garden and orchard plants were part of the flock, the souls which they tended. We cannot be certain of their views about heaven above, but they surely sought paradise on earth.

Early in the twentieth century, another kind of earthly paradise with religious foundations was under consideration. This is what is now named "Neot Kedumim", a 222 hectare arid land revegetation and cultural project located in Israel between Tel Aviv and Jerusalem and bordering Palestine. Established by Jewish academic botanists and their entrepreneurial son, Nogah Hareuveni, Neot Kedumim is designed to recreate the history of the Jewish people in their quest for the land "promised" them by Yaweh (God) in the *Old Testament*. It is both a horticultural and a cultural park which through planting, land use structures, events and documents retells, the *Old Testament* story of the migrations of the Jewish people, their occupancy, cultivation and care for land. Deeply regretting, as we must, the bitter controversy over the Jewish settlement, or resettlement, of these disputed lands, we can also admire the energy and creativity of the settlers of modern Israel in their management of waterless arid and semi-arid lands. Neot Kedumim symbolizes, for me, not disputed occupancy but the ability and persistence of people, faced with adversity, to call on nature for their sustenance, survival even, yet in so doing to use the land sustainably, conserve it, and so study it as to draw lessons of value for land users everywhere.

On the matter of lessons, we continue in this chapter, reflecting on important changes introduced into Australian – and English, American and European – schools from the middle to late in the nineteenth century. What we know as "nature study" was emblematic of a turn in educational thought, policy and practice towards recognition of the child, not just as a cipher, an object of adult authority and manipulation but a person with interests, needs and the urge to discover, create and act in the world. That world was encapsulated in the idea of experiences of and learning about the natural world of plants and animals, understanding the earth and its forces, and helping to prevent damage and restoring it when damage has been done. In a modest case study, of policy changes and reforms in Australian primary schooling at the beginning of the twentieth century, one of the most important, but often disregarded, strands of the modern environmental movement, school nature study, is opened for further investigation.

We move finally, in this chapter of multiple episodes of caring, of hope and achievement, to another ancient culture, that of the Aboriginal peoples of Australia. In the current surge of political claims being made by Aboriginal leaders – land rights, constitutional recognition and equitable access to the nation's resources – reference is commonly made to the antiquity of the culture and to 60,000 years or more of continuous occupancy of the land. Against this backdrop, it is important to consider what is now meant by "Aboriginal culture." Traditionally, and in survivals of that tradition, Aboriginal culture ranks high among the world's conservers of the land and its creatures. The land is material; its products sustain life for the creatures and the people alike. It is no less sacred and its care is the unending responsibility of the people. A deeply spiritual culture, that of indigenous Australians, was for long misunderstood. There is much to learn from it in the management today of the land of the Australian nation.

Chapter 17 completes the book: *Reaching an Understanding and Love of Nature*. Themes that have been taken up in earlier chapters are briefly recapitulated, but the main purpose is to reinforce the arguments that underline the title and subtitle of the book: the need to care for and to save the earth, and the need to do whatever we can to protect it from harm, to ensure its continued life as a means of our happiness, enjoyment and survival as a species. But not just our survival and enjoyment are at stake; no less are that of our fellow creatures. Environmentally conscious people, feeling helpless in face of the destruction now so widespread and well aware of the looming dangers to the life and health of the earth, ask what as individuals they can do. There is much – from avoiding or reducing unnecessary travel, exercising care in choice of material goods, joining environmental groups and movements, to enacting more radical changes in personal lifestyle, political action and so forth. As argued in this book, attitudes and beliefs do indeed matter and knowledge and learning are fundamentally important. Loving feelings for the natural world, if widespread throughout the community, are transformative. They are the basis of understanding, not only of what we ought to do but of what we freely wish to do and of our ability to act as natural beings, friends and companions, sharers and partners – as once we thought the resources and beauties of the world were ours exclusively by gift and by right.

Notes

[1] Singer, P. (1975) *Animal Liberation. A New Ethics for our Treatment of Animals*. New York: New York Review/Random House; Regan, T. (1983) *The Case for Animal Rights*. Berkeley and Los Angeles: University of California Press.

[2] Harari, Y. N. (2011) *Sapiens. A Brief History of Humankind*. London: Vintage Books.

[3] Angus, M. (1975) *The World of Olegas Truchanas*. Hobart: O.B.M. Pty. Ltd. p. 51, quoting from O. Truchanas at Art Exhibition, Richmond, Tasmania. 19.11.71.

[4] *Ibid.*

Full Details for Wikimedia Commons Images

1.2 Bison near a hot spring in Yellowstone. https://commons.wikimedia.org/wiki/File:Bison_near_a_hot_spring_in_Yellowstone.JPG Attribution: Daniel Mayer, CC BY-SA 3.0 https://creativecommons.org/licenses/by-sa/3.0, via Wikimedia Commons.

1.3 Timur granting audience on the occasion of his accession. https://www.flickr.com/photos/hopkinsarchives/8220594313/ Attribution: Bihzād, CC BY-SA 2.0 https://creativecommons.org/licenses/by-sa/2.0, via Wikimedia Commons.

Chapter 2
A Ravaged Land: The Mining City of Broken Hill

We travel in this and the next chapter to the mining city of Broken Hill, in the semi-arid lands of far western New South Wales. It is the site of one of Australia's most impressive projects of land restoration and conservation, following large scale destruction of the land, mainly through mining but also pastoralism. That destruction of the western land is the theme of this chapter. Conservation follows, in Chapter 3.

2.1 From Tented Encampment to Heritage City

The city of Broken Hill marks the location of a fabulously rich ore-bearing body – silver, lead and zinc among many other minerals. Originally a rough encampment, as mining declined, other riches have emerged. The city is now an architectural gem, an artistic and tourist centre, a base for eco-tourism into the country's arid lands, deserts and pastoral leases, and for film-making, for visits to Aboriginal rock art, for botanical, zoological, geological, archaeological excursions, and for travel into a vast and starkly beautiful inland landscape. The innovative "School of the Air", which provides schooling at a distance to children in remote homesteads, has its headquarters there. It is a base for the nation's Royal Flying Doctor Service, once part of a Christian Inland Mission.

Broken Hill has developed from a rough mining-rush encampment late in the nineteenth century into a colourful city of wide, tree-lined streets, parks, gardens, sportsgrounds, fine public buildings, country style hotels faced with imposing verandas, simple cottages, modern shops and historical monuments. The most poignant of these monuments is dedicated to the hundreds killed in the course of a century and more of mining operations.

In 2015, Broken Hill received national heritage listing, the first Australian city to be so honoured. The citation included reference to mining, of course, but also to the fine heritage of public buildings and private dwellings and there is

M. Skilbeck, *Loving and Studying Nature*, International Explorations in Outdoor and Environmental Education 10, https://doi.org/10.1007/978-3-030-80751-1_2

acknowledgement as well of the restoration and renewal of the natural environment upon which mining has had such a heavy and damaging impact.[1]

Great wealth has over many decades accrued to the city, to state and federal governments and to investors both within Australia and abroad. Together with the pastoralism of the western lands, Broken Hill mining has contributed substantially to the growth of the Australian economy – in the case of mining - through royalties and taxation, employment, technical innovation, scientific research, international trade, transportation, financial management and the creation of capital. The development of a distinctive Australian identity and culture, often attributed to military exploits of the First World War, the "Anzacs," is no less an outcome of these two great industries of mining and pastoralism and other elements of rural and industrial life, as recorded in the poems and stories of Henry Lawson, Banjo Paterson and Judith Wright among many others.

Some of the world's leading mining companies have their roots in the Broken Hill mines: BHP-Billiton, Rio Tinto and Pasminco. Innovations in the extracting and processing of mineral ore were paralleled throughout the twentieth century by pioneering achievements in setting health and safety standards and in providing amenities for mine employees and their families. Not least, mining, together with pastoralism, played an important role in the development of the country's industrial relations and award system.[2]

Thus, both the mining in and around Broken Hill and pastoralism across the western lands have been major factors in Australia's development, not only economic and industrial but also political, social and cultural. However, all this has come at a substantial cost in human terms and in a devastating impact on fragile lands. Destruction and seemingly irreparable damage, the result of ignorance and carelessness rather than ill intent, are the main themes of this chapter. In the following chapter we turn to the long process of recovery, through the combined efforts of the three main actors in this story who shared a vision and collaborated in its realization: the mining companies and their employees, the amateur botanists and nature lovers, Albert and Margaret Morris, and the Barrier Field Naturalist Club and its tireless band of community-minded members.

2.2 Dust

Writer, explorer and traveller, Ion Idriess, wrote of the fierce dust storms he experienced as a child growing up in Broken Hill and travelling in the western lands:

> Redder grew the sky, redder still until a ball of fire was the face of the sun. From the horizons to the heaven now was red haze, rapidly growing denser, redder, as clouds of fine red sand were lifted to the skies ... The storm lasted for seven days and nights. It arose from the edge of the Simpson Desert, gathered momentum in the desolation north of Lake Eyre ... It raced south, south-west and south-east until its vanguard covered a thousand miles wide of country ... swept on over the harsh lands and plunged Broken Hill into twilight.[3]

Fig. 2.1 Dust, a threat alike to mining assets and the comfort and health of mine workers and residents. Dust storm, Broken Hill. Photograph c.1890. 1/5/14/1/43. CRA Collection, University of Melbourne Archives. Courtesy CRA

Figure 2.1 indicates the scale and power of dust storms. Like Ion Idriess, but decades later, I spent my early years in Broken Hill and like him felt the impact of huge dust storms that turned day into night, piled dust over fences and walls, and pushed a fine, silky carpet under doors and around window frames. I experienced fierce, gritty winds that howled along the streets and over the low hills of the Barrier Range, on one of which the family house was firmly set. And I felt the general relief when temperatures dropped, thunder cracked, rain plopped down and muddy streams coursed along the unmade street verges.

Dust is the key to understanding the grand and successful re-vegetation experiment, dating from the 1930s, which gives to Broken Hill the status of a world leader in arid land restoration. Experience of the effects of dust in burying high fences around his family domain was to set a pioneer ecologist, metallurgist and mine assayer, Albert Morris, on a quest for solving "the dust problem". The threat of dust damage to newly erected buildings and expensive equipment was enough to stimulate a mine manager, A.J. Keast, to talk with Morris about a partnership for immediate, large scale action. Their meeting, in 1936, marked the beginning of a far-sighted scheme to re-vegetate the barren lands around Broken Hill in a bid to ameliorate the effects of the wind-blown dust, which we address in Chap. 3.

Dust storms have not disappeared from inland Australia and at times in recent decades, Sydney has been blanketed by tonnes of red particles driven across the whole of New South Wales from the arid lands of dry Lake Torrens far to the west in South Australia. The Broken Hill re-vegetation project beginning in the 1930s did, however, point the way towards an understanding of how, through knowledge and skilful use of plants, the dust menace might at least be ameliorated if not brought fully under control. The revegetation project also threw up a challenge, yet to be adequately addressed, to land use in the vast semi-arid interior of the continent. It is an inspiring example of how well-grounded, experimental, collaborative, community action can produce far-reaching results.

2.3 Inland Living

Dust notwithstanding, growing up in Broken Hill in the late 1930s was in many ways similar to the experience of childhood in other Australian inland towns and indeed to that in coastal towns and the outer suburbs of the state capitals. Children had the freedom of the streets, the public swimming pool swarmed with lean young bodies, Saturday afternoon was for the flicks at the town centre picture house. There was weekly Sunday school or Mass followed by the habitual roast dinner or summer salad. On weekday mornings we trudged to school, or if someone was the fortunate possessor of a bike, it was an easy ride along wide streets largely free of traffic. During the long, drawn-out summer holidays, we either endured the stifling atmosphere and, bare-footed, the burning bitumen, or joined the moneyed classes at one of Adelaide's seaside boarding houses, hotels or tented caravan parks.

If Broken Hill in these ways resembled other regional centres and outer city suburbs across the country, there were also differences. For a start, what was the reason for a thriving city of some 27,000 people and more, in the remote drylands of NSW, 713 miles (1148 kilometres) from the state capital, Sydney, and 320 miles (515 kilometres) from the South Australian state capital, Adelaide? For well over a century after mineral-bearing ore was discovered in the 1880s, Broken Hill existed for mining, and evidence of that era is plain to see today across the city.

Beneath the semi-arid, rocky and tumbling terrain of the Barrier Range, on which Broken Hill sits, for millennia there lay an immense deposit of rich ores: silver, lead, zinc and hundreds of other minerals, discovered late in the nineteenth century by German-born stockman and boundary rider, Charles Rasp (Fig. 2.2). According to

Fig. 2.2 Charles Rasp, the knowledgeable boundary rider who discovered Broken Hill's mineralized lode, pegged a site and set in motion one of the world's great mining industries. Photograph. 1/5/14/1/139 cropped. CRA Collection. University of Melbourne Archives. Courtesy CRA

legend, still current during my childhood, Rasp's horse quite by chance one day in 1883 stumbled on an outcrop of rather interesting quartz-bearing rocks. The fact of the matter is that Rasp, initially trained in Germany as an edible-oil technologist, had something of a scientific background. Moreover, mining had already been under way in the region of the Barrier Range with the discovery in the 1870s of silver some 36 km west of the site of Broken Hill. By 1882, Silverton, the first mining township on the Barrier Range, had been established.

Although rich ore discoveries were at times made by chance, prospecting was a geologically-informed, deliberate and frequently heart-breaking affair. Charles Rasp's horse was mounted by a sharp-eyed observer whose curiosity about a mass of dark mineralized ore thrusting through the rocky surface of the hilly land resulted in a fortune for himself and his partners, from the minerals in the claim that was quickly pegged out. The assayer's report on a collection of these interesting stones, despatched to Melbourne, was the beginning of one of the world's great stories of mineral exploration and exploitation continuing, although much lessened, to the present day.

In Broken Hill, it is never possible to be long out of sight of the looming manifestation of the mining boom that ensued on Rasp's discovery. Dividing the city is a kilometres-long, dark mass of rock and spoil brought up from the depths by the winding gear on the poppet heads of the mining companies. Initially thrusting through the surface rock and soil, the lode over time had to be mined far below and mineral re-extraction from the debris has been a periodic and technically challenging enterprise (Fig.2.3). The great slash of mine tailings burying the rocky hilltop across the skyline, known as the "Line of Lode", is a stark brooding presence, a reminder of wealth extracted through skill and sweaty labour, and of many hundreds of lives lost. No less, it tells of entrepreneurship, industrial conflict, technological innovation and the birth of Australian industry and global corporations.

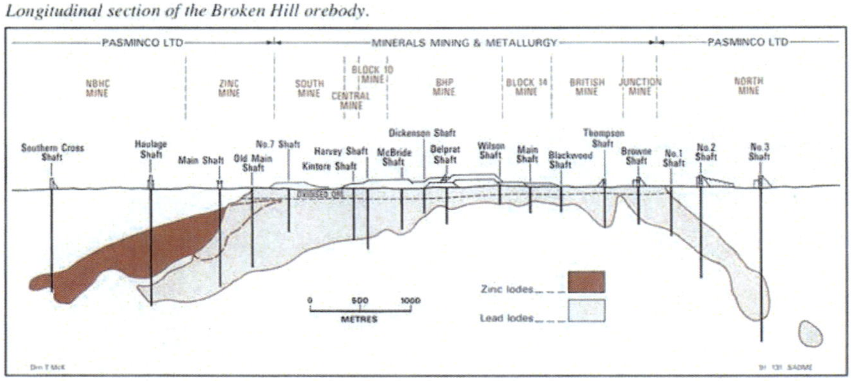

Fig. 2.3 The up-ended boomerang shape of the "Broken Hill" ore body – the "Line of Lode", from first surface mining (centre) to the depths later mining reached to, including the Zinc mine during Keast's management. Mining companies operating in 1990 shown at top of diagram. Longitudinal section of the Broken Hill orebody. Courtesy Department for Energy and Mining, South Australia

Fig. 2.4 Line of Lode, Broken Hill today, modern memorial and tourist buildings at top. Courtesy Stephen Yarrow

Visitors can make their way to the top of the Line of Lode which divides the city into North and South Broken Hill. They can and do marvel at the riches it represents, reflect on the long list of names of miners who died, and admire the striking metal sculptures. And they may well contrast their expensive lunch at the skyline restaurant now topping the huge mullock heap with the sandwiches which the miners took underground with them in their tin lunch boxes, their crib (Fig. 2.4).

The closest friend I had in my Broken Hill childhood was the son of a miner who lived with his family in a nearby cottage. My father was the manager of the blue metal quarry on one of the mine leaseholds and thus on the margins of the managerial mining community, on friendly terms with bank managers, solicitors and others in the city's higher social echelons. Did this infer a cosy family niche in a tightly structured, class-directed society? Despite the bitter and protracted strikes of the miners culminating in the greatest of the confrontations, in 1919–20, from a child's standpoint it was shared rather than separated interests and experiences that generally prevailed. All the town's children attended the state primary schools or their Catholic equivalents (and that was a divide), and there was but one state high school with secondary classes for girls in a Catholic convent. There was some residential stratification, notably for mining executives and other professionals, but our comparatively substantial house, for instance, was in a neighbourhood of mostly modest dwellings.

In contrast to this comfortable, child-centred view, in the adult world was a profound division between capital and labour which underpinned the growth of the town's trade unionism and fuelled bitter, periodic strikes. In a unique form of worker dominance of civic affairs and a pervading atmosphere of industrial militancy, the powerful Barrier Industrial Council (BIC), a combination of the local unions, was formally inaugurated in 1924. Giving Broken Hill a unique place in Australian industrial history and a role in the evolution of radical politics, the BIC has also written a chapter in local government history through its decades of dominance of the City Council.

During visits, respectively in 1898 and 1902, by British trade union leaders, Ben Tillett and Tom Mann (who remained in Australia as a union organiser and industrial activist until 1910), the community, or the mine-workers' part of it, was given instruction in the ideas and ideals of the British Socialists, the International Workers of the World (the so-called "Wobblies") and Karl Marx's *Capital*. In the course of the great strike of 1909–10, Mann was jailed for a short time but continued his attacks on the capitalist system. The bitter disputes between the unions and the mine managers eventually subsided due in part to more intelligent, cooperative leadership on both sides but also to improved (yet still to a degree dangerous and unhealthy) work conditions, better markets, a realisation of the gradual decline of economically recoverable mineral resources and developments in Australian arbitration and conciliation systems. For whatever particular reason, by the 1930s the era of massive confrontation had passed.

So class antagonism and structural division did exist, if increasingly overlaid by many shared leisure and recreational activities and a strong community-wide sense of identification with the mines and Broken Hill itself. The one distinction I recall my English-born father making on the topic of "shared community values" related to a nearby Italian family. This was at a time when Italy was being drawn by Mussolini and the Fascists into the German Nazi Axis. The threat of another European war, which was to become global, was in the air. After relating a pleasant visit to the nearby cottage of a newly-made friend from an Italian family, I was told that this association should not continue. I do still savour the memory of the thick slices of white bread smeared with dripping that my new friend's mother gave us! But I don't recall any other attitudes of the like being expressed by my parents. Indeed when a passing band of people, described as "gypsies", knocked on the door asking to gather olives for pickling from a large garden tree overhanging our high iron fence, consent was cheerfully given. We, however, had no such understanding of this Mediterranean culinary tradition any more than of the comestible uses of the flesh of the native quandong tree. Its hard, pock-marked stones were, however, a favourite for conker games similar to those with horse chestnuts threaded on strings, in England (Fig. 2.5).

Most Australians now live in urban concentrations on or near the coast where beach culture has flourished from the early years of the last century. By contrast, Broken Hill's impact on the life of children, as of adults, is that it was (and remains) a place relatively isolated in the vast semi-arid and arid space of the inland. No great excursion from the straggling town boundaries was then, or is now, required to experience a kind of continental emptiness. For children, a short bike ride or a ramble quickly led to a world of silence, remote from streets, houses, vehicles and people. "Nature" was, in this respect, less the presence of abundant plant and animal life than absence of people and the paraphernalia of human settlement. It seemed to be a boundless land either flat or with low hills forming a soft purple line along a distant horizon and with a patch of dull green trees along a dry creek bed. Under a great arc of pale blue sky, sparse, scrubby vegetation was scattered across red soil. Littered elsewhere with hard gibbers, the stony surface was stripped by wind. We

Fig. 2.5 The Quandong – *Eucarya acuminata*. The native quandong tree, with edible (albeit thin) flesh and nut was, like practically all native vegetation, little understood and less used (timber excepted) by the settlers. Courtesy Barrier Field Naturalists Club Inc. and sourced from Broken Hill Outback Archive

were to learn later that it was not only wind that denuded the soil. Human factors in many places were no less significant.

Water was the big issue for Broken Hill, even after 1932 when a pipeline was laid from the Darling River, a distance of 61.75 miles (99.4 km). It was not until the 1950s that substantial water storage dams on the Darling at the small settlement of Menindee ensured a permanent supply, threatened in more recent times through huge water extraction by cotton farmers further upstream. The mean annual rainfall, 1891–1944, was 8.87 inches (225 mm), dropping in the leanest years almost to zero. With extremely high evaporation rates, and no nearby lakes or rivers to draw on, town and mines alike were highly vulnerable to drought. Figures in the Barrier Field Naturalists Club archive in the Broken Hill City Library indicate that on 26 of the years between 1889 and 1950, the rainfall was less than 8 inches per annum (203 mm), with as little as 3.83 inches (97 mm) in 1927 and 2.28 inches (58 mm) in 1940. There was a record fall of 16.72 inches (425 mm) in 1917, a rare inundation. Such falls, flooding creeks and the flowing Darling River system, led on occasions to a fresh green mantle on the land, bringing in great quantities of birds, animals and wildflowers.

However, the latter half of the 1930s in western New South Wales was marked by one of the many long, intense and enervating droughts, a familiar hazard for all inland dwellers. With my father, on one occasion I walked across the sun-cracked bed of one of the city's two reservoirs, Umberumberka. Twenty two years later, just before the new pipeline brought water from the Darling River, emus were observed on the dry bed at the town's other reservoir, Stephen's Creek.

Trains at times were required to bring water from the Darling River and from South Australia. There was for many years a running battle between the mining companies together with the City Council on the one hand, and the NSW State Government on the other, which was only finally resolved with the construction of the Menindee Lakes major water storage, pumping and reticulation system. By contrast with the procrastination of the New South Wales authorities, the Western Australian government, through the Eastern Goldfields Water Supply Scheme, had in 1903, half a century earlier than the Menindee Lakes scheme, run a pipeline taking water from Perth to the gold mining city of Kalgoorlie, a much greater distance of 353 miles (568 kms).

But even in times of drought, staff houses on the mine leaseholds did seem to have a supply of water adequate for gardening. One was occupied by the family of a friend of my father. It had a flourishing vegetable garden organised in well defined beds with firm edges and laced with a network of paths. Our family usually left after a visit with a fine haul of its produce. It was wonder at that garden and admiration of its most capable garden master that was to enlist me lifelong in the ranks of home food producers and ornamental gardeners.

If water was the key - for the more favoured citizens – to successful domestic gardening, its absence helped to explain the barrenness then of the streets and the hard, dry, plots of bare earth in which most houses were set. For much of the time, the air was dry, as was the soil. Winds blew scorching hot or shiveringly cold from the deserts to the west. Nearby creek beds were rivers of sand while often the only "water" to be seen in any quantity was in the shimmering landscape of a tantalizing, distant mirage.

The owner of our rented house had clearly been a special kind of gardener. With the same surname as one of the leading lights of the Barrier Field Naturalists Club, he may well have been that member. At any rate, the spacious sloping hillside on which our house was set had once been planted with shrubs and trees, most if not all of them presumably indigenous. Alongside their dry, broken stems and trunks, earthenware pipes of the kind used in sewage systems jutted out of the dry, rocky ground. Here was evidence of what was required to (re)vegetate an arid hillside. Our family at the time lacked both the knowledge and interest – to say nothing of an ample water supply – to fill these pipes. If the plants were not already dead by the time we took up occupancy, it was not long before they were, thanks to boisterous childhood games played on that hillside.

It was against this backdrop of dawning horticultural awareness that in the late 1930s as a seven-year old, I heard 1 day from my father about a new project on one of the mine leaseholds. There was to be a programme of tree and shrub planting on barren land where, under the recently appointed mine superintendent (later manager), Asdruebal James ["Jim"] Keast, the Zinc Corporation was extending its mining operations. This kindled my curiosity and indeed was a source of a lifelong personal interest in trees and plant life generally, their cultivation, conservation and uses.

The Broken Hill planting project began in the 1930s as an unlikely partnership between the mine manager, A.J. Keast and Albert Morris, an amateur botanist, horticulturalist and Secretary of the Barrier Field Naturalists Club (and professionally an assayer at another mine). Initially a modest, experimental plantation of trees and shrubs on a few hectares of the Zinc Corporation leasehold, re-vegetation work was extended by stages to large areas of the Willyama (Broken Hill town) Common. The Keast-Morris alliance was the starting point of one of Australia's - indeed the world's - great pioneering achievements in ecologically sound restoration of land degraded by human activity. Today, there is a dense, grey-green kilometer-wide girdle of shrubs, trees, perennial plants and grasses stretching across the Willyama Common and encircling the city. The story, to which we return in Chap. 3, is one of close and successful collaboration between groups often in confrontation with one another: environmentalists and miners.

For several years, beyond the first generation of initiators, interest in the re-vegetation project seemed to flag. The impact of the second World War was followed by a long period when various proposals were made for use of the common lands surrounding the town: off road vehicle driving, agistment, horse-riding, a golf course and a drive-in cinema. For a variety of reasons, fortunately none of these proposals materialized and recognition of the value of the re-vegetated common land has steadily grown, to the point where it is now formally acknowledged in the city's National Heritage listing.

Knowledge and understanding of the value of the Broken Hill revegetation program, although still strangely limited in conservation circles, has been spread in a variety of ways. The publication in 1992 of a well-researched study by former school teacher, Horace Webber,[4] led to a renewal of interest and wider recognition of the environmental significance of early achievements. In 2001, prominent Sydney botanist, Barbara Briggs (and niece of Albert Morris's wife Margaret), spoke to a national audience on the ABC's *Ockham's Razor* program on the achievements of Alfred Morris.[5] The Sydney Powerhouse Museum took to the road in a travelling exhibition "Greening the Silver City: seeds of bush regeneration".[6] Launched in Broken Hill in collaboration with the City Council, it attracted audiences of adults and children in regional centres across NSW. The exhibition organiser called on people in these places to bring forward examples of local environmental initiatives. On the outskirts of Broken Hill, the establishment of The Living Desert Flora and Fauna Sanctuary in 1992 has contributed to a wider public appreciation of the restored natural environment. To reach the dramatic sculptures which are part of this initiative, visitors pass through thickly vegetated tracts with seasonally rich wildflowers. Since 2009, landscape architect, David Jones, in a series of papers has analysed the underlying designs, principles and planting strategies pioneered by Albert Morris, and together with many others, argued for full heritage recognition of the overall project.[7] In these and other ways, both public and professional awareness of the significance of the Broken Hill re-vegetation project has grown. Heritage listing will doubtless further help in this regard. However, this is to anticipate details of the revegetation project to which we return in Chap. 3.

2.4 Discovery, Settlement and Destruction of Western Lands: From Charles Sturt, Explorer, to the Spread of Pastoralism

The destructive forces that mining – and pastoralism – brought to bear on the general living conditions and environment in Broken Hill and its surroundings form a not insignificant part of the overall history of settlement and land use in Australia. Scrutiny of them will help in explaining actions needed and eventually taken to revegetate the Broken Hill townland and especially the unique contribution made by Albert Morris in developing an ecological perspective on restoration of damaged land.

First, and well before the time when Charles Rasp puzzled over those lumps of quartz-flaked rock on the "broken hill", which led to a rush of settlement and the land degradation that quickly followed, there is the question of the land in its more or less pristine state. It is necessary to qualify the term "pristine". Knowing little and understanding less of Aboriginal culture, early settlers and observers were commonly of the view that the land, because unsettled and uncultivated in their terms, was in a "state of nature". Moreover, it was "unpeopled" in that the Aboriginal inhabitants were seen to be of a different and lower order of humanity. Hence, the convenient colonial fiction of "terra nullius" (land belonging to nobody) was conceived and was only finally overturned by the High Court of Australia ruling in 1992 (Chap. 16). It is now recognised that the whole of the continent was peopled, if sparsely, by Aboriginal groups, and that vegetation in particular had been transformed by the ways Aboriginals had hunted, foraged and fished over tens of thousands of years. Research by archaeologists, anthropologists, botanists, geographers and historians has led to the abandonment of the idea of a pristine environment, an Australian "state of nature," free of significant human intervention.[8]

No Aboriginal written records exist and little enough effort is known to have been made at the time of first European settlement to gain from Aborigines a clear picture of the nature of the land and its management prior to European arrival. In addition to the findings of scholars and Aboriginal oral traditions and legends, it is mainly to the accounts of the early explorers, settlers and travellers that we must turn for an understanding of conditions prior to the inland expansion of European settlement which occurred from the middle of the nineteenth century onwards. The records that were officially required to be kept by explorers, including Captain Charles Sturt, in the epic journeys he and his party made in the 1840s into central Australia, are a primary source, not only of observations and discoveries but also of dreams and of hopes frustrated. Another important source is the diary kept by Ludwig Leichhardt in his 1840s journey from southeast Queensland to the northern coast of Arnhem Land (Chap. 15).

Whether or not Albert and Margaret Morris had the Sturt journals, other stories by explorers or many books on inland Australia in their Broken Hill home library, is not known. But there were available popular versions and magazine articles on the inland that would have provided a backdrop to their own travels out from Broken Hill, enabling them to form a mental picture of the land prior to the arrival of pastoralists and miners. Effective land restoration requires understanding of conditions prior to its destruction and the Morrises certainly studied the land and its vegetation around Broken Hill.

For the Broken Hill region, the journal kept by Sturt is of great value. There are also other records, including the sketches, paintings and field notes of Ludwig Becker who was part of the Richard O'Hara Burke and W.J. Wills epic transcontinental exploration. On their fatal quest in 1860–61 to cross the continent from Melbourne to the shores of Carpentaria, Burke and Wills passed within some 30 miles (48 kms) of the future site of Broken Hill. One of Becker's paintings is of the Darling River, which Sturt had named in traversing the Murray in his 1828–29 inland expedition (Figs. 2.6 and 2.7).

Charles Sturt had a definite mission, which was to seek to determine the existence of a great inland sea, in which he fervently believed. The absence of significant rivers flowing into the ocean from an island continent whose vast size had been fairly well determined by the 1830s was puzzling. On the alluring example of America, Africa and Europe, it was widely believed that a continental-scale land mass the size of Australia must have a mountainous interior and extensive flows of water. Rivers on the eastern side of those mountains that were known to exist, most conspicuously the north-south spine of the Great Dividing Range, flowed to the ocean. Perhaps – probably - there were inward flowing rivers, comparable with those of the other continents, flowing to a great inland sea? However, the early explorations westward in New South Wales by John Oxley (1817–1818) and

Fig. 2.6 In 1861, Ludwig Becker, artist, naturalist and geologist with the ill-fated Burke and Wills expedition, painted *Nord. End of the Pass through the Muntanie Ranges* [now known as Bynguano Range], an area now in the Mutawintji National Park, showing country relatively typical of the Broken Hill region prior to European settlement, with Callitris pines, low shrubs and tussock grasses. State Library of Victoria Pictures Collection. Ludwig Becker Sketchbook. H16486. Public domain

Fig. 2.7 Ludwig Becker. *River Darling and the mouth of the Bamamero Creek, at sunset with the anti twi-light. Decb. 191,860. Darling Depot. L. Becker*. State Library of Victoria Pictures Collection. Ludwig Becker Sketchbook. H16486. Public domain

Thomas Mitchell (1835) raised serious doubts, as rivers trickled into swamps or dribbled into scattered, drying pools. In South Australia, E.J. Eyre (1838–40) encountered, not the hoped for mountains and rivers to the northwest of Adelaide, but seemingly endless flat land, salt pans and daunting desert. Nevertheless, the belief in an inland sea persisted, supported by analogies and dubious deductions, just as the early geographers "deduced" a great southern continent to "balance" Europe, Asia and North America in the northern hemisphere (Chap. 15).

The explorers brought to the vast, open spaces of inland Australia expectations as well as hopes and these are reflected in their observations and records – of distances traversed, topography, soil, river and creek flows, climate, vegetation, animal life and encounters with Aborigines. There were both scientific interests and commercial expectations: land for crops and pasture especially. And the explorers also expressed feelings – often of great discomfort but also comments on the beauty of the landscape. Overriding all this was a determination to uncover the unknown and in their journals to record for their colonial fellows and European audiences, a newly revealed world. E.T. Giles was, together with Sturt, one of those explorers whose journals added substantially to knowledge of the "dead centre". They "constitute a record of inner experience as well as outward observation".[9] The outward observation included plant life since on his expedition Giles had commissions from the government botanist in Melbourne, Ferdinand von Mueller.

Sturt, one of the greatest explorers of Australia, in his expedition of 1844–46 from Adelaide, made the first European crossing of the Barrier Range, close to the site of the present city of Broken Hill. The elusive inland sea was widely believed to be in the distant, remote north. Sturt had predicted its existence partly on theoretical and comparative grounds and partly on the basis of deductions from the direction of flights of birds that he had observed in his earlier (1828–29) expedition. In the course of this expedition, along the Murray and the Darling Rivers, Sturt determined great stretches of the inland river system of the continent. That must have seemed encouraging to his belief in the inland sea. At any rate, the cavalcade leaving Adelaide on tenth August 1844 included a boat, visible in the celebratory painting by S.T. Gill which now hangs in the Art Gallery of South Australia, not far from where the party set forth.

The boat notwithstanding, Sturt's 1844–46 expedition coincided with a river-depleting drought which was to prove almost fatal to the whole expedition. In his journal, Sturt gave a vivid account of the topography, climatic conditions, plant and animal life, as well as making observations of fleeting encounters with small scattered groups of Aboriginal people. At the same time, and in measured language that scarcely conceals anguished disappointment on failing to achieve his goal, he recorded the ferocity of the heat and the dryness, the terrible hardships endured and the loss of life both human and animal. Finally, he was forced to abandon his quest for the inland sea. Finding neither sea nor flowing rivers, and discarding his boat at Depot Glen far to the north of the site of Broken Hill, Sturt spent almost 6 months stranded in a desert dug-out between extremely sparse water sources neither of which he could reach, in a parched, flat landscape. It was only in 1888 that John McDouall Stuart, who had been a member of Sturt's 1844–46 expedition (and

B.H. Babbage in a separate expedition) disproved the inland sea hypothesis. Nevertheless, the periodic flooding of Lake Eyre and grandiose schemes to flood the interior kept the dream of an inland sea alive.

In a poignant contrast with the "dead heart" experience that ended his quest, earlier in his journal, Sturt had admired the relative lushness of vegetation in the immediate vicinity of the Darling River:

> We passed flat upon flat of the most vivid green, ornamented by clumps of trees, sufficiently far apart to give a most picturesque finish to the landscape. Trees of denser foliage and deeper shade drooped over the river, forming long dark avenues, and the banks of the river, grassed to the water, had the appearance of being made so by art.[10]

By contrast, in his earlier (1828–1831) explorations, he had been disappointed in the saltiness of the river, finding the land devoid of promise:

> There is no life upon its surface, if I may so express myself: but the stillness of death reigns in its brushes and over its plains. It cannot be doubted, however, that we visited the interior during a most unfavourable season. Probably in ordinary times it wears a different appearance, but its deserts are of great extent, and its productions are of little value.[11]

These two passages illustrate several aspects of the environment with which early settlers had to come to terms: a land of periodic, lengthy droughts and (more rarely) floods; the unreliability of creek and river flows; and the general low fertility and aridity of the interior lands.

In places up to and somewhat beyond the Barrier Range, Sturt commented favourably, even admiringly, on the trees, shrubs and grasses, native pines (*Callitris*), *Casuarina*, box (Eucalyptus), wattle (*Acacia*), *Hakea*, *Rhagodia*, saltbush (*Atriplex*) and others. "There is no impediment to travelling," he wrote, "save the want of water. There is an abundance of feed everywhere."[12] He collected seed of indigenous plants, as later mentioned in the report by Phillip Chauncy of a visit he made in 1848 to a Mr. Robinson (commonly known as Little Robinson), who was a squatter, botanist and horticulturalist in Australia Felix (the rich pastoral lands), inland from coastal Portland: "He gave me seeds of different valuable plants – some latterly brought by Capt. Sturt from the Centre of Australia."[13] Sturt's name is justly celebrated in the emblematic Sturt's Desert Pea (*Swainsona formosa*; formerly called *Clianthus formosus*). The creeping grey-green foliage and brilliant red (also pink or white) and contrasting black of these startlingly beautiful flowers now carpet parts of Broken Hill's Willyama Common, after rain.

What Sturt presented in his account of this, the first part of his journey to the interior, as of his earlier expedition, was to serve as a prelude and indeed incentive to the taking up of pastoral leaseholds and settlement. In England there was an eager public for the 1848 publication of Sturt's findings.[14] His account of great tracts of vegetated land opened vistas to prospective pastoralists, his reports of fierce dusty gales, intense, debilitating heat (128 degrees Fahrenheit, 53 degrees Centigrade, and above) and a bleak interior all notwithstanding.

The reports of Sturt's expedition, those of the early explorers of the inland rivers, John Oxley and Allen Cunningham, the expedition of Ludwig Leichhardt in northern Australia and that of Burke and Wills across the continent led, from the 1850s onward, to the advance of intrepid pastoralists into semi-arid and arid regions of the vast triangle bordering southwest Queensland, northwest NSW and northeast South Australia. They had penetrated the land west of Menindee on the Darling River by 1850 and into the far northwest corner of New South Wales by 1878.

What was the explanation for the interest taken by pastoralists and investors in the so-challenging western lands and why were the practices they adopted and supported to prove so unfavourable, both to the land itself and to their own interests? The answer to the first question takes us back to the early years of settlement and the success of the pioneer NSW pastoralists, John and Elizabeth Macarthur, who with their family in the 1820s developed a flourishing merino stud just west of Sydney.

Due to the relative narrowness of the plains between the coast and the Great Dividing Range – the twisting spine running from northern mountainous and upland Queensland to Western Victoria – almost from the time of the first European settlement at Sydney Cove, a westward movement was seen as the road to survival, with dreams of wealth for a growing population. Thus, encouragement and a measure of faith, and financial support, private as well as public, prompted exploration, for much of the nineteenth century.

As for the semi-arid lands of distant north-western NSW, Queensland and northern South Australia, the lure was reports (in good seasons) by explorers of extensive native grasslands stretching west and far beyond the Great Dividing Range, and a ready, profitable market in England. The spinning and weaving machinery of England's expanding industrial midlands and north provided an outlet for the high quality fleeces that resulted from cross breeding Spanish merino sheep, one of the many interests Sir Joseph Banks had taken in the colonial economy from its foundations (Chap. 15). Leasehold land in the interior was relatively cheap and excessively large capital outlays were not required. However, "the outstanding difficulty in settling the Far West was lack of water".[15]

By the time of the great mineral discoveries along the Barrier Range in the 1880s and 90s, pastoralists had taken up huge leaseholds in western NSW. Bore water was discovered in the 1860s and dams and open tanks (many to be silted later) were constructed. The "pastoral frontier" had encircled the whole of Australia by 1880, leading to the process of grazing which, along with mining, was very shortly to result in a blasted landscape.[16] The stocking ratios were in terms of acres per single sheep and these acres multiplied in the years of drought. It was a portion of this devastated land that Keast, Morris and the Barrier Naturalists set out to restore in and around Broken Hill in the 1930s (Fig. 2.8).

Charles Sturt's expeditions had paved the way for take-up of the western lands. His traverse of the Barrier Range in the 1840s proved to be of particular interest to the Barrier Field Naturalists Club almost three quarters of a century later. The Naturalists were responsible for the erection of a bust of the explorer, one of the City's many fine historical monuments. Mining executive, Maurice Mawby, one of the Barrier Naturalists, alluded to Sturt when dedicating the entrance gates of the Zinc Corporation planting to the memory of Albert Morris. He linked Sturt, whose exploration opened the district to the pioneers (and to the ravaging of the land), with Morris who showed the way to its restoration and recovery.

Fig. 2.8 Broken Hill Common, 1936. The surface of the western land, denuded of vegetation by rabbits, sheep, mines and households. Photograph. 1/5/14/1/248. CRA Collection University of Melbourne Archives. Courtesy CRA

During the era of the great inland explorations, there emerged a promise - or at any rate hope - of successful agricultural and pastoral settlement far beyond the crowded Eastern and South-eastern coastal strip of Australia. But there were massive obstacles which were inadequately appreciated by the settlers, governments and the general public. Most notable was – and remains – the climate, characterized by long, periodic droughts, intensely dry atmosphere, severe wind storms and shorter, irregular periods of heavy, flooding rains. Daytime temperatures range from some of the hottest in the world to freezing night-time conditions and heavy frosts. Many of the early pastoralists were unduly optimistic about the prospects of success. For a variety of reasons, not all of their own making, many succumbed to the temptation to over-stock and over-graze the fragile western lands.[17]

With hindsight, it is all too easy to condemn the pastoralists. Some were careless, willful and in quest of quick returns, but others were possessed of a pioneering spirit and a determination to establish a viable, enduring industry. Legislation, regulations and government policy affecting leasing arrangements, size of holdings and stocking rates were factors that at times led to poor management practices. Fluctuations in wool prices were another factor, as were access to distant markets and transportation difficulties. Among these man-made considerations was inexperience and misunderstanding of the overall environmental conditions - the nature of the land and the climate.[18] In the drama of pastoral settlement of the semi-arid interior acute tensions arose: between human desire to settle the land for productive purposes and financial gain and the forces of nature which could not accommodate those desires and needs; between the experiences of everyday life in a remote, lonely setting and the city-based apparatus of legislation, regulations and bureaucracy; between the uncertain rhythms of sheep farming in a harsh land and the dynamics of northern hemisphere markets.

Through a combination of inadequate understanding, carelessness and rash experimentation in this context of a demanding, unfamiliar climate, challenging topographical and soil conditions, market fluctuations and public policy failures, pastoralists quickly reduced the semi-arid lands to a level of impoverishment from which, it seemed, they could never fully recover. Degradation accentuated climatic extremes, and floods eroded denuded creek banks and scoured the bare land. Immense losses of surface soil and widespread destruction of vegetation, loss of indigenous plant species and infestation by alien animal and plant species occurred in just a few decades from the middle of the nineteenth century.

2.5 The Western Lands Commission of 1900–1901

In the 1890s, severe drought, economic depression, distress and hardship among the pastoralist settlers led to the establishment by the New South Wales Legislative Assembly of a Royal Commission, reporting in 1901. The evidence documented by the Commission comprises an extraordinarily searching appraisal – by the pastoralists themselves, land agents and inspectors, forest experts and others - of just how much deterioration of the land had occurred, and why, in the relatively few years of European pastoralism. Devastating as were the Commission's findings and summary of evidence, alarm bells had been rung before. For example, in Victoria, sand drift in the Wimmera in the northwest of the state had been identified in 1878.[19] Reports in Adelaide and Melbourne in the middle of the nineteenth century spoke of dust in the air, in the houses and on the streets, turning to axle-deep mud after rain.

Conditions reported earlier in South Australia by explorers, Giles, Eyre and others and in Queensland, at the time of the Broken Hill revegetation project, underlined the impact of climatic extremes. It was written of inland Queensland, "the drought was so bad that even the snakes were scarce and the goannas were protected because they kept the rabbits in check by raiding their burrows. All the soil that had not blown away was nearly pure sand with clay beneath it. The only trees worth the name were cypress pine and a few wilgas. The gum trees … were called tumble-down gums, because there was nothing their roots could get a hold in and high winds brought them tumbling down."[20]

The government botanist in Melbourne, Ferdinand von Mueller, who had commissioned travellers and collectors to furnish his Botanic Garden with native plant seeds and specimens, had warned that forest clearing would lead to soil sterility and affirmed the usefulness of indigenous vegetation. In 1892, the Hon. A.C. Hamilton wrote on the effect which settlement in Australia had produced upon indigenous vegetation, while in the same year pastoralist, Samuel Dixon, wrote about adverse effects of settlement and pastoral occupation in Australia upon the indigenous vegetation.[21] Together, these and other articles documented damaging forest clearing, species loss, soil degradation, overstocking, weeds, rabbits – and drought. Moreover, from the first years of European settlement and exploration, scientific observations and descriptions of the distinctive qualities of the natural environment had drawn meaningful contrasts with the very different British and European conditions. Presumably, neither government officials nor pastoralists and their backers were ready readers of literature of this kind.

The New South Wales Royal Commission was established in 1890 to inquire into the depressed economic and social conditions of the western Crown land tenants. It reflected concern being publicly expressed about poor land management and an unsatisfactory legislative and regulatory framework in the context of the 1884 Crown Lands Act.[22] Severe drought and economic recession were exacerbating factors. The Commission's succinct, terse Report, and even more so the voluminous evidence, graphically spelt out the consequences of these interrelated factors. While more recent appraisals have to some extent moderated – but not eliminated - these concerns, the report of the Commission and its legislative and regulatory aftermath mark a watershed not only in policies affecting the management and uses of the western lands of New South Wales but in awareness and understanding of just how fragile the land is and of the importance of conservation and restoration policies and programmes nation-wide. Immense damage had been done, was well known to have been done, and was continuing at the time the Broken Hill re-vegetation project was launched, some 35 years later.

In reviewing the principal factors identified then, and reinforced since, in the use and abuse of the land, the Western Lands Commission report of 1901 anticipated the kind of ecological awareness and understanding that only in quite recent times has become – or is becoming - the norm. It was the interaction of several factors, not any one of them in isolation, that accounted for the destruction and degradation of the soil and of its vegetation.

"The causes of the present depression," in the order adopted by the Commission, were:

- Low rainfall and frequency of droughts: "The primary and most constant cause of the difficulties which beset the western grazier." (*Report* p.135). Charles Sturt was cited on: "a heartless desert"; there also was "the seductive influence of a short run of good seasons" (*Report* p.136). But the resulting "distress" had been converted into "disaster" by the huge infestation of rabbits.
- Under drought conditions competition between sheep and rabbits had become acute: "The rabbits devoted themselves to ringbarking edible shrubs and eating them out by the roots". Even this might not have led to disaster had overstocking not resulted from, "the much too favourable early view of the carrying capacity

of the land". When favourable seasons were the rule, the graziers had mistakenly believed that heavily overgrazed land would quickly regenerate after rain. It did not, at least not for long. The problem of overstocking was exacerbated by higher rents as properties changed hands and stock sale prices declined. Moreover, rabbit eradication was costly as well as difficult or well nigh impossible.

- A further factor in the decline was seen to be the "immense" damage resulting from sandstorms – loss of surface soil and the mounding of sand over vegetation and fences with sheep buried alive.
- Also identified were the consequent growth of inedible scrub, the overall decline in prices of pastoral products, the inadequate size of land holdings and short tenure.

Witnesses compared the conditions in 1880–81 before stocking - when "natural grasses", saltbush and cotton bush, were abundant on soft, spongy, absorbent ground - with the destruction wrought by rabbits and the spread of noxious weeds. They noted the "moving mass" of sand, silted water tanks, poor water conservation practice and indiscriminate ring-barking of valuable hardwoods. That issue, incidentally, was to be taken up by the Barrier Field Naturalists Club when in 1935, members protested against the destruction of trees in the Broken Hill district by local pastoralists.

In his evidence to the Royal Commission, R.W. Peacock, Manager of the Government Experimental Farm at Coolabah, claimed that timber reserves would benefit both climate and surrounding vegetation. Then, in something of an anticipation of the 1930s project at Broken Hill, he envisaged a regeneration zone which Morris or Keast, or both, may well have read about or heard reference to:

> If I had a large area of country which had been eaten out – as a lot of the Western country has been – and if the pastures were ruined by drought and overstocking, and the saltbush had disappeared, I would go in principally for the conservation of edible shrubs and saltbush as my principal standby, and I would run a secondary fence right across my country from another fence. I would keep the stock completely out of that area … In that way we would get saltbushes and seedlings of edible trees and that would provide a seeding ground for saltbushes and other plants, which could be distributed …

Another witness, A. Smith, a station manager from the Broken Hill district, noted that wire netting had been effective in controlling vermin, as we see in the following chapter. In the Keast-Morris re-vegetation project, the first planting of indigenous species was to be within a well fenced area including wire netting.

2.6 Aftermath. A Partial Diagnosis and Continued Failings

Overall, there was very little disagreement among the Royal Commission witnesses, either about the factors at work in the degradation of the land or the need for improved management and landcare practices. This, and the urgency of the situation, explain the passing of the Western Lands Act (1901) which, inter alia, established a unified Western Lands Board - whose first chairman was C.J. McMaster chair of the Royal Commission - in place of a plethora of local boards. Land tenure legislation was overhauled in response to criticism of the link between overstocking and the size and turnover of leaseholdings although it did not resolve a fundamental conflict of interest between large corporations (such as that of the Kidman family) seeking consolidation, and individual settlers. Perhaps of greatest long-term significance was the (eventual) emergence of a more thoughtful and intelligent attitude to the arid and semi-arid lands of Australia, to be seen henceforth as a fragile resource – a rich, diverse and complex ecological system requiring knowledgeable and sympathetic care. Nevertheless, decades later than the publication – and publicising - of the Western Lands Commission report, boosterism of the economic potential of the arid inland continued, as for example in the attacks on Sydney University geography professor, Griffith Taylor, who, in the 1930s, questioned the continual economic exploitation of arid lands.[23]

The science and management of the interior has progressed quite substantially and for the arid and semi-arid inland as for the encircling semi-arid regions, the message is much the same: "Of all the resources of the central region of Australia, plants are among the most ubiquitous and, in the long term, the most vital ... Man and his animals, wildlife, and the very stability of the landscape itself depend on their continual existence."[24]

Yet, as reported in *The Guardian Weekly*, November 1, 2019, clearance of woody vegetation for large-scale industrial farming continues today in northwest New South Wales and other parts of Australia, despite its known negative impact on the warming climate.

This reference to "man" in the inland raises an issue which the 1901 Report of the Royal Commission did not address, yet it was and remains a vital human, social and cultural component of a toxic ecological mix. Not only were the lands and livelihoods of the European occupants of the Western lands threatened with irreparable damage and loss. As well, but apparently much less noticed at the time, was the annihilation of an ancient culture, that of the Barkindji people. In two powerful books, *Lament for the Barkindji: The Vanished Tribes of the Darling River Region* (1976) and *West of the Darling* (1969), Bobbie Hardy, noting the focus by the report of the Royal Commission on the land and its utilisation by pastoralists, deplored its inattention to the original human inhabitants. She held the NSW Government partly responsible for land use, through its encouragement of overstocking when wool prices were booming and its poor policies and procedures governing water extraction from the Great Artesian Basin. Precious and already scarce surface water resources were seriously depleted by wind-blown loamy topsoil. "Sturt's Parnari (the Gorge) on Stephens Creek (on the outskirts of present day Broken Hill), the rockholes of Yancowinna, the deep shaded pools at Depot Glen – were now made shallow by silt, and laid open to the fierce evaporation of the summer sun. Much of it also went to augment the celebrated dust-storms and sandstorms that became such a feature of western life in the years following the early nineties."[25] These storms were known ironically in the 1870s as "Darling (River) showers".

Of the impact on the Barkindji people of westward expansion and pastoral exploitation of the land and its resources, Bobbie Hardy recounts a story of even more devastating effect, from which there was no recovery. Probably in their thousands, she wrote, people were starved and hunted from their land. A further destructive effect she attributed to mining, "As the colossus of Broken Hill began to ravage and begrime the Barrier Range". Deterioration of the land and collapse of ancient culture followed fast on settlement: "the advance of the rabbit coincided with the bid for closer settlement [following the 1884 Lands Act] … under their combined impact the fine natural balance of the soil, the seasons, the plant life and its harvesting was thrown awry, and this spelt disaster for black man and white man alike. The attack on the vegetation and its very roots caused the first disruption of the ancient harmony. As the mulga and bluebush disappeared from the hills, the saltbush and perennial grasses from the plains, their nurturing seedbed was left vulnerable to the scorching gales of summer so it was whipped up and scattered in blinding red sandstorms of a magnitude unknown to the Barkindji in their tribal days."[26] (Chap. 16 further discusses pre-settlement Aboriginal culture.)

The public, at least in Sydney and parts of NSW, were made aware, if not of the blight on Aboriginal culture and life, then of the devastation of the land. A young reporter on the *Sydney Morning Herald*, C.E.W. Bean (later Australia's official historian of World War 1) was sent west to report for the newspaper on the state of the wool industry. His book *On the Wool Track*, subsequently published in 1909, was reprinted in 1925, 1945 and 1963. In recounting the Royal Commission's findings on the destruction consequent on pastoralism and settlement, Bean made a point about national identity; an image of "the Australian back country" from which was drawn "a considerable fraction of the Anzac forces in the First World War." His concern, however, at the time of his tour, was the devastating combination of overstocking and rabbits. Newspaper readers in Sydney were presented with stories of sheep buried, as were fences, under great drifts of sand. While alive, and under drought conditions, the sheep had "so trampled and trodden, and powdered the face of the country that it was blown clean away and piled itself up behind the tree clumps and over fences and old stock-yards where we saw it and could have driven a buggy over it, fences and all."[27]

As we shall see in the chapter that follows, it was this mounding of dust over fences that in the 1920s set Albert Morris on his path toward re-vegetating the Broken Hill lands. What Bean also commented on was the remarkable but quite tenuous plant growth which followed rain. It was this regeneration, fragile as it was in the succeeding heat and dryness, that encouraged over-optimism about the stock-carrying capacity of the western lands. It was also a trigger to Albert Morris's understanding of ecology and the capacity of the land, under specific favourable conditions, to recover.

Francis Ratcliffe, an English scientist and one-time pupil of eminent biologist (and first head of UNESCO) Julian Huxley, was in the 1930s invited to prepare a report for the then Council for Scientific and Industrial Research (now Commonwealth Scientific and Industrial Research Organization (CSIRO)) on Australian land use. Scientifically-trained and more dispassionate than Hardy, Ratcliffe reached similar conclusions to hers and to those of witnesses at the Royal Commission, on the impact of pastoralism on the western lands. "Wind erosion is exacerbated by the destruction of vegetation so if overgrazing is avoided vegetation cover remains even during prolonged drought. Since this dried vegetation remains nutritious and because it is widely available through perennials like salt bush (*Atriplex*) and blue bush (*Kolchia*, [now *Maireana*]), it is often made available to stock. This increases the risk of overgrazing, hence of wind erosion, resulting in a long-term destructive cycle."[28]

In a personal communication, Sydney botanist Barbara Briggs adds: "Soil stability in arid areas generally also depends importantly on a thin surface crust formed by bacteria, algae and lichens. Hoofed livestock break this crust exposing the soil, so that wind generates dust. The importance of this was not known in Morris's time. As vegetation regenerates the soil crust will also regenerate." Significant as were Ratcliffe's research and the regeneration work of the Morrises, the field naturalists and the miners, land restoration on a vast scale would be required to ameliorate the damage already done, and still being done.

Governments, local authorities and an array of public agencies and voluntary bodies have become more conscious of the need to continue building knowledge and understanding of the land. Conferences and workshops, groups such as Landcare, newsletters, leaflets, radio and television reinforce and spread both scientific findings and good practice; for example, advice regarding salination and tillage, vermin control, re-vegetation, erosion, etc. But the challenge is great and continuous. An instance of this is the construction and maintenance of just one central measure - the erection and maintenance in the inland of hundreds of kilometres of wild dog fences.

As for rabbits, Ratcliffe commented that even careful control does not lead to eradication due to their fecundity, "which defeats that great and effective hunter the fox". The fox, of course, is no less an enemy in the eyes of the pastoralist due to depredations at lambing time. The introduction by CSIRO in recent decades of viral weapons has not resulted in the elimination of the rabbit so it remains a threat to the land, its vegetation and indigenous wildlife, as do other feral and introduced species such as camels, goats, pigs, cats, dogs and foxes.

Despite changes following the 1901 NSW Royal Commission report, Ratcliffe in the 1930s found conditions in the north east and north west of South Australia, adjacent to the NSW Western Region, to be as bad as those uncovered by the Commission decades earlier. He observed sand banking over fences and houses up to the tops of trees, land stripped of vegetation, dust storms and abandoned sheep runs and homesteads. It is as if recollection of loss and destruction at their worst during the periodic droughts is no less periodically effaced by the land itself springing again to life following rain. Thus do the dreams and ambitions of one generation triumph over hard won knowledge and understanding of its predecessors.

The western lands inland story has been re-told many times, perhaps most graphically by the Broken Hill-born explorer, prospector, adventurer and popular writer, Ion Lloyd Idriess. In a book scanning the human and pre-human history of the arid lands north of Lake Eyre and into far western New South Wales and south east Queensland, *The Great Boomerang* (1941), Idriess recapitulated the desolation of the arid and semi-arid environment, the loss of life and livelihood and the fearfulness of Ratcliffe's shifting sands as they encroached on pasture and arable alike.

Idriess did not, however, confine his attention to the findings of science, improvements in stock management and restoration of devastated lands. In the footsteps of the distinguished Queensland-born engineer, Dr. John J.C. Bradfield, designer of the Sydney Harbour Bridge, who campaigned to divert several of Queensland's coastal rivers to irrigate Queensland's western plains, Idriess urged a much grander scheme to divert water from the whole of the continent's eastern and northern coastal river system in order to restore thousands of square miles of ancient inland rivers and lakes. This dream of recreating Sturt's inland sea, though of course never realised, tells of the shock felt by a widely experienced and travelled observer, at witnessing the combined impact of natural forces, human settlement and exotic animals on Australia's arid lands. No less does it tell of the persistence of hope and ambition generated by the very harshness of the inland.

By contrast, Ratcliffe's conclusions and recommendations envisaged no bold nor grandiose engineering feats of the kind imagined by Bradfield and Idriess but largely repeated those of the 1901 Royal Commission. In his view, "the pastoralist's existence will always be a gamble in inland Australia".[29] He doubted whether the gentle treatment of the land through a knowledgeable balance between vegetation and stock would become the norm in practice. This sombre conclusion has been backed up numerous times, from the reckless practice of arable farming beyond the "Goyder Line" – the state surveyor's assessment of the northern limits of agriculture in South Australia - to the huge, evaporation-prone dams of cotton growers and the over-optimistic irrigation projects of Australia's "food basin", in the cradle and watersheds of the Murray and Murrumbidgee rivers. Clearance of tracts of indigenous vegetation continues in the interest of large-scale farming. The widely reported death in 2018–2019 of millions of native fish in the depleted waters of the Darling River is testament both to drought and inadequate management of inland hydrology.

By the 1930s, there had been in the western lands of NSW an amalgam, on the one hand, of costly experience and lessons insufficiently taken and, on the other, a range of initiatives by way of legislation, regulations, the establishment of new public agencies, better co-ordinated land use policy and collaborative moves towards improved pastoral practice. This was an equilibrium of sorts, with an eye to economic returns but not one that was to ensure the balance Ratcliffe among others believed to be required. For wiser policy and practice to prevail, more ecologically-grounded science is needed, more intensive and well-focused education at all levels is required, together with wider public appreciation and understanding of the land.

As we see in the following chapter, the practical visionaries, Jim Keast, Albert Morris and the Barrier Field Naturalists, were the principal actors in developing something of this kind of equilibrium in and around Broken Hill. This was not to be between well-managed stock, vegetation, landscape, climate and economic markets, needed as they are, but a balance of the historical and continuing impact of mining and pastoralism on the one hand and, on the other, the intelligent application of re-vegetation practices grounded in careful observation, horticultural experience, the insights of field naturalists, botanical knowledge and clear-sighted leadership.

The setting was not huge sheep runs but the outskirts of a thriving inland city which was nevertheless suffering from a barren environment, encircling degraded lands and public policy shortcomings and failures.

Morris was reaching for an opportunity, after years of campaigning as Secretary of the Barrier Field Naturalists Club, to undertake a large scale project based on his botanical interest and growing knowledge and success in cultivating indigenous flora in his own backyard. For Keast, the regeneration project had potential benefits for the mining community and for himself as a modernising mine manager, a pioneer in improving living conditions for his workforce and their families. He was also someone who loved trees. The Field Naturalists added their knowledge, experience and concern for community well-being. Although Morris had visualized a revegetation project from the 1920s, it required the interest, the financial commitment and the managerial capability of the mining industry to bring the scheme to fruition. This is the subject of the following chapter.

Notes

[1] "Broken Hill, the 'city in the desert' is significant for its unique mix of architecture and mining infrastructure, set in a vast arid landscape The City of Broken Hill is of outstanding heritage value to the nation for its significant role in the development of Australia as a modern and prosperous country. The listing recognises the significance of over 130 years of continuous mining operations, its contribution to technical developments in the field of mining, its pioneering role in the development of occupational and health standards, and its early practice of regenerating the environment in and around the mining operations..." Australian Government Department of Environment and Energy. National Heritage List Inscription 28 January 2015. The heritage values are elaborated in documents commissioned by the City Council notably in a well-illustrated text by consultant architect Vines, E. (2010) *Broken Hill. A Guide to the Silver City.* Broken Hill: Broken Hill City Council. Other cultural and environmental heritage studies commissioned by the City Council in the years prior to heritage listing include: R.W. Corkery and Co. Pty. Ltd. (2003) *Willyama Common Land Assessment* and Appleton, J. (2003) *Willyama Common.* Broken Hill, New South Wales. Such expert studies of the geological and cultural history of the land were not available to the pioneers who in the 1930s set out to restore the damage done by mining and pastoralism. See Chap. 3.

[2] Solomon, R.J. (1988/ 2008) *The Richest Lode. Broken Hill 1883–1988.* Sydney: Austral Books (reprint of 1988 edition Sydney: Hale & Iremonger). - a general historical, geographic, economic and social source which includes a superb array of historical photographs and diagrams.

[3] Idriess, I. (1941) *The Great Boomerang.* Sydney and London: Angus & Robertson Ltd. Ch 1 Water is Life. Idriess returned to the impact of the great dust storms of his youth in other books on the western region of NSW: *The Silver City* and *Lightning Ridge.*

[4] Webber, H. (1992) *The Greening of the Hill. Revegetation around Broken Hill in the 1930s.* Melbourne: Hyland House.

[5] Briggs, B. (2001) "Albert Morris, Pioneer environmentalist from Broken Hill." Australian Broadcasting Commission. Ockham's Razor. April.

[6] Powerhouse Museum, Sydney. "Greening the Silver City. Seeds of bush regeneration." Travelling Exhibition on the Broken Hill revegetation project. Curator Ms. Annie Turnbull. Opened in 2000.

[7] Jones, D. (2011a) "Re-greening 'The Hill'. Albert Morris and the transformation of the Broken Hill landscape." *History of Gardens and Designed Landscapes.* 31:3. pp. 181–195.; Jones, D. (2011b) "Green planning heritage/ Industrial nature: the plant lands or "Regeneration Reserves

of Broken Hill." Paper presented in Track 7 (Planning History) at the third World Planning Schools Congress. Perth (WA) July 2011. Mimeo.

[8] Gammage, W.L. (2011) *The Biggest Estate on Earth: How Aborigines Made Australia.* Crows Nest, NSW: Allen and Unwin. That "making" of the landscape includes the eel and fish trap site (Budj Bim National Heritage Landscape) in Western Victoria, dating back 6000 years. It was added to the National Heritage list in 2004 and to the World Heritage list in 2019.

[9] Green, L. "Giles, Ernest (1835–1897)" *Australian Dictionary of Biography*. Canberra. Australian National University National Centre of Biography. https://adb.anu.edu.au/biography/giles-ernest-3611. (accessed 10-3-2021).

[10] Scott, E. (ed) (1929) *Australian Discovery by Land.* London and Toronto: J.M. Dent and Sons Ltd., p. 281.

[11] Sturt, C. (1834) *Two Expeditions into the Interior of Australia During the Years 1828, 1829, 1830 and 1831. With Observations on the Soil, Climate and General Seasons of the Colony of New South Wales.* 2 volumes. London Cornhill: Smith, Elder & Co. second ed. Vol.1 p. 108.

[12] Davis, R.C. (ed) (2002) *The Central Australian Expedition 1844–46. The Journals of Charles Sturt.* London: The Hakluyt Society. p. 96.

[13] Martin, S. (1993) *A New Land. European Perceptions of Australia 1788–1850.* Sydney: Allen and Unwin/ State Library of New South Wales. p. 79.

[14] Sturt, C. (1848) *Narrative of an Expedition to Central Australia 1844–46.* London: T and V. Booth.

[15] Jervis, J. (1948) "The West Darling Country: its Exploration and Development. Part III" *Journal of the Royal Australian Historical Society*. Vol.XXXIV p. 169.

[16] Bolton, G. (1981) *Spoils and Spoilers. Australians Make Their Environment 1788–1980.* Sydney: Allen and Unwin.

[17] Solomon, R.J. (1988/ 2008) *The Richest Lode. Broken Hill 1883–1988.* Sydney: Austral Books (reprint of 1988 edition. Sydney: Hale & Iremonger). pp. 354 ff.

[18] Blainey, G. (1966) *The Tyranny of Distance: How Distance Shaped Australia's History.* Melbourne: Sun Books.

[19] *Victorian Naturalist* IX 1892 p. 32.

[20] Holthouse, H. (1975) *The First 150 Years of Queensland Schools.* Brisbane: Queensland Department of Education. p. 47.

[21] Kynaston, E. (1981) *A Man on the Edge. A Life of Sir Ferdinand von Mueller* Melbourne: Allen Lane/ Penguin; Heathcote, R. (1972) "The visions of Australia 1770–1970" in Rappaport, A. (ed) *Australia as a Human Setting.* Sydney: Angus and Robertson; Young, Ann (2000) *Environmental Change in Australia Since 1778.* Melbourne: Oxford University Press. second edition.

[22] Western Division of New South Wales Royal Commission to Inquire into the Condition of Crown Tenants, appointed by the Legislative Assembly 11 August 1900. The Report and Summary of Evidence were published in Sydney by William Applegate Gullick, Government Printer, in 1901.

[23] Hutton, D. and Connors, L. (1999) *A History of the Australian Environmental Movement.* Cambridge: Cambridge University Press. p. 58

[24] Friedel, M.A. and Squires, V.R. (1983) "Plants as a resource." In Crook, G. (ed) *Man in the Centre*. Melbourne: Commonwealth Scientific and Industrial Research Organisation. p. 103.

[25] Hardy, B. (1969/1981) *West of the Darling*. Milton, Queensland: Jacaranda Press. p. 198.

[26] Hardy, B. (1976) *Lament for the Barkindji. The Vanished Tribes of the Darling River Region.* Adelaide: Rigby. p. 152.

[27] Bean, C.E.W. (1963) *On the Wool Track.* Sydney: Angus and Robertson Ltd. / Sirius Books. p.vi. and p. 7.

[28] Ratcliffe, F. (1938) *Flying Fox and Drifting Sand.* London: Chatto and Windus. p. 201.

[29] Ratcliffe *op.cit.*, p. 338.

Chapter 3
Recovery

"Jim Keast at the Zinc mine has run a wire netting fence across the land. It's to keep down the dust," my father said one day at lunch.
"How's that?" I asked.
"The wind raises the dust and blows the dead plants along. The dust and the plants hit the fence and drop down," was his response.
"Oh," I said.
"As the wind blows the plants, their seeds are scattered about. The fence keeps out the stock and the rabbits. It rains and the seeds grow," he continued.
"So that's how it's done?" I responded, thinking, "so they say, but it sounds funny to me."
Clearly neither Jim Keast nor Albert Morris thought the idea crazy although there were many in Broken Hill at the time who did. Or perhaps they were of the opinion that it would be wasted effort as the mineral resources ran out and the town dwindled away.

3.1 A Mining Turnaround

With the steady, inevitable decline and occasional peaks of mining following the boom years of the first three quarters of a century from the 1880s, Broken Hill might have tracked other mining camps and settlements – like nearby Silverton and many others around the world – into decline and abandonment. This did not happen. Although the population, now, of some seventeen thousand is well below the peak in the 1920s, which bordered on thirty thousand, Broken Hill has become a vital inland centre drawing on its rich mining heritage and developing new industries – tourism and the arts among others. Perhaps the existence of well developed infrastructure, modern amenities, (relatively) assured water supply and a legacy of fine public and commercial buildings, churches, houses, sporting grounds and parks seemed just too solid to be abandoned. Or perhaps the city's much remarked upon

M. Skilbeck, *Loving and Studying Nature*, International Explorations in Outdoor and Environmental Education 10, https://doi.org/10.1007/978-3-030-80751-1_3

spirit and pride, with a grounding in great industrial battles bitterly fought, has led to an obstinate determination to stay put. There is a commonplace saying that confirms the locals in a deep sense of pride in place; of them, it is said that they are either, "on the Hill, up the river (Menindee), t'Adelaide - or away". Clearly, Broken Hill is the place to be.

Broken Hill has boomed again, the pride of its citizens. As we saw in Chapter 2, it is now a popular tourist centre and focal point for outback excursions, painting, photography and sculpture. Its handsome built heritage has been carefully studied, restored, architecturally embellished and heritage-colour-painted. While much of this is the result of the work of public authorities and commercial interests during the past few decades, in preparation for heritage listing, its beginning lies much further back – first in the erection of a cluster of fine stone and brick civic as well as commercial buildings that from as early as 1890 began to replace the hastily erected structures of a mining camp and then in a resurgence in the 1930s when a group of mine managers, field naturalists and concerned citizens began to map a new future for the town. The rich photographic record clearly shows the progressive transformation of what was in danger of becoming an environmental wasteland (Fig. 3.1).

A comprehensive programme in the 1930s of modernizing mining and improving the amenities of working and civic life, which notably included restoring lands degraded by mining and pastoralism, started with the knowledge, ideas and determination of a very small band of people. A leading figure in the drive to ameliorate working and everyday living conditions was the newly appointed superintendent, then manager, of the Zinc Corporation mine, A.J. Keast. While his name was familiar to me as a child, it was only many years later that I came to understand his key

Fig. 3.1 West side of the "broken hill" c.1890. Miners' cottages in the foreground. Stripped vegetation and surface work of early mines and the hilltop over the underground mineral deposits, the "line of lode," beyond. Photograph. BHP Archives, Melbourne

role in the revegetation project together with mine assayer and amateur botanist, Albert Morris. It was later still before I came to appreciate the role of Morris's Quaker wife, Margaret, of the network of local naturalists in the Barrier Field Naturalists Club and of other mining executives and staff. Their part in the story is developed below.

To return to my first awareness and the puzzling remarks of my father: "It was," he said, "a matter of fencing", that the erection of iron fencing posts (from a disused rail track) and wire netting on the leasehold land of the Zinc Corporation mine was supposed, in time, substantially to reduce one of the scourges of town life – dust storms. How could this be? My father's understanding of the fencing was that it would help cushion the impact of wind on the surface of the dry soil, catching the spinning plant debris which would in turn help form a seed bed and with rain lead to natural re-vegetation of the parched, partially vegetated earth. That was one element of a more complex, later story but it was the key to the whole enterprise.

My father's interest, I suspect, was more in the ingenuity of the technique than in the plant growth that was to be fostered thereby. He belonged to the hunter or free-shooter category of nature lovers - of whom US President Theodore Roosevelt was a notable leader - who flourished as Britain and other colonial powers spread their tentacles around the globe.[1] I recall one early morning episode when a crow was cawing noisily high up in the large gum tree in our front yard. Emerging from the house, in his pyjamas, revolver in hand, my father took aim, fired and brought the bird to the ground, its black wings slowly flapping. On another occasion, with my brother and me in the back seat of a car being driven furiously by one of his friends along a bumpy, dusty road beyond the town, he raised his rifle and brought down a racing emu.

It was not such petty – and shocking - dramas as these, but rather the fascination of seeds springing to life that appealed to my childish imagination. Reading the memoirs of Jim Keast now, I warm to his sympathies and interests which aligned more with the gatherers and their families than the hunters. Through a series of indigenous plantations, both in Broken Hill and later at the Port Pirie smelting works, then at the new and controversial uranium mining town of Mary Kathleen, Keast was responsible for planting parklands, vegetable gardens and orchards and trees. He combined engineering and management skills with a concern for fostering diverse forms of plant growth and improved living conditions for his fellow mine workers and their families.

3.2 The Leadership of A.J. Keast and H.S. Robinson

The achievement of Jim Keast as an ambitious and successful mine executive is a story of a highly successful career from quite unpropitious beginnings. He recounted it modestly in a self-published autobiographical sketch and in an interview with Horace Webber.[2] In a combination of enlightened health, safety and welfare measures and a keen environmental awareness, Keast pre-figured much of the best thought and practice in the modern mining industry. These qualities are apparent in the personal part he played, together with several other mine authorities, in launching the scheme to re-vegetate barren land, working with Morris, and the Barrier Field Naturalists Club. Without Keast's leadership, the zone of regenerated indigenous vegetation that now encircles Broken Hill would not exist (Fig. 3.2). Albert Morris's role as botanist, horticulturalist and environmentalist was, as we shall see, vital but Keast, together with the Melbourne-based managing director of the Zinc Corporation, William S. Robinson, ensured the initial financial and human resources, steered the decision making and managed the overall project. Without this drive and commitment, the scientific expertise and horticultural experience that Morris and the Barrier Field Naturalists were to bring to bear could never have been mobilised. It was Albert Morris who had the idea of a re-vegetation zone; its initial realization was largely Keast's (and Robinson's) achievement.

Broken Hill today is a city of plaques and other monuments to its past. The work of Morris and others – such as May Harding, artist, botany teacher at the local Technical College and long-serving member of the Field Naturalists, and that of doctor and naturalist Dr W.D.K. MacGillivray, the first President of the Barrier Field Naturalists - are memorialized in various ways with busts, plaques, gates, and park and street names.[3] Keast appears to have, as a visual reminder of his role, just a small park named for him together with a modest plaque with a brief text. The Zinc Corporation has long since moved on. Robinson Community College (a not-for-profit community organization funded by the NSW Department of Education and Training to deliver Vocational Education and Training programs) is a fine tribute to the head of the Zinc Corporation, who encouraged and warmly supported Keast's efforts, but one wonders how many Broken Hill residents and students at the College know anything about the role Robinson and Keast played, ultimately on their behalf.

Fig. 3.2 A.J.(Jim) Keast, 1938. Zinc Corporation mine manager and pioneering leader of the project to "re-green the hill." *ZC Conveyor*, 1938. Broken Hill Outback Archive

An appropriate monument to the environmental impulse of the Broken Hill mining companies are the land restoration and community outreach programs of the most enlightened mining companies operating today elsewhere in Australia, as are the changes they helped to effect in and around Broken Hill from the late 1930s. It is due in no small measure to the combined efforts of Keast and Robinson and other mining figures of the time that a new environmental consciousness has developed in recent decades in the Australian mining industry. Awareness of all this is needed to balance the generally (and often deservedly) negative image of mining in contemporary environmental circles.

Financial contributions to the re-vegetation project as it progressed were made not only by the Zinc Corporation but also by other Broken Hill mining companies. Personnel of the Zinc Corporation other than Keast and Robinson also played important parts – notably mill manager, Maurice Mawby who, as a member of the Barrier Field Naturalists Club, introduced Keast to Morris and Jack Scougall, a foreman of works at the mine, who led the practical planting and erection of fencing and was also crucial in the early management of the plantation. However, it was Keast who brought their ideas and participation together and who with Morris ensured that what might have been agreeable speculation about a possible future became a reality in the rich carpet of indigenous vegetation that now dominates Broken Hill's once denuded and barren Willyama town common. Who was this man Jim Keast, and what led him to blend mining management with environmental concern?

From an impoverished background, first in rural Victoria then in Queensland, Keast's parents, together with their daughter and Jim, moved to Broken Hill when he was 15 years old, in search of work. Their first house was a three-room iron structure with a fireplace but no stove, no bathroom and an outside "dunny" (Fig. 3.3). With only the rudiments of a peripatetic, formal school education, capped by training as an assayer in the town's technical college, Keast obtained a succession of mining positions and worked his way up from office boy to technical trainee. Subsequent service in the Australian and then the British Army in the First World War led to British Government support to study for a Bachelor of Science degree at the Michigan (USA) School (later College) of Mining, where he was a top student. From this, he embarked on a mining management career in Canada.

It was on the basis of his reputation as an outstanding mining engineer there that Keast was recruited on the advice of Andrew Fairweather, Broken Hill South Mine manager (and later leader of the Association of Mine Managers and fair-minded industrial negotiator), to superintend the Zinc Corporation's expanding operations in Broken Hill, taking up post early in 1936. In negotiations with Board Chairman, W.S. Robinson, Keast stated that he required, in close collaboration with Robinson, a clear hand to bring about "first of all, modern working conditions on the Zinc Corporation, and follow through with good living conditions in the city itself"[4].

Keast's appointment was instrumental in an ambitious plan to expand and upgrade the Zinc Corporation's mining operations which was being spearheaded by Melbourne Managing Director, Robinson. A new mine shaft was to be sunk

Fig. 3.3 The kind of iron-clad house eventual mine manager and conservationist Jim Keast lived in as a boy when his family moved to Broken Hill. Photograph. 1/5/14/1/38. CRA Collection, Melbourne University Archives. Courtesy CRA

following a favourable survey of the ore body. New buildings were to be erected, advanced machinery and equipment installed and innovatory processes introduced.

These innovations included the progressive refinement of a technique of ore extraction known as the flotation process. From the oxidized ore close to the surface lead, zinc and silver could be readily procured through smelting. As the surface lode dipped deeper and shafts were sunk for underground mining, the minerals, in an unoxidised sulphate form, could not be extracted economically. This was known in the industry as the "sulphide problem" (recalled, incidentally, in the naming of one of Broken Hill's many mining-related street names, Sulphide Street). For several years from the beginning of the twentieth century, there were trials with a variety of techniques including magnetic separation and, in the technique that eventually triumphed, the use of gravity separation with acid, oil and violent aeration. Numerous patents were taken out in a wave of metallurgical and industrial developments of world-wide significance. One of these was adoption of the electrolysis process of treating zinc concentrates on the foreshore of the Derwent River at Risdon near Hobart. Through experimentation and technological innovation, processing of the immense quantities of dumped, low grade material brought from underground and forming a high barricade (the Line of Lode) fronting the town could get under way. Employment increased with output but under working conditions of heat, poison, dust, and fumes, poor lighting – and high accident rates. The Zinc Corporation (whose initial co-managing director in 1905 was mining engineer and future U.S. President, Herbert Hoover), saw an opportunity to treat the large tonnage of tailings in an effective and economic way. Keast's appointment in 1936 and the charge given him and readily accepted, to modernise the mine and improve working and living conditions, was thus in line with technical innovation and changed working practices which had been under way for several decades, although still falling short of needs and expectations.[5]

Keast was seen to possess the necessary engineering and managerial skills to lead the projected expansion and technical upgrading at the Broken Hill mine. More than this, the project of land restoration which followed his appointment was to be the forerunner of a comprehensive programme affecting health and safety and general working conditions, housing, recreation and civic life. The assessment made of Keast at the time of his appointment was fulfilled, and handsomely acknowledged by Robinson: "He was one of the most notable figures I have ever been associated with; he had determination, courage, imagination, and a capacity for hard work rarely equalled but never excelled, though as he himself freely admitted, he did not always get on well with his staff."[6]

Conditions, if improving at the time, were often poor in mines and mining towns in Australia and around the world. Safety standards were low; the impact of mining processes on the health of miners and their families even when known was often ill-considered in practical arrangements and housing could be crude and makeshift. Poor hygiene was a factor in recurring diseases. In his recollections, Keast made no bones about where responsibility lay: "the objective of all the mining companies was to make as much money as they could as quickly as possible. Very little was done to help the townsfolk and mine employees in their homes and little enough was done by the New South Wales Government."[7] Even dogs, he said, suffered poisoning from the dust of the lead-impregnated spoil dumps.

The concerns Keast expressed were shared by the managing director of the company. From a position first as a fruit grower then as a journalist on the Melbourne newspaper *The Age*, Robinson rose to positions of prominence and power in the Anglo-Australian mining industry, but not as the ruthless captain of industry and thoughtless exploiter of human and material resources portrayed from time to time in Broken Hill's vociferous newspaper *The Barrier Truth*, organ of the Barrier Industrial Council, the consortium of trade unions. "When I entered the industry in 1914 I was struck by the care devoted to inanimate power and the carelessness displayed to man power ... whether in selection for employment or in working conditions. No attention was paid to housing, or to transport to and from work, or to feeding or hospitalization, or educational facilities for a man's children or amenities for his wife ... As soon as possible I introduced the slogan 'At least as much care for the man as for the machine.'"[8]

Robinson first visited Broken Hill in 1905, when employed as a journalist. He found a drab settlement lacking in care for people and place: "Most of the houses were small iron cottages without even the vestige of a lawn or a shade tree. Very little rain fell near the town, and not much had been done to collect what did fall. The same lack of forethought marked the relations of mine-owners to men, and more effort was put into fighting one another than into fighting the environment."[9]

In his memoir, edited by Geoffrey Blainey, Robinson pointed out just how important Broken Hill mine investments were to become in Australia's industrial growth and economic development through their revenues (rents and patents), direct and indirect employment, and extension beyond the mining industry itself with BHP probably the most obvious example: "The directors of the Zinc corporation," he claimed, "were the first to recognize that Broken Hill was not just another mining camp but rather the heart of a great group of industrial enterprises."[10] Not least of the economic contributions has been the training and up-skilling of engineers, technicians, managers, scientists and a wide range of craftspeople.

3.3 Extending the Vision

Civic amenities in Broken Hill in the 1930s were still relatively modest, even if leisure activities were varied and animated. The bitter strikes and lockouts which punctuated industrial relations from the 1890s to the 1920s lingered in memory, a source of mistrust of whatever projects the mine managers might conjure up. Governments only gradually seized hold of their responsibility and while content to receive revenues from the mining industry, were slow to acknowledge the infrastructural, social and cultural needs of such remote settlements as Broken Hill. In a telling epigram, New South Wales premier, John Story, had in 1920 located Broken Hill as "on the edge of sundown".

Keast, Robinson and other directors of the Zinc Corporation felt a need to enlarge the perspective of their mining enterprise. Extending and upgrading mining operations could be concurrent with embracing the wider life interests of the miners and their families, using the wealth generated by their efforts not only for the capitalist principle of benefit to investors but for the betterment of conditions of the whole working force and their dependents. In this respect, while maintaining a conservative political ethic, the mine managers and directors appear to have shared many of the ideals and may have emulated the practice of enlightened industrial pioneers in Great Britain in the course of the nineteenth century. These included the Socialist mill owner, Robert Owen in his New Lanark, Scotland, textile mill; the Quaker Cadbury and Rowntree enterprises, respectively the model village, Bournville, south of Birmingham, and New Earswick, near York; and the Lever factories, housing and social amenities at Port Sunlight. All of these enterprises gave attention to housing, civic amenity and working conditions. In view of the mineral processing operations in the Whyalla-Port Pirie region in South Australia and the numerous links with Adelaide, it is likely that the town planning culture dating from its foundation in 1836 also had an impact on the Broken Hill mining executives. Possibly, too, word of Ebenezer Howard's garden city vision had reached them, and they would have been familiar with the "green ideas" expressed in the plans of Walter Burley Griffin for Canberra and then the New South Wales country towns of Griffith and Leeton. Such endeavours by industry and civic authorities arose from a variety of motives, self-interest and economic advantage among them. Nevertheless, there were discernible benefits for the whole of the mining workforce and their families in Broken Hill, while significant contributions were made by the mining industry to town planning, civic amenity, architecture and design, construction, modern arbitration and dispute resolution practices.

In this context, the planting of trees and shrubs, first on the Zinc Corporation leasehold, then on the town common and in the city's streets and parks were all of a piece with a general reformist strategy of increasing the efficiency of mining operations and ameliorating living conditions for mine workers and their families. With respect to naturalist Morris's role in partnership with the miners, it has been suggested that the parkland encircling Adelaide's city centre and the generally advanced town planning ideas current in both the Adelaide city council and the state government were a likely influence.[11] It was with South Australia along with Melbourne, rather than Sydney, that the mining industry had closest links and it was with South Australia that the Morrises had closest family and professional ties.

This South Australian influence was a likely source since Albert Morris and his wife and botanical partner, Margaret, maintained connections with family and friends there. However, Albert's experience of growing up in Broken Hill, his personal encounters with great drifts of sand piling up and over his garden fence and his great interest in indigenous plants and their soil retentive capacity were likely to be more powerful forces in his thinking, as indicated below.

Among the changes being instituted by the mining industry in the 1930s were new industrial agreements, including the introduction of the famous lead bonus which in addition to financial benefits gave miners a stake in the success of mining operations and improved productivity, together with strengthened employees' sickness funds. Co-operative housing schemes, kindergartens, clubs, sports fields, parks and gardens, and hospital and dental services were all supported. The Zinc Corporation, which became the model mine, was not, however, alone in these efforts. The mining companies as a whole joined in helping to finance existing amenities – hospitals, schools, water supply, sewerage and power supply. Despite his great admiration for Keast, Robinson in writing of the re-vegetation project identified Morris as the key player. Surprisingly, there was no mention of Keast who may have been seen as merely the agent of company policy: "Perhaps the most miraculous change of all at Broken Hill was the landscape. No man did more for Broken Hill than Albert Morris, the great botanist."[12] Horace Webber, however, acknowledges the significance of the part Keast played. There is sufficient evidence in his close personal interest in the developments at Broken Hill and later appointments in other mining centres to warrant placing him alongside Morris.

Keast's appointment and the mission he was charged with in superintending the mine was in the context of the favourable survey mentioned above which disclosed substantial untapped mineral resources within the Zinc Corporation's leasehold. At the same time, attention of the Corporation's Melbourne-based proprietors and of Keast on the ground in Broken Hill was turning towards their civic responsibilities. The goal was to develop the best mine in Australia in terms of health and safety as well as overall manner of operation and community responsibility.

3.4 Re-vegetation Starts

In a chance meeting early in 1936 with the Melbourne-based managing director of the Zinc Corporation, William Robinson, Morris outlined his ideas about re-vegetating tracts of land around Broken Hill. Subsequently, in May of that year, Zinc mine mill manager and Barrier Field Naturalists Club member, (Sir) Maurice Mawby, arranged a meeting between Morris and Keast to discuss fencing an area of the leasehold for a future plantation in place of the dusty, barren expanse of land which had resulted from mining operations and town settlement. A re-vegetation plan was prepared following this meeting and a planting program drawn up. In the first instance, this was for establishing a plant nursery, fencing some 22 acres (approx. 9 ha.) of the Zinc mine leasehold, and planting and care of trees and shrubs on this land, adjacent to the area of the mine's new buildings and mine shaft. Morris visualized a larger sanctuary and it fell to Keast to procure the gazetting of a by-law in 1937 to protect the native flora on 412 ha. Morris was appointed a "ranger" under the gazette[13] and this ensured his active engagement as consultant and adviser working in close association with the mining personnel.

Initially, and overseen by mine foreman, Jack Scougall, 2,800 trees were planted with protection to the windward by Old Man saltbush hedges (*Atriplex nummularia*). This tough arid land plant was able to keep growing even when mounded over with sand, but having been eaten out even to the roots by rabbits and stock, it had almost disappeared from the region. The plantation was fenced with wire netting and iron sheets supported by iron posts. Fifteen to eighteen species (accounts vary) were planted, including *Eucalyptus, Acacia, Callitris* and *Leptospermum* (Figs. 3.4 and 3.5). Over time, the plantation was extended further into the Willyama (Town) Common and the species count rose. As an aside, I note that the fencing idea was quite ancient. The oasis gardens of the ancient Persians were fenced to check drifting sand and keep out marauding stock. The Broken Hill pioneers lacked the Persian (Iranian) amenity of fresh spring water, using instead waste water from the mine.

In a handwritten essay dated June 20, 1942 by an unnamed author and possibly a student of Technical College Lecturer and Librarian of the Field Naturalists Club, May Harding, some forty species in the plantation were identified. In addition to Old Man Saltbush (*Atriplex nummularia*), they included Western boobiala (*Myoporum montanum*), River Red Gum (*Eucalyptus camaldulensis*, formerly *E. rostata*), Curly Mallee (*E. gillii*), Black Box (*E. largiflorens*), *Acacia lignata*, Shoestring acacia (*A. stenophylla*), Cooba (*A. salicina*), and many small species formerly in *Cassia*, now in *Senna*. Soon after the area was fenced, *Sida* and the perennial grass *Stipa variabilis* emerged, together with Mulga and Sturt's Desert Pea (*Swainsona formosa*). Not all plants were from the region; for example, *Eucalyptus dundasi* from Western Australia and, from Iron Knob in South Australia, the Western Myall (*Acacia sowdeni*) and the Flinders Wattle (*A. notabilis*). Exotics were not eschewed; they included mesquite from Arizona and tamarisk. The California tamarisk, or athel, was also part of an extensive plantation at the smelting

The top. photo., taken soon after the first red gum and acacia seedlings were planted, gives some idea of the desolate nature of the original site. In the centre photo. the scene is beginning to change; salt bush and myoporum in foreground and eucalypts inside the fence. The lower picture is barely recognisable as the same view.

Fig. 3.4 The importance of preparing for re-vegetation within defined areas protected from the ravages of stock, rabbits and other vermin is shown in these illustrations of early, then mature plants. They demonstrate the stark contrast between revegetated land and the ravaged land. Photos show Plantation No.1 in June successively in 1937, 1938 and 1939. *ZC Conveyor*, 1939. CRA Collection, Melbourne University Archives. Courtesy CRA

Fig. 3.5 Two years after planting, 1938, trees beginning to screen buildings beside Plantation 1. *ZC Conveyor*, 1939. CRA Collection, University of Melbourne Archives. Courtesy CRA

works in Whyalla, presumably a result of Morris' visit at the invitation of BHP executive, Essington Lewis, shortly before the commencement of the Broken Hill plantation. Morris's horticultural skill was to the fore, as on his advice, "about 1600 young gum trees were brought in from the edge of the reservoir. The roots and tops were suitably pruned, and they were then planted in kerosene tins in the nursery. When the warmer days of spring arrived, the trees planted in the kerosene tins put out tiny red shoots from the dry looking sticks. About 95% of them grew."[14] Anyone who has tried transplanting eucalyptus seedlings from a patch of bushland will know what a feat this was.

Morris concluded his account of the project: "The whole scheme is a marvellous example of what has been made possible by the co-operation of the three Broken Hill mining companies and their generous financial assistance. As the years roll by we feel sure that it may point the way to others to improve their lot in greater or lesser degree. What Broken Hill can do, others can do also."[15]

Early scientific appraisals were positive, as evidenced in a report by Sydney University botanists.[16] Prepared only two years after the initial planting, the report found that while fencing had had variable effects, dependent upon the prior condition of the land, it was demonstrably an effective technique. At the same time, it threw up a firm barrier to the sceptics and scoffers. Assessments over a longer time span have been made, affirming the effectiveness of the horticultural designs and methods and the management system that was adopted. Both have been widely taken forward in the mining industry.[17] In discussions, Broken Hill public authorities testified to me the environmental and civic value of the re-vegetation zone (Fig. 3.6).

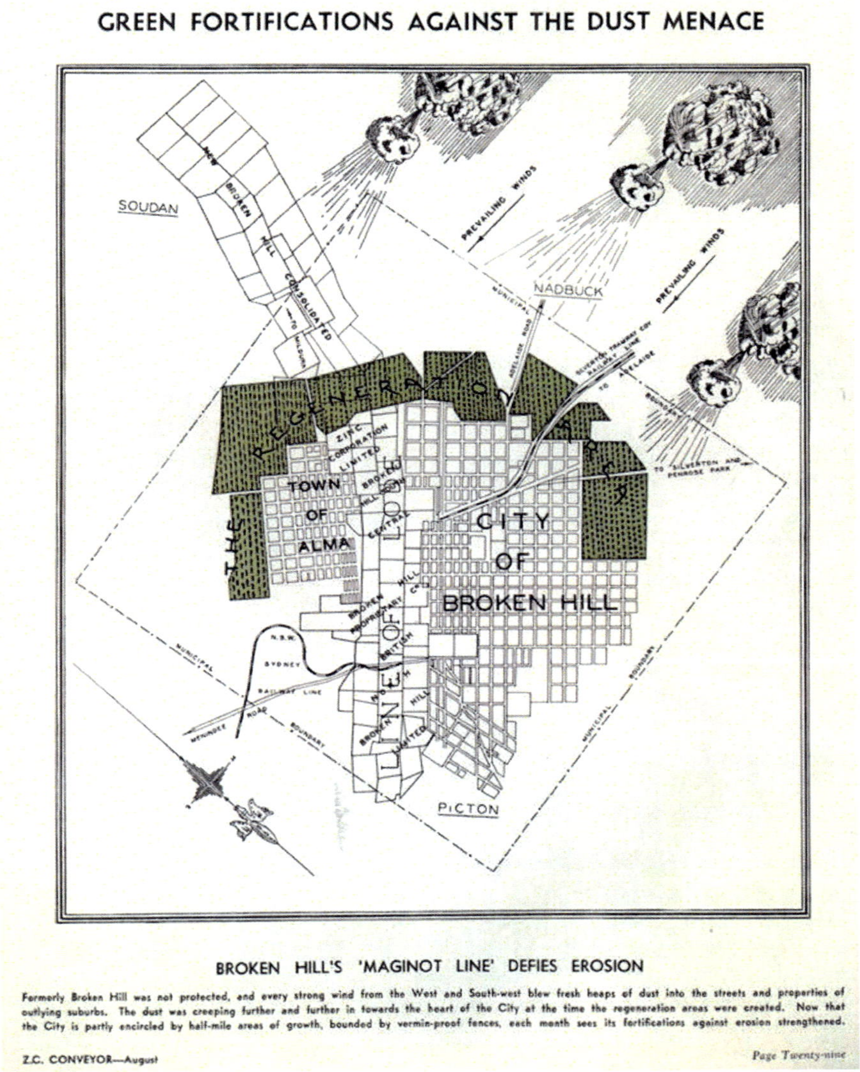

Fig. 3.6 Regeneration areas extending from north to southeast in an arc a kilometer or more wide, ca 1939. Today the city is encircled. *ZC Conveyor* August 1939 p.27. CRA Collection, Melbourne University Archives. Courtesy CRA

3.5 Nature's Friends: Albert and Margaret Morris

"Albert and Margaret Morris laid the foundation – their work can be studied in the Broken Hill area – and let us hope their simple hobby with its spectacular results will be studied with increasing interest by trained scientists, pastoralists and town planners. They showed the way ... His memory will live forever in his great work of the regeneration of native flora and tree planting in the Broken Hill district."[18] (Figs. 3.7 and 3.8).

Trained as an assayer and metallurgist, self taught as a botanist and horticulturalist, Albert Morris, had been a gardener from his childhood, growing up in a botanically-minded household. He cultivated exotic as well as indigenous species in his Broken Hill home garden and experimented with ways of controlling drifting sand.

Fig. 3.7 Margaret Morris, Quaker, naturalist, partner of Albert Morris who shared the vision and the work required for its realisation. Courtesy Barrier Field Naturalists Club Inc. and sourced from Broken Hill Outback Archive

Fig. 3.8 Albert Morris, botanist, naturalist, metallurgist, Secretary of the Barrier Field Naturalists who nurtured a vision and worked for its realisation. Recognised by the Zinc Corporation at his death in 1939. Photograph 1/5/14/12/13. CRA Collection, Melbourne University Archives. Courtesy CRA

In choosing exotics, his principle was to draw on plants, whether American or African, from similar climate zones. His conclusion was that while the exotics were often adaptable, they generally required water over and above rainfall. They could be used in gardens and parks, including the Zinc Corporation park later named in his honour, but for the extended re-vegetation zones, out into the town common, it was the indigenous species that were required.

Morris played the key role in planting plans for the regeneration zone. With his Barrier Field Naturalists colleagues and the mine staff, he established conditions for successful plant growth and the progressive extension of the original planting. He drew on experience and ideas developed over many years of botanising in the region and experimentation with exotic plants such as yuccas and cactus and indigenous species whose seeds he and his wife, Margaret, collected on local excursions.

If the Zinc Corporation executives felt a necessity to act and had the resources to do so, within the staff ranks were committed and competent artificers who sowed and planted. On the other hand, Albert and Margaret Morris and the Barrier Field Naturalists were the landscape architects. They did indeed dip their hands in the red soil but their principal role was that of the imaginers and testers of practical ideas for revitalized and recharged indigenous vegetation. Their purpose included the utilitarian and important protection of machines and property. But their primary concern was the restoration of a beloved landscape and the enhancement of civic amenity. The recovery of a botanical Eden destroyed by human ignorance and hubris and by the imperatives of economic production was a practical outcome of their love of nature.

For the Morrises, intertwined with their botanical studies and interests was a powerful sense of social responsibility. Dr Barbara Briggs, the niece of Margaret Morris, reminded me of this in a personal note. From a family background in a sharing community at Crafers in the Adelaide Hills, where her parents had considered joining William Lane's (unsuccessful) socialist venture in Paraguay, Margaret had grown up in Broken Hill imbued with strong Quaker (Religious Society of Friends) principles. Her mother had worked actively there for the local hospital. Her sister, Edna, as a school student (later to become the first female physics graduate of Sydney University and mother of Barbara Briggs) volunteered with a range of charities for the poor and for Aboriginal people. Some time after their marriage, Albert, who had been an active Anglican, joined Margaret in membership of the Religious Society of Friends. (Chapter 12.6 further discusses the Society and its attitudes toward nature).

In their field naturalist roles and activities, the Morrises showed a strong sense of civic responsibility. Tree planting, marking historic sites, campaigning for conservation of indigenous vegetation within the town boundaries and public education were among the means they adopted to enhance the quality of life in a remote mining community.

Albert and Margaret Morris, together with their fellow naturalists, gave impetus to a growing appreciation of the semi-arid landscape of Australia's interior. This required a re-orientation of settler values and ways of seeing the vegetation. The

Barrier Range struck many observers as quite inhospitable and it still seems so to those who do not attempt to study and understand it sympathetically.

Early settlers were unfamiliar with the character of the indigenous vegetation, comprising mostly low growing annual and perennial grasses, shrubs and stands of often stunted trees. In classical European terms it was a "wilderness" to be exploited according to human aspirations and need. Lacking built structures and visible evidence of cultivation – the appurtenances of civilised life – to most people, it was scrub of no intrinsic interest or value. For pastoralists, it was stock fodder; for miners, it was scrub to be cleared for buildings, roads, tunnel props and fuel for smelters. For householders, it was fuel for fireplaces, kitchen stoves and water heaters. Only for a tiny minority was it a botanical treasure trove, a source of cultural understanding and aesthetic enjoyment.

Over many centuries and during the first decades of European settlement of Australia, the wilderness was widely conceived as a barrier to be overcome, for human occupancy of the land. Trees stood in the way of agriculture; in the European tradition, forests housed wild animals, outlaws, goblins and ghosts. Forest thickets fuelled terror in fertile imaginations and spawned tales of monstrous creatures. As late as 1896, Lewis Carroll invoked the fearsome Jabberwocky in the forest wilderness and Tennyson, in his *Juvenilia*, portrayed the mysterious monster, the Kraken, slumbering in the depths of the ocean wilderness, until, "In roaring he shall rise".

Although there were elements of an appreciative understanding of the land and the life forms it supported, there was much concern among settlers and travellers about its dangers, and there was fearfulness of Indigenous people, loneliness, lack of water, prolonged drought and snakes and spiders.[19] Lingering notions of wilderness from a European past, and the immediate presence of deserts to cross, of swamps and rocky terrains to negotiate, compounded the climatic challenges presented by the interior of the country. While some people endeavoured to understand and to value the distinctive features of the Australian inland, many were either overwhelmed or fought bitterly against them. It has fallen to artists, writers, botanists, zoologists, geographers, students and lovers of nature and to the more insightful and intelligent settlers to find beauty and deep, abiding interest in what others have found harsh, forbidding and monstrous.[20] Thus it was the Morrises and the Field Naturalists who brought to the partnership with the miners' values, experiences and ideals which, if in many ways from a different world view, were ones that were to mesh closely in practice with the interests of the mining fraternity. Morris himself was, after all, employed as a metallurgist and assayer.

As a person, Albert Morris seems to have been reserved and quietly spoken, seldom working in isolation but thriving in the company of fellow field naturalists and of young people. He and his wife were childless; they entertained young teachers and nurses in weekly music and social nights in their home on the outskirts of the city. Many of these joined in the field excursions and in visits to sites of special interest in Albert's car.

Morris came to Broken Hill as a child, moving, like Keast, with his family who were searching for work. He was born in Bridgetown, South Australia in 1886, his father a stone-cutter from England, his mother Australian born. With his family, he travelled to NSW, first to Thackaringa, one of the early western prospecting sites, then to the main mining fields of Broken Hill. Completing his schooling, he studied metallurgy, achieving a diploma and a fellowship at the Broken Hill Technical College. He spent his whole career, until his premature death in 1939, as an assayer on the South Mine, rising to the position of chief metallurgist. Nothing appears to have been publicly reported about his performance in that role, but his integrity, reliability, methodical work as founding Secretary of the Barrier Field Naturalists, (recorded in minutes of the regular Club meetings) fine eye for detail and interest in the geology and the natural environment generally, suggest that he would have been highly regarded in that occupation as he most certainly was for his horticultural expertise, botanical researches and field naturalist activities.

Albert was clearly very well supported by his wife, Margaret Ellen Sayce, whom he married in 1909. "Support" does not adequately express her role since in both their joint botanical investigations and in the Barrier Field Naturalists Club, she became an equal partner. It was she, for example, who coloured the large collection of glass slides of indigenous vegetation that were prepared following their photographic expeditions into the surrounding country. (Many of them, now faded, are held in the Broken Hill City Library Archive). After Albert's death, Margaret and fellow naturalists brought his field notes together to be published as the first comprehensive scientific botany of the West Darling region of NSW, *Plant Life of the West Darling* (Fig. 3.9). Moreover, she continued for many years as a committed and highly active member of the Barrier Field Naturalists Club, becoming Secretary in 1940. It is therefore fitting that Albert and Margaret should be treated – and acknowledged – together as a highly effective partnership. Indeed, as will emerge in the following pages and by contrast with the story of the part played by the Zinc Corporation, in the industry line as it were, in the botanical-field naturalist line, women played a major part.

Childhood interest in plants and gardening grew in Albert Morris to a passion for botanising. Aided by the noted Adelaide naturalist and Quaker, Edwin Ashby, of Wittunga Botanic Garden fame (Blackwood, South Australia),[21] Albert needed to develop systematic botanical knowledge for the identification, collection, drying, recording and distribution of plant specimens. These, he, his wife and fellow members of the Barrier Field Naturalists Club, collected on field trips across the district and beyond. His expertise came to be recognised through international exchanges and the Morris herbarium of some 7,000 specimens was eventually acquired by the Waite Institute in South Australia. In her Ockham's Razor talk for ABC radio in 2001, Barbara Briggs mentioned that a collection of some 500 species is held in the data base of the herbarium at the Royal Botanic Gardens Sydney. For many years, Morris's *Plant Life of the West Darling* was the only substantial scientific source and although since supplanted, served both as a significant expert resource and a valuable field guide to the region's flora. This book provides a botanical compendium as well as Morris's field notes on various excursions – to the Darling River, Mt. Robe and Cockburn on the South Australian border, and an article "Broken Hill fights sand drift" first published in 1938 in the journal *Walkabout*. It is evident from his field notes, numerous lectures and regular botanical exhibitions and in the

Fig. 3.9 Cover of the first scientific botany of the West Darling Region of New South Wales (1920s – 1966), designed by fellow naturalist, May Harding. Courtesy Barrier Field Naturalists Club Inc. and sourced from Broken Hill Outback Archive

meetings of the Field Naturalists that Albert Morris had a highly developed ecological understanding of the land, its plant life and people. As a skilled horticulturalist, he gave practical advice on seed germination, care of young plants, economic uses (timber construction, tan bark) and the garden uses of exotic arid-land species as well as the indigenous flora. He appreciated not only the beauty of the local flora, animals, bird life and landscapes - while seeking their conservation and rehabilitation – but also uses for grazing, amenity and, not least, scientific inquiry.

Through his energetic work as Secretary of the Barrier Field Naturalists from 1920 until his death, his classes for students, writing, lecturing, field work and the great re-vegetation project, the impact of Albert Morris on his fellow citizens was wide, and profound. Fellow field naturalist, May Harding, paid tribute to his many qualities as botanist and teacher:

> When on excursions he was surrounded by enthusiasts and he did not overlook the juniors, tapering his descriptions in terms of their understanding, illustrating the finer points with his hand lens.
>
> Always a teacher, his knowledge was at the disposal of all who sought it, young or old, professional or amateur, the man on the mine interested in plants or the overseas expert. In study groups when flowers were plentiful, he revealed his technique as a botanist. There were no microscopes and adjustable lights then, but only hand lens and Black's Flora ... they were memorable hours, redolent of wildflowers. The complexities of ovaries, embryos, cotyledons were studied. Exquisite details on parts just visible to the eye were looked for and seen.[22]

According to Margaret Morris, Albert's passion for gardening took a turn during his adolescence when, among other ventures as a student, he raised funds for study by selling peppercorn plants (*Schinus aeria var. molle*). These old time country favourites from South America are still found gracing farmsteads of rural properties, school yards and country towns, several in my own country garden in southern Victoria. As an adult on his own plot of land, Morris developed a widely admired collection of dry land plants: "Bush house, fish pond and nursery made the garden a source of interest to botany students and other visitors to whom Albert Morris was tireless in introducing his treasures."[23] He enjoyed music and reading and had on his shelves at home books on travel, biography and natural history as well as botany. Regrettably, the authors and titles are not recorded, with the exception of a handful of mainly technical botanical publications. We do not know whether he was familiar with the ideas of the distinguished eighteenth and nineteenth century botanists and naturalists and the litany of nature writers who popularised natural history. There are reasons to consider him a "natural" in that his childhood plant interest and his empirical approach to mitigating mounding dust and wind blast over the stone of the family home on the west side of Broken Hill fostered an interest in growing a desert garden. And it was this, together with his field studies of indigenous flora, that lay at the heart of his designs for the major re-vegetation project.

The musical, reading aloud, and social evenings that Albert and Margaret regularly conducted for young teachers and nurses were a source of pleasure for both of them. It is not difficult to imagine how their warmth and enthusiasm sparked and fed interest in an arid land environment which for many of these young women, many from the east of the state, would have been quite unfamiliar and perhaps forbidding.

In Broken Hill, the Morrises - and Albert in particular - had almost heroic stature. It is not surprising that a prominent memorial to him stands in front of the centrally placed Technical college in the city's main thoroughfare, Argent Street. On the back of an imposing block of polished marble, housing a fountain, is a bronze tablet bearing the simple text: "Nature's friend". Ornamental lamps on the four corners of the block were donated by A.J. Keast who, in paying the warmest of tributes following

Albert's death, said of him, "There goes to his rest one of the finest citizens of any community in which it has been my lot to live."[24]

The impact that Morris, Keast and his colleagues and the Field Naturalists had through the re-vegetation project is not confined to Broken Hill and its surroundings. Nor is it to be observed only in those regional towns and mine sites where tree planting, restoration of degraded land and an enhanced environmental sensibility have become more than just a matter of law and regulation. Horace Webber gives a number of examples of the spread of the re-vegetation idea, several under Keast's watch as he took up a range of mining posts on leaving Broken Hill not long after Morris's death.[25]

In addition to instances of direct transfer of the principles and practices of the Broken Hill project, other mining companies have adopted a similar approach. Mt Morgan copper mine in Queensland, for example, under the leadership of Adam Boyd in the 1930s, adopted similar environmental and health and welfare approaches. Eventually closing about the year 2000 with, it has been claimed, little adverse visible trace of its physical impact on the land, Mt Morgan is one of many mines where advanced environmental restoration processes were and are being adopted. In 1990 Alcoa Australia Ltd was entered on the United Nations environmental Protection Agency's Global 500 Roll of Honour for its rehabilitation of mined areas and environmental care in alumina refining operations in Western Australia. It was the first Australian company and the first mining company in the world to be so recognised.[26] This recognition indicates the kinds of changes that occurred following on, and sometimes directly a result of, the Broken Hill project.

Albert Morris influenced many individuals, colleagues as well as classes of students and study groups. Students of his regular botany classes presented him with Ralph Tate's *A Handbook of the Flora of Extratropical Australia* (1890), the only book identifiably from his library held in the archive of the Barrier Field Naturalists Club in the Broken Hill City Library. One of his acolytes, Thistle Y. Harris, was to become a noted environmentalist, and later in life, the wife of the prominent naturalist David T. Stead (father of the distinguished novelist, Christina Stead). For some eighteen months, Thistle Y. Harris, environmentalist, educator, author of a popular and frequently re-issued book, *Australian Wildflowers*, was a teacher of biology at Broken Hill High School. While there, she joined the Barrier Field Naturalists Club and became Assistant to Albert as Secretary. In the oral history records held in the National Library of Australia in Canberra, Thistle says that it was from Albert Morris she learnt to be an ecologist.[27] It seems that the approach he and the field naturalists took in seeing the plant life of the region in its interrelationships and its distinctive environmental setting was something of a revelation. Harris' popular wildflower book, written shortly before her Broken Hill period, is a set of short texts on individual flowers painted by botanical artist, Adam Forster.[28] There is no suggestion in this "pre-Morris" book of an ecological awareness of plants as communities in symbiotic relationship with soil, aspect, climate, landform - or people. That realisation, it seems, came through Morris.

There is nothing in what Morris wrote or is known to have said that suggests he articulated the re-vegetation project in the most advanced ecological terms. For example, little if any explicit attention was given to the contribution re-vegetation might make to depleted indigenous animal and insect life. Yet it is not too fanciful to suggest that those responsible for the project had a deep if unformulated sense of both biological and social ecology. They recognized how the relationship between the degraded physical environment, pastoralism, the mining industry, the mines and the people of Broken Hill could be transformed. Through concerted action, there could be – and were – immediate improvements and these in time reverberated through the whole community in a wider transformation of the city and its culture.

3.6 The Barrier Field Naturalists Club

"The most influential green movement in all Australia in the 1930s was inspired by the field naturalists club in Broken Hill".[29]

If Albert Morris and A.J. Keast, together with a handful of mining executives and colleagues, formed the two principal dynamic forces in the creation of the re-vegetation project, it was the Barrier Field Naturalists Club that provided the indispensable cultural/community resource for their activities (Fig. 3.10).

The Barrier Field Naturalists Club, a community of like-minded amateur naturalists, provided a forum in which Maurice Mawby, the mine mill manager, and other mining personnel together with teachers, and others, mainly professional people, were able to share enjoyment of natural history, widen their knowledge and develop plans for the betterment of their own environment. The scale and condition of the extensive re-generation zone which today holds Broken Hill in its embrace is a

Fig. 3.10 "Not all field trips were on warm days!" Barrier Field Naturalists field trip, undated. Courtesy Barrier Field Naturalists Club Inc. and sourced from Broken Hill Outback Archive

tribute to all of these energetic, capable and far-sighted ecological pioneers, not just the prominent leading figures.

The immediate concerns and diverse local interests of the Barrier Field Naturalists grew from strong, indeed quite ancient, roots. The achievements of Keast, the other mine workers involved in the re-vegetation project, the Morrises and the Barrier Field Naturalists Club connect with a movement whose antecedents could be said to stretch back to the Ancient Greek and Mediaeval Arab philosophers and students of plant and animal life. Early naturalists laid the foundations of a scientific approach to medicine, nutrition, animal care, horticulture and agriculture. While there appears to be no clear, direct line of descent to the modern Field Naturalists movement from these ancient and mediaeval antecedents, through to the beginnings of modern science in seventeenth century Europe, there are many connections and echoes. Traces of these remain despite the significant divergences which emerged in the nineteenth century: between amateur and professional science; science and the humanities; mathematical and experimental methods and field observation.

With the professionalisation of science and the underpinning of observation and inquiry by powerful theories and mathematical methods, the resulting specialised approaches resulted in time in a bifurcation of "science" and "nature study." It led people with a broad, general interest in the natural world – the amateurs, the nature lovers - to set up clubs and societies of their own. Whereas, until the mid to late nineteenth century, for example in the Royal Society of London and the Australian colonies' Royal Societies, membership comprised a broad church of interest in and engagement with antiquities, natural history and inquiry into natural phenomena, specialisation and professionalisation gradually led to the emergence of two different if still related cultures. Science, in becoming academic, arcane and specialized, separated and elevated physics, chemistry and biology, with their specific methodologies, theories, academic and professional institutions, fellowships, journals and career structures. Nature study, or love of nature, on the other hand, became a more generalised, multi-disciplinary, observational, field-based, literary, artistic, conservationist and broadly educational activity, a forerunner of the modern "green movement". (Chapter 16.)

Professional scientists did, nevertheless, join and become active in the societies and clubs established by the naturalists and wrote for their magazines, doing so as members of these more open, interest-based communities of nature lovers (amateurs). While both the diverging paths remained at first under the broad umbrellas loosely defined as natural history/natural philosophy, progressively divisions sharpened. This was notably so from the late nineteenth century onwards in the establishment of the research structure of the European and American universities, botanical gardens, museums and specialist institutes, where a transformation of subject matter study and methods of inquiry occurred. The orientation of scientific societies, journals, exploration and voyages of discovery moved from 'general enlightenment' for the educated classes and enthusiasts toward expert knowledge systems. Results of the sciences could be – and were – popularised but their methodologies and intellectual frontiers were the province of discrete specialist categories of knowledge workers.

The importance attached now to formal academic qualifications, professional recognition through gatekeeper bodies and the badge of "scientist" are historically of relatively recent origin. For much of the nineteenth century in England and Australia this was not the case; Charles Darwin had a modest divinity degree from Cambridge and incomplete medical studies at Edinburgh. He held no academic posts. Thomas Henry Huxley, his great champion, had three years of medical training in London and became prominent as an essayist and public intellectual. Alfred Wallace, after meagre schooling and apprenticeship as a surveyor, through years of field work in the Amazon and Southeast Asia, developed a theory of evolution virtually identical with Darwin's. Initial scientific education of these great pioneers was largely through personal interest in the natural phenomena of their childhood surroundings and the great voyages undertaken in early adulthood, in Her Majesty's ocean-charting ships, *HMS Beagle* (Darwin) and *HMS Rattlesnake* (Huxley). These were followed by years of wide reading, travel, private study, inquiry, reflection, analysis and international exchange – and hands-on field work.[30]

The empirical research and field data that Darwin used, and handsomely acknowledged in his *The Origin of Species* (1859), were drawn from the work of specialist scientists and amateurs alike – and included the breeding practices of pigeon fanciers. For long, those biologists who wrote and published scientific papers were content to rely on diverse sources of knowledge and notably systematic observation and descriptive field, garden, river, shoreline and ocean findings. Amateurs, such as the English fossil collector, Mary Anning, were also responsible for pioneering zoological and botanical studies which met the scientific criteria of the time - less exacting than now. Albert Morris had no formal training as a botanist, yet *The Plant Life of the West Darling* was an original contribution to botanical science. There was still, in the 1930s, considerable overlap and interplay between the two traditions with lines much less sharply drawn than now.

In the English-speaking world, the field naturalist whose work is most often seen as the greatest milestone, if not the foundation, is the Reverend Gilbert White (Chapter 13). His *The Natural History and Antiquities of Selborne* and *Garden Kalendar* are diaries of acute observations of daily weather patterns, plant life, animal behaviour and other domains of everyday observation and comment in a rural environment, made in White's own backyard and neighbourhood. *The Natural History and Antiquities of Selborne* has remained in print since first published in 1789; it is the bible of field naturalist writers in English, of whom there has been a distinguished succession, in the living tradition of field naturalist writing.[31] One of White's observations, on earthworms, set Darwin off on a study in his own garden.

The development of the field naturalist movement is a large and complex topic. The branch with which we are concerned here is that of the clubs and associations in Australia which were established from the late nineteenth and into the early twentieth century. The naturalist movement, throughout these formative years, combined in a single organization the amateurs: bird lovers, collectors of butterflies, insects, fossils and minerals, flower and wildlife artists, photographers and writers, botanists, ethnologists, anthropologists and antiquarians. They also numbered in their ranks professional scientists; in the Victorian Field Naturalists Club, for example,

Baldwin Spencer, English naturalist became a leading scholar in Australia of Aboriginal culture. Even if specialists in different branches of natural history – as many were – members were happy to cross the boundaries, in delivering lectures or providing audiences to lectures and lantern slide shows. They were ready to join excursions, reading and discussion groups. In a nineteenth century tradition of continuing, lifelong learning, they took pride in a general culture of "nature" or, more precisely, in curiosity about the natural world extending to certain aspects of the past and to cultures and civilisations other than their own. Following establishment in 1880 of the Victorian Field Naturalists Club (Victoria's longest established conservation body), the Australian field naturalists movement spread to other states.

With a highly regarded local doctor as President (W.D.K. MacGillivray) the no less respected Albert Morris as Secretary, and Margaret, her mother and Albert's sisters among its founding members, the Barrier Field Naturalists Club was established in Broken Hill in 1920. The rules of the Field Naturalists' section of the Royal Society of South Australia were adopted. "The objects of the Club," then as now, "are to encourage the study of natural history." Study, in this case, has also entailed direct action, whether in the form of protests about environmental threats or action to restore damaged lands, tree planting, and recognition of achievement in the form of monuments, cairns, avenues and public events.

For many years, Broken Hill maintained a most active club but the heyday seems to have been in the decades of the 1920s and 1930s when Morris was Secretary. It was during this period, too, that the "mother" club in Victoria was most vigorous. In Broken Hill, of the total initial membership of seventy four in 1920, forty five were teachers; forty six of the members were female, including four of the five juniors (presumably school students). During its first forty years, four of the twelve presidents were teachers. The Club at the outset was numerically dominated by teachers and their students, and women were in an overall majority. This may reflect the popularity and acceptability of botany, wild flower painting, gardening and other natural history pursuits among women throughout the nineteenth and continuing into the twentieth century. It is likely also to relate to the establishment of nature study as part of the Victorian and NSW primary school syllabi in the early nineteen hundreds – for which many teachers were ill-prepared (Chapter 16.3). Since the female teacher members were single (once married, if working in public schools, they were required by the state department of education to resign), it is likely that membership of a field naturalist club would be an attractive outlet in a mining town which lacked a rail link to Sydney until 1927 and where many recreational, sporting and leisure activities were dominated by men. According to the *Centenary History of Alma (South Broken Hill) Public School*, accommodation was a big problem for female teachers.[32] Hotels were expensive and in any case did not welcome women. The hospitality shown them by Margaret and Albert Morris and active involvement with the field naturalists would have been much appreciated.

Membership of the Club rapidly grew, to one hundred and seventy nine by 1924. It fell away during the Second World War and reached a peak of one hundred and ninety one in 1963. By 2011, the membership had shrunk to a handful and included no teachers.

At their monthly meetings from 1920, the Broken Hill naturalists were informed and entertained by lectures on such topics as volcanoes, astronomy, exploration, reptiles, fossils, geology, sea birds, mosses, natural variations in Sturt's desert pea, foreign travel and "Einstein's theory". In 1932, Albert Morris delivered a lecture on "Nature's garden" i.e. the indigenous plants of the region. In 1936, his topic was "Ecology", while in 1937 Dr McGillivray spoke on "Erosion west of the Darling" and in 1940, Margaret Morris followed with "Broken Hill, Maginot Line Against Erosion". Botany Professor, Eric Ashby, joint author of the first scientific assessment of the re-vegetation project, as noted above, lectured in 1946 on "Education in Russia".

The Club's field excursions extended beyond botany and bird-watching to Aboriginal rock art sites, distinctive geological formations and explorers' feats, among others. Trees were donated and planted; for example, in the town cemetery and hospital grounds. The naturalists in their monthly meetings exhibited local plant species, stones, shells, insects, birds, reptiles, corals, opals and wood specimens. In an effort to attract a wider public interest, for some years they managed to place a number of their lectures in the columns of *The Barrier Miner*, the less politically radical of the city's two newspapers.

As conservationists, the Naturalists protested not only against the destruction of the remaining sparse stands of indigenous trees and shrubs within the town boundaries and the continuing adverse effects of the pastoral industry. From the late 1920s, they agitated for fencing vegetated areas of the townlands against stock and traffic. They planted monuments as well as trees, notably in memory of Charles Sturt's 1844-6 expedition. They wrote letters, made appeals and pressed both the state and local authorities to show a more enlightened awareness of the fragility of the environment and the needs of the community.

An example of their tactics was a motion proposed by a member of the Club on April 29, 1938, to protest to the public authorities about the felling of living trees by pastoralists. This motion was withdrawn and replaced by one to connect with the Pastoralists Association, "because we are out to assist in every way and not to antagonise". The original motion was replaced with the suggestion that a minute as follows be put to the Pastoralists and jointly forwarded to the authorities: "We believe the sand menace around Broken Hill could in time be somewhat checked if an area only a mile wide was fenced to prevent access of stock, and natural regeneration would result, which could be assisted by our Society by scattering native seeds, such as various kinds of Salt-bush, shrubs and trees indigenous to this part of the country. The worst areas near the town are on the South-West and Nor-West sides." Eventually it fell to the mines, not the public authorities, to join with the Naturalists in this grand endeavor, even though in a 1936 visit, the State Erosion Committee and the Surveyor-General were made aware of the chronic sand problem.

Concern for vandalism in the galleries of Aboriginal engravings and rock art in the gorges of the Bynguano Range was expressed by the Barrier Naturalists in a 1925 petition to the New South Wales Government to declare the whole area a reserve. In 1927, Mootwingee was declared crown land for the protection of Aboriginal sites. This was a victory but, without an adequate management strategy, vandalism continued, as reported from time to time at the Club's meetings. Mootwingee was declared a national park in 1982 and, in 1998, as Mutawintji National Park, was handed to the original owners.[33] (Fig. 2.6)

To foster an appreciation of wildflowers and reduce plant destruction, in their first years the Broken Hill Naturalists organised wildflower displays including painting exhibitions. The first prize winner in a painting competition in 1922 was sixteen year old school student, May Harding, who joined the Club in 1925. Later, as teacher of art and then botany at the Technical College, and throughout her life, she maintained active and responsible membership of the Club. Another flower painter, Ebenezer Gostelow, was one of those who, with MacGillivray and Morris, proposed establishment of the Club.

Headmaster of Alma Public School in South Broken Hill, Gostelow was joined by several teachers there in initial membership of the Field Naturalists. A fine flower painter, in both a decorative and an accurate botanical style, Gostelow held an exhibition of his local wildflower paintings in 1920 shortly before taking up school appointments elsewhere in the state. A collection of his paintings, about one third from the Broken Hill region, is now held in the National Library of Australia in Canberra which has published a selection of them.[34] The student, May Harding, and the headmaster, Ebenezer Gostelow, created and publicly displayed pictures in what has, from the time of first European settlement, become a rich tradition of botanical illustration and wildflower painting in and of Australia, both amateur and professional (Frontispiece, Figs. 3.11 and 3.12).

The Field Naturalists in the 1920s and 30s, if not evidently given to scholarly study of the history of ideas, were well embedded in a deeply rooted and living tradition of inquiry and practical action. The carefully kept records of the Club in the archives of the Broken Hill City Library record interests and achievements that have been quite central to, and indeed form part of an evolving culture of diverse naturalist activity, creativity and expression.

In the inquiries, field excursions, reports and discussions of the Barrier Field Naturalists Club several features of this knowledgeable, active environmentalist culture stand out:

- Curiosity about the natural world; interest in educational travel, exploration, especially in places distant or separate from settled and urban areas; and interest in aspects of exotic or non-European culture (in this instance mainly Aboriginal rock art);
- Close and systematic observation of natural phenomena: plant and animal life, minerals and geological formations and processes, climate, weather and 'remarkable' natural events such as volcanoes, earthquakes, storms and astronomical occurrences;
- Enjoyment and the direct experience of 'nature', conceived as either 'wilderness' or 'countryside', beyond the paraphernalia of close settlement and industrial development - as distinct from villages, hamlets and farmsteads (or in the case of the Broken Hill naturalists, neighbouring sheep stations and pastoral runs);

Fig. 3.11 Desert Rose. E.E. Gostelow. *Gossypium sturtianum J.H.Willis, var. sturtianum*, family Malvaceae, Broken Hill region, New South Wales, 1920. National Library of Australia, nla-obj-135347405. Public domain

- Appreciation of artistic and literary renditions of nature, as in landscape and flower painting, wildflower displays, nature poetry, travelogues, stories of exploration on sea and land, diaries, notes and accounts of excursions, in particular those of fellow naturalists and prominent, heroic figures, such as Sturt in Central Australia, and Scott in the Antarctic …
- Engagement with public authorities, landholders, industries, developers to ensure protection and preservation of areas of great interest botanically, zoologically, geologically, scenically and historically; hence, also, in the erection of monuments, construction or preservation of nature walks, reserves and areas threatened by mismanagement or destructive natural forces;

Fig. 3.12 Barrier Field Naturalists 1922 award to May Harding. Courtesy Barrier Field Naturalists Club Inc. and sourced from Broken Hill Outback Archive

- The education of children and the wider public to establish a groundswell of public opinion and well-informed decision-making by authorities, including the establishment of museums and aquaria, animal and bird sanctuaries; organising of campaigns, producing leaflets and articles for school magazines, proposing and making submissions to inquiries and undertaking or promoting follow-up action.

Many of the above have come into much greater prominence since the 1930s, through the strengthening or emergence of a variety of journals, magazines, radio, television and film, and conservation and sustainability organisations and campaigns; traces of all of them are to be found in earlier writing or campaigns, some of which have become classics. In the records of the Barrier Field Naturalists Club, this wide range of interests and concerns is well in evidence.

Due to the dispersion of the Barrier Field Naturalists book collection, established in 1923, the sparseness of details of books in the Morris household and the lack of biographical details of members, it is not possible to say whether any of the field naturalist 'classic' writings were familiar to the naturalists in Broken Hill. The remains of the Naturalists' book collection, in The Broken Hill City Library archive, tell us something but are disappointingly thin and miscellaneous. They include: *The Emu*, official organ of the Royal Australian Ornithological Union (presumably reflecting the interest in birds of founding President, Dr W.D.K. MacGillivray); *The Outline of Science* by J. Arthur Thompson (ed); an issue of *Antiquity*, an English quarterly review of archaeology; *Nature, a Weekly Illustrated Journal of Science*; a single year's run of *The Gum Tree* (journal); *The South Australian Naturalist and Wildlife* and W. Gillies and R. Hull *Nature Studies in Australia* (a text for teachers of nature study in schools). The latter is marked *ex libris* Dr MacGillivray. The Naturalists' collection at its inception also included the journal *Walkabout* and Sturt's 1848 *Expedition into Central Australia*. Minutes of the Club in 1932 record worry over increased journal costs, viz *The Geographer Magazine*, and a decision to discontinue subscriptions to *The Victorian Naturalist, The South Australian Naturalist* and *The Gum Tree*, since practically no use was being made of them. In 1932, the Botany class which Albert Morris taught presented him with a copy of Professor Ewart's *Flora of Victoria*. Tantalisingly, Margaret Morris in her biographical sketch, tells us that Albert was a keen reader but not what he read. From the admittedly flimsy evidence and absent details of their libraries and bookish interests, it does not appear that reading and discussing the classic naturalist authors was a priority of the members. Their interest in nature was practical rather than literary.

The Broken Hill naturalists were, however, mostly well educated and if not directly acquainted with them, then they shared many of the insights, values and understandings of Gilbert White and his successors in the literary genre of nature writing. Their familiarity or otherwise with the wider naturalist movement may seem of minor importance compared with the re-vegetation project and the serious environmental campaign the Barrier Field Naturalists Club conducted. The Keast-Morris project, in the context of the meetings and activities of the Broken Hill Field Naturalists Club, is firmly within the wider naturalist tradition and a pioneering contribution to environmental consciousness and action.

3.7 Impact and Legacy

The starting point for this account of the Broken Hill re-generation project was the impact of its beginning on a child growing up in Broken Hill in the 1930s and 40s. That impact, while vivid and strong, was entirely personal; it was of no significance to the general life of the town or to the actors directly involved. A snatch of conversation between father and son could quickly have faded into a fuzzy past. That it did not may be of some interest in considering the uses we make of our childhood experiences and the effect of striking impressions from that time on later life interests and activities. Wordsworth made the point in *My Heart Leaps Up When I Behold:* "The child is father of the man."

The interest I, like many other students of nature, have had throughout life in plants and their cultivation, their place in history, art, literature and scientific inquiry and in nature more generally is not that of the trained specialist. Rather it is of the amateur (the lover) who takes pleasure in the kinds of both subjective and scientific knowledge and activities which open up and illuminate appreciation and enjoyment as much as understanding of the natural world. The limitations of such an approach are manifest. They include both fragmentary, partial and idiosyncratic rather than objectively structured knowledge and systematic inquiry, and the kind of emotional responses and preferences which can cloud judgement and blanket fruitful lines of inquiry. At the same time, an emotional response, if well informed, can itself be enormously satisfying and serve as a stimulus to thought and action. This is particularly so in the fields of art and literature where nature is a primary subject and source of profound feelings, powerfully expressed. The amateur has also played a leading role in conservation movements.

Recalling the field naturalist agenda, and the achievements of the Morrises, Keast, Mawby, Robinson, Scougall, Harding, Gostelow and that crowd of distant figures who came together in the 1920s and 30s in the physical and cultural landscape of Broken Hill can serve several purposes. From the "greening of the Hill" we may take the conviction that optimistic, determined practical action is required to restore damage and destruction. It is possible, and at times needful, to turn back the clock and reverse what may at first seem to be inexorable processes and irredeemable damage (Figs 3.13 and 3.14). In the close interest the Morrises took in species

Fig. 3.13 Broken Hill Willyama Common before and after bush regeneration. Photograph by Albert Morris, hand coloured by Margaret Morris, 1938. Courtesy Barrier Field Naturalists Club Inc. and sourced from Broken Hill Outback Archive

Fig. 3.14 Panel mapping contemporary regeneration "green belt" areas around Broken Hill, Riddiford Arboretum, Broken Hill. Photo Helen Connell, 2011

adaptation to environment and particularly their understanding of their own semi-arid environment, there is an appreciation of land and landscape and diverse organic life, often regarded as dull, unremarkable, devoid of interest and aesthetically barren. The point here is appreciative understanding, which comes from the effort to study and learn. An open, inquiring mind is really all that is required to learn to love what is otherwise passed by indifferently or dismissed as "uninteresting" or regretted as devastation beyond repair. Limited formal training and no academic qualifications in botany, earth sciences, horticulture and land management were no impediments to the Morrises and their fellow naturalists who, in respect of their interest in these fields, belong to a rich and productive autodidactic, amateur and self-help tradition. They used their relations with like-minded people within and beyond Broken Hill to test, verify and extend their developing knowledge and understanding.

Those interested in the uses of indigenous plants both in restoration projects and in planting parks, footpaths, schools and other institutional grounds can learn much from the experiments the Morrises, the field naturalists and the mine employees made with seed collection and sowing, vegetative propagation, growing on, transplanting, protecting and monitoring young plants. Since the 1930s and especially since the 1960s, the Australian plant movement has found champions; for example, the Society for Growing Native Plants, Landcare, local societies of Field Naturalists among them. The Broken Hill project was a major pioneer of practices now widely adopted and successfully deployed across the country, but still in need of further support and extension (Fig. 3.14 and 3.15).

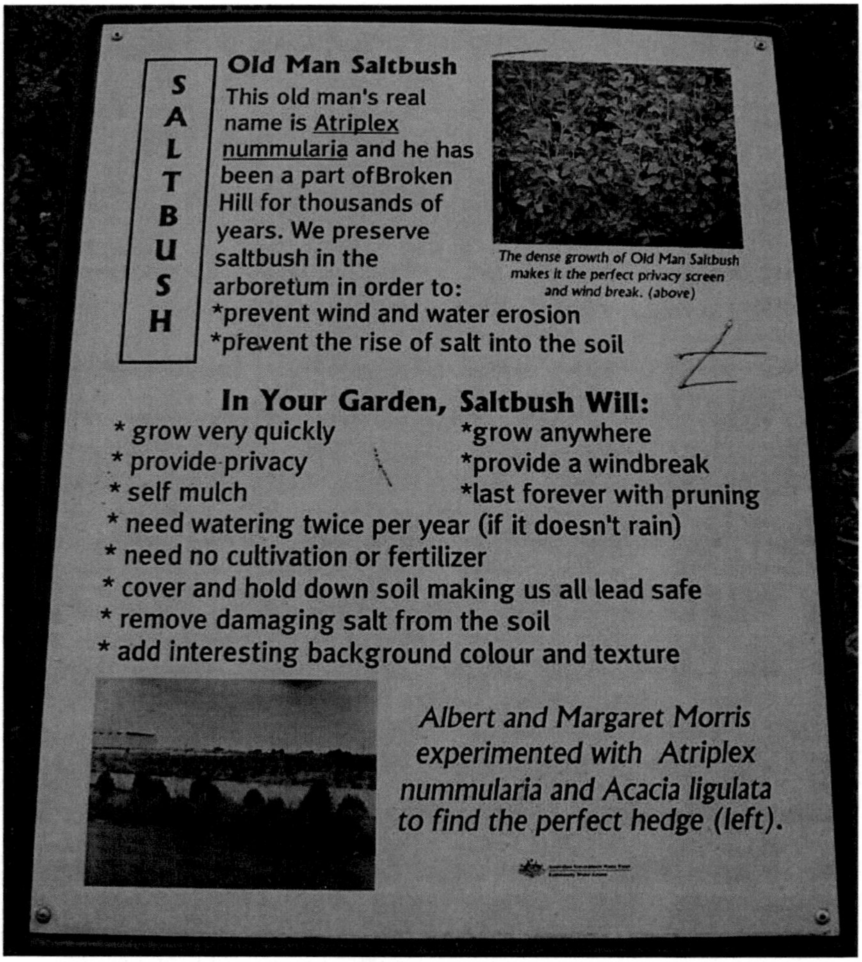

Fig. 3.15 Panel promoting use of saltbush in home gardens, Riddiford Arboretum, Broken Hill. Photo Helen Connell, 2011

Whether the Morrises and other Field Naturalists were active in growing vegetables and fruit as part of their vision of a richer and fuller town life is not recorded. However, an orchard, including a citrus grove was established and flourished on the Zinc Corporation leasehold, with both Morris and Keast originally involved. The idea was in part to give guidance to townfolk on backyard fruit growing and in part a decorative intent. It was fed with wastewater and processed sewage from the mine. The meetings of naturalists, according to their records, indicate no interest in horticulture and gardening for self sufficiency. Inadequate rainfall and the high cost of water purchase would have been significant deterrents and it is only in more recent times that in the drought-prone areas of Australia methods of efficient, low water use for these purposes have received much attention. Edwin Ashby, a family relative of Margaret Morris, as an Adelaide nurseryman and naturalist, had experimented with what became known as the "Ashby deep penetration system" of watering but this was for shrubs and trees and his interest was mainly in indigenous plants.

The Zinc Corporation had facilitated home gardening, at least for its senior employees, through access to water on its leasehold. As mentioned above, I was in childhood first alerted to the possibility of household self-sufficiency in vegetables through family visits to the large and impressive vegetable plot tended by one of these employees. The Morrises, however, appear not to have had this interest or, if they did, we have no records. Indeed the re-vegetation project had no direct connection with those productive uses of the land, such as forestry, firewood gathering, cropping and grazing since it was such activities that had brought about the devastation they set about reversing. The step beyond preservation and restoration is adoption of an intelligent and sustainable approach to the economy of land use and Morris certainly believed in this. That is a step which while much discussed, advocated and analysed in recent years, has yet to be fully taken.

For the Morrises and the Field Naturalists, it can be surmised that they would have been warmly receptive to the contemporary notion that land is a precious resource and that its uses and management must be highly respectful of natural processes, avoiding destruction and fostering regenerative and sustainable uses.

For their pioneering work, the Morrises are credited with providing the "foundations of international arid zone rehabilitation science."[35] The townscape of Broken Hill, including the regeneration reserves, now has national heritage recognition. Consequently, the reserves are, for the foreseeable future, protected from the threat of inappropriate development that has surfaced from time to time. The re-vegetation project is, however, an unfinished story. One puzzle in the project as recorded and described is the absence of reference to indigenous animals, birds and insects. There was, among the Field Naturalists Club members, considerable interest in all three kingdoms – animals and minerals as well as plants. It is curious, therefore, that in their developing ecological perspective, the Morrises do not appear to have seen the project as a restoration of other life forms than plants, which it undoubtedly was.

It is a happy coincidence, for me, that with a few words from my father lodged in memory, when living in Canberra many years later, I made a professional visit to Broken Hill, read Horace Webber's book and took photographs of two sides of the regeneration zone fence. The striking contrast between bare dry earth on one side and dense five metre high vegetation on the other symbolized a huge achievement. More recent visits have re-ignited long dormant interests and stirred a personal inquiry into the wider reaches of the naturalist movement. This story is taken up in the chapters that follow.

Notes

[1] Griffiths, T. (1996) *Hunters and Gatherers. The Antiquarian Imagination in Australia.* Cambridge: Cambridge University Press.

[2] Keast, A.J. (1944) "Straws in the Wind." Mimeo. Broken Hill City Library. (and self-published in 1977: Canterbury, Victoria. A. Keast); Webber, H. (1992) *The Greening of the Hill. Revegetation around Broken Hill in the 1930s*. Melbourne: Hyland House. Chapter 4 Assessment; ch.5 The Spreading of the Green.

[3] It was an elderly May Harding who designed the cover of Morris's *Plant Life of the West Darling*.

[4] Keast, A.J. *op.cit.* p. 95.

[5] Blainey, G. (1963) *The Rush That Never Ended.* Melbourne: Melbourne University Press; Blainey, G. (1968) *The Rise of Broken Hill*. Melbourne: Macmillan; Solomon, R.I. (1988/ 2008) *The Richest Lode. Broken Hill 1883–1988.* Sydney: Austral Books (reprint of 1988 edition Sydney: Hale and Iremonger).

[6] Robinson, W.S. (1967) *If I Remember Rightly. The Memoirs of W.S. Robinson 1876–1963.* Edited by Blainey, G.. Melbourne: Cheshire. p. 156.

[7] Keast, A.J. *op.cit.* p. 6.

[8] Robinson, W.S. *op.cit.* p. 159.

[9] Robinson, W.S. *op.cit.* p. 30.

[10] Robinson, W.S. *op.cit.* p. 161.

[11] Jones, D. (2011) "Re-greening the Hill: Albert Morris and the transformation of the Broken Hill landscape." *History of Gardens and Designed Landscapes* 31:3 pp. 181–195.

[12] Robinson, W.S. *op.cit.* p. 162.

[13] Jones, D. (2011) "Regreening the Hill" *art. cit.*

[14] Morris, A. (1975) *Plant Life of the West Darling*. Compiled by the Barrier Field Naturalist Club. Revised edition. p. 45.

[15] Morris, A. *op.cit.* pp. 49–50.

[16] Pidgeon, I. M. and Ashby, E. (1940) "Studies in Applied Ecology 1: A statistical analysis of regeneration following protection from grazing." *Proceedings of the Linnaean Society of New South Wales*. 65. pp. 123–143.

[17] Jones, D. (2011) "Green Planning Heritage/ Industrial Nature: Regeneration zones of Broken Hill." Paper presented to Track 7 (Planning History) at the 3rd World Planning Schools Congress. Perth, W.A. July. Mimeo.

[18] Mawby, M. (1975) Foreword to Albert Morris *op.cit.* "Tribute to Albert Morris at time of his death. Barrier Field Naturalist Club."

[19] Martin, S. (1993) *A New Land. European Perceptions of Australia 1788–1850.* Sydney: Allen and Unwin/ State Library of New South Wales.

[20] European scientists, plant lovers and naturalists were among those excited by the novel fauna and flora of the colonies: Mabberley, D.J. (2019) *Botanical Revelation. European encounters with Australian plants before Darwin.* Sydney: University of New South Wales.

[21] Jones, D. and Morley, B. (2002) "Ashby, Edwin" in Aitken, R. and Looker, M. (eds) *The Oxford Companion to Australian Gardens*. South Melbourne: Oxford University Press. pp. 33–34.

[22] Harding, M. (n.d.) "The Albert Morris Story." Barrier Field Naturalist Club Archive. Broken Hill City Library. Mimeo.

[23] Broken Hill City Library Archive: Miscellaneous references on Albert Morris.

[24] Keast, A.J. (1939) as reported in *Barrier Daily Truth* January 10.

[25] Webber, H. (1992) *The Greening of the Hill. Re-vegetation around Broken Hill in the 1930s*. Melbourne: Hyland House. ch.5 "The Spreading of the Green."

[26] Parbo, A. (1992) *Down Under. Mineral Heritage in Australasia.* Parkville, Victoria: The Australian Institute of Mining and Metallurgy. pp. 67 and 78.

[27] Webb, J. (1998) *Thistle Harris. A Biography of Thistle Yolette Stead.* Chipping Norton, Surrey: Beatty and Sons.

[28] Harris, T. Y. (1966) *Wild Flowers of Australia.* Sydney. Angus and Robertson.

[29] Blainey, G. (2001) *2001 Boyer Lectures: This Land is all Horizons – Australian Fears and Visions.* Sydney: ABC Books for Australian Broadcasting Commission.

[30] McCalman, I. (2010/2009) *Darwin's Armada. Four Voyagers to the Southern Oceans and Their Battle for the Theory of Evolution.* London: Pocket Books.

[31] Mabey, R. ed (1995) *The Oxford Book of Nature Writing.* Oxford. Oxford University Press.

[32] Fletcher, I. et al (1989) "A History of Alma School and District." Broken Hill City Library Archive. Mimeo.

[33] River Junction Research (2006) "Unincorporated Area of New South Wales heritage Study." Mimeo.

[34] Mattingley, Christobel (2010) *For the Love of Nature: E.E. Gostelow's Birds and Flowers.* Canberra. The National Library of Australia.

[35] Jones, David (2011) "Regreening The Hill" *art.cit.*, p. 181.

Chapter 4
Bison and Beyond – Artists Encounter Animals

The Middle Ages attributed to the Holy Ghost the composition of two books. The first was, as is well known, the Bible; the second, the whole world, whose creatures had locked up in them moral teachings. In order to explain these teachings, Physiologi, or Bestiaries, were compiled in which accounts of birds and beasts and fishes were laid over with allegorical applications.

(Fastitocalon by Jorge Luis Borges)[1]

4.1 Ancient Southern European Cave Art

It is a very long gaze back into the pre-historic past from the nature conservationists and land restorers in the twentieth century to the possibly 17,000 year old cave paintings of animals in the north of Spain and the south of France. Yet there are elements in each which point to a common thread of values, beliefs and understanding of the natural world. Does not our common humanity rest upon a grasp and an acceptance of our shared earthly existence, the mysterious natural ground of our being, over aeons of time?

4.1.1 Reading Ancient Cave Art

Since late in the nineteenth century, discoveries of magnificent animal paintings on the walls and ceilings of caves and rock faces in Europe, Africa, America and Australia have changed our understanding of early human mental capabilities while continuing to raise questions about sources of our beliefs and values. At the same time, these revelations invite reflection on what meanings and attitudes the galleries of ancient animal art suggest we might well carry into the present day. Moreover, they serve as a sharp reminder that many of the species portrayed no longer exist or, if they do, only in the museum condition of zoos and small reserves.

Following the pioneering work late in the nineteenth century of French palaeontologist, Emile Carthailhac, and pre-historian abbé, Henri Breuil, in recent decades

M. Skilbeck, *Loving and Studying Nature*, International Explorations in Outdoor and Environmental Education 10, https://doi.org/10.1007/978-3-030-80751-1_4

there has been a considerable amount of scholarly and popular writing on the meaning and significance of the cave art of South-western Europe and beyond. Some of it has been highly speculative and wide-ranging, such as Georges Bataille's essays on people's relations with the animal world. He has used the paintings in the Lascaux cave as a vehicle for views about ancient concepts of death and attribution of semi-divinity to animals. South African anthropologist and ethnologist, David Lewis-Williams, in reviewing different theoretical perspectives, proffered a shamanistic explanation. He argued for visionary states and altered consciousness in the artists and their choice to locate the paintings in dark, inaccessible chambers below the earth's surface. American journalist and author, Gregory Curtis, himself excited by the richness of interpretation that the cave art has inspired, wished to bring this "interpretative landscape" to a wider public audience. Other scholars focus less on imputed meaning, more on techniques of execution: engraved lines, painted outlines, dots and infill painting in colours, the use of flat and rounded surfaces, protruberance, rock cracks and so on.[2]

These and other speculative, scholarly and popular works, including films, are important and valuable in maintaining a vital interest in the nature of humanity itself. They testify to continuing and fresh interest in the lives of our ancient forebears, in the origins and significance of artistic expression and in the material and physical life conditions of human groups – the world of ancient man. In this chapter, however, the purpose is somewhat different. It is to offer a personal reading, in the context of my overall study of the ways in which love and understanding of nature have been variously expressed at different periods of time and in different social and cultural settings. The cave paintings represent one such episode from which there is something of value to learn for life now.

The chapter thus has a threefold aim: first to recapture some of the excitement and mystery associated with the late nineteenth century discoveries of ancient cave art in South-western Europe; second, to use one of the splendid animal forms depicted in this art, the bison, as a metaphor of habitat loss and species destruction; and finally, to suggest that the genre of animal art and representation has continued in historical times as an enduring way to enjoy, respect and celebrate the lives of animals.

4.1.2 Discovery – Recovery

The bison cave art story began when a father and daughter accustomed themselves to the dim light in the entrance to Altamira cave, located a few kilometres from the north coast of Spain. Let us try to visualise a scene of excitement and amazed wonder: "Look, Papa – bulls," exclaims eight year old Maria, pointing up to what looked to her like animal forms scattered across a wall of rock just above their heads some thirty metres from the cave entrance. Her father, a wealthy local landowner, has been peering into the shadowy depths of the cave they have just entered. He spins around. "Where?" he demands, craning his neck and holding a lamp above his head to where Maria is excitedly pointing. In the feeble light he makes out a great herd of animals – bison – outlined in the darkest black, their muscular bodies tinted in glowing shades of red, orange and violet. The brilliant painted scene pulses with life and raw energy, even those animals that seem to be in repose (Figs. 4.1 and 4.2).

Father and daughter had clambered to this cave on a limestone ridge, coming from the small town of Santander in northern Spain, some thirty kilometres to the east. Nearby was the even smaller township of Santillana-del-Mar. The land was dry and hot under the summer sun, the cave a cool, shady retreat.

The year was 1879 just a decade after the cave had been stumbled into by a local share-farmer. As far as anyone knew, it was only another cave, containing signs of ancient occupation but of little interest except to adventurous boys and those who enjoyed poking about in fissures in the limestone, collecting chunks of flint, rocks or wildflowers. There were, however, some who were intrigued by the relatively new and baffling scientific subject of palaeontology. One of these, Maria's father, by the name of Marcelino Sanz de Sautuola, enjoyed the hobby of searching for insects, minerals and fossils. In the tradition of the plant-loving nineteenth century field naturalists and their patrons who filled their glass-fronted cabinets with assorted

Fig. 4.1 Maria Sanz de Sautuola c.1877, the girl who discovered the cave paintings of Altamira. Santander Prehistory Museum, Public domain, via Wikimedia Commons

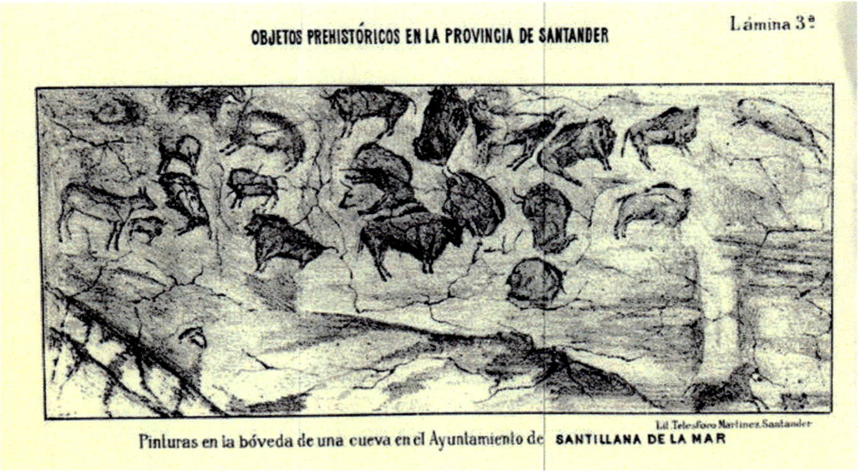

Fig. 4.2 Marcelino Sanz de Sautuola's black and white drawing of Altamira's polychrome ceiling in his 1880 publication. Marcelino Sanz de Sautuola. Fotografía por el Museo de Altamira, CC BY-SA 3.0 via Wikimedia Commons

specimens – animal, vegetable and mineral – he is credited with being the first to cultivate on the Iberian Peninsula species of the now ubiquitous Australian eucalyptus tree.

First attracted to the cave in 1875, Marcelino had found a scattering of small bones and flint implements. In the hope of finding more, while introducing Maria to the delights of cave exploration, he had returned on that summer's day in 1879. Thrilled by the paintings on the ceiling and walls, shortly afterwards they came again, this time accompanied by Vilanova y Piera, a professor at the University of Madrid.

Convinced that the cave was a site of great importance, Vilanova in 1880 reported the finds, which included bones engraved with animal figures, to an international congress in Lisbon. None of the delegates, however, accepted his invitation to visit. In the same year, Marcelino produced an illustrated booklet modestly titled *Breves apuntes sobre algunos objetes prehistoricos de la provincia de Santander (Brief notes on some prehistoric objects in the Santander province).*

Although, elsewhere, cave art was being uncovered by French and English archaeologists – many of them amateurs – no rock paintings of anything like such grandeur had come to light. At first, and for several years, response by experts to the Altamira cave was disappointing. The animal paintings were derided, as either the scribblings of Roman soldiers fighting in the Cantabrian war late in the first century BCE or the work of frauds. This latter assertion was not at all implausible as in the early years of palaeontology several notorious frauds had deceived the experts. Even into the twentieth century an assemblage of human bones found in Sussex, known as "Piltdown Man" and purporting to come from as much as a million years ago, was shown to have been tampered with and to be of much more recent origin.

Scientific frauds continue to the present day so it is understandable that eminent palaeontologists joined the chorus challenging the authenticity of the Altamira cave paintings. Surely art of this quality, if genuine, was from a much higher order of civilisation than the ancient domain of "primitive" cave dwellers? However, in the Dordogne region of France, where evidence of Palaeolithic cave paintings was also being uncovered around this time, small carved objects and splendid rock paintings, including bison, were found at Combarelles in Les Eyzies and at Font-de-Gaume nearby. These findings were instrumental in swaying opinion in favour of the Altamira bison paintings.

Together with young prehistorian, abbé Henri Breuil, the distinguished, elderly palaeontologist, Emile Carthailhac, visited the Spanish cave in 1902. Breuil made beautiful copies of the paintings which in 1908 were published by the Institut de Paléontologue Humain (with a second edition in English and Spanish published in 1935 by the Real Academia de la Historia). Meanwhile, Carthailhac, with suitable humility, had published an article in 1902, "Me culpa d'un scéptique", in which he retracted his earlier dismissal of the Altamira paintings.

Just when the caves of Southern Europe were painted has been the subject of warm debate and scientific analysis. It seems that they date from the time of the end of the last great Ice Age, somewhere between thirty or more and ten thousand years ago. The master-works at Altamira have been dated to some 14,500 years ago. Various motives for such works, often in the darkest depth of the caves, have been attributed to the artists and their kinsfolk; they are discussed below.

Chance favoured several of the great rock art discoveries in Southern Europe. Just as the Altamira paintings had been brought to light by the circumstance of keen-sighted young Maria accompanying her father, the famous Lascaux cave network in South-west France was uncovered, in 1940, by a group of youths and a dog. Following a severe storm, a gaping sink-hole which emerged from an uprooted oak tree attracted the dog. Stories differ – the dog fell in and the boys went to its rescue, or one of the boys peered down and, observing a chamber, resolved to come back and explore it. Either way, with three friends he did return, entered through the sink hole and uncovered one of the greatest of all pre-historic rock art galleries.

The Lascaux cave was soon closed to the public due to the harmful effects of the exhalations of visitors, humidity and fungal growth. Today, a reproduction in concrete is open, and many superbly illustrated books provide glimpses of the numerous ancient masterworks spread through the rock chambers. Another set of rock paintings, discovered only in 1994 in the Chauvet-Pont-d'Arc cave in the Ardeche region of southern France, was the subject of a film in 2011 – Werner Herzog's *Cave of Forgotten Dreams*. This film offers a rich vicarious experience of ancient cave art on the walls and ceilings of chambers far from the cave entrance and in eerie total darkness save for flashes of light from the film maker's apparatus. Lit by burning flames, this is perhaps how the cave may have appeared to the painters assembled to their work. Dating from some thirty six thousand years ago, Chauvet is a treasure trove of animal art – approximately four hundred and forty representations of bison, buffalo, rhino, horse, ibex, auroch, and a remarkable figure of a bison-woman.

While Lascaux, Chauvet-Pont-d'Arc and other French caves may be better known to the public, given the publicity they have received and the great volume of tourists that every summer floods into the south of France, the bison of Altamira have a unique appeal as the climax of a remarkable expression of pre-historic artistic genius; "one of the grandest expressions of cave art brought to its peak."[3] They not only serve today to symbolise some of the earliest efforts by humans to create art as a way to express what may be taken to be love and understanding of nature but also may be seen as metaphors for the ways humans and animals have interacted, to the marked detriment of the latter. This is not to suggest that there are not other species whose decline and loss is any less shocking. Nor is the south European rock art the oldest of its kind. That of the Australian Aborigines in Arnhem Land, thought to date from possibly sixty thousand years ago, may well be older. Other artistic achievements of comparable interest include those of the Indigenous Americans, the Tassili frescoes of the Sahara and the San rock paintings in Southern Africa.

Thus the animals of the European caves by no means stand alone. Moreover, not all of the world's subterranean caves and rock overhangs have been explored and documented. Fresh discoveries continued to be made throughout the twentieth century; there are likely to be further revelations in the future. These other works of ancient artists notwithstanding and whatever discoveries may lie ahead, the character of the Altamira bison paintings, even in reproduction, has a suggestive magnificence which gives them claim to be among the greatest ever of all animal paintings. This is not to say that controversies over their purpose and meaning have been laid to rest, as we shall see in the following section.

4.1.3 Interpretations

Geoffrey Grigson, English poet, essayist, critic and naturalist, made an extensive tour of the south of France caves before their closure. He wrote most insightfully of the cave animals that they, "dictate, insist upon themselves", in an art that emphasises and combines, "elemental traits of life".[4] Another English poet, the influential art theorist and critic, Herbert Read, had in the 1930s defended the thesis that ancient cave art was not just a form of magic, and was something more than "photographic realism". As creative images in the mind of the artist "the paintings reveal a scale of values, and these values are not those of verisimilitude, but of vitality, vividness and emotive power, which are precisely aesthetic qualities."[5] The paintings themselves, Read argued, showed in their organic sensuousness that the artist enjoyed creative activity and pursued it for its own sake.

Reflection on the fate of the great bison family as it was portrayed in caves in Europe, during the Stone Age – and as it had become by the time when the paintings were discovered late in the nineteenth century – is both poignant and deeply troubling for modern nature consciousness. These animal paintings are significant not only for palaeontologists and historians of ancient art. They are symbolic of the continuing and fateful contest for survival between humans and other species.

But let us consider, first, the paintings as works of art. Just what is it about the Altamira cave bison that warrants such claims as, "the Sistine chapel of Quaternary art"[6]? Likewise, what does gazing at artistic representations do for our understanding and appreciation of the life of animals that is not done more comprehensively, effectively and, as many believe, with so much greater verisimilitude by the scientific accounts of modern zoologists and the dazzling achievements of modern photography of living species?

In addressing these questions, we of course must acknowledge the immense achievements of the scientific study and photography of animals, and that, in the domain of painting representational or figurative, art has to a considerable extent been overtaken by abstract and other non-representational forms. Are ancient rock paintings of long-since departed animal herds of any more than fusty antiquarian interest? As we shall see below, they are far from that. At any rate, it was the ancient big game hunters or artistic kinsfolk living in their midst who, in their animal portrayals, first created Western naturalistic art, a long line extending to the present day and a powerful invocation of the fragility and beauty of nature.

What became known as "the Gallery of the Bisons" at Altamira (now called the "Polychrome Ceiling") stretches over a floor space of some 18 by 9 metres, with a height from the floor of 1.15 to 2.65 metres, much the size of the living-dining room in which I am now writing, but with a considerably lower ceiling. Close to the cave entrance, the paintings were long protected from damage and deterioration by an ancient rock fall. On the low ceiling and walls are some 150 figures. In addition to bison, there are wild boar, hyena, a hind, together with human hand-prints and their outlinings. Delineated in black charcoal, the bodies of the animals glow in rich ochre tonings. The abbé Breuil, who would have studied them closely for the drawings he made, gave the name "conceptual realism" to the style commonly used by the cave painters. It was, he said, the species, not the single animal that was represented (Fig. 4.3).

Today we can visualise these animals as herds which long ago grazed the generally stunted vegetation of northern Spain (and southern France) at a time when the last of the great glacial inundations was retreating northwards. The time might be thirty, or more likely twenty or even ten thousand years ago. Into the north of Europe, as recently as 20,000 years ago, stretched great ice-fields and continuous blankets of snow. Permafrost extended for thousands of kilometres across central and deep into the mountainous regions of southern Europe. Herds of horse, deer, auroch, mammoth, lion, bear, hippopotamus and rhinoceros had yet to retreat, some to Africa, others to Northern Europe and others again to become extinct. Temperatures were low, winters freezing. Icy winds swept across the plains and through upland valleys. Tundra vegetation prevailed in regions of perma-frost and bog with mosses, sedges, scaly lichen, and small tough-leaf shrubs, perhaps including conifers and dwarf willow. In warmer regions, plants flowered in damp or marshy soil above the permafrost.

Fig. 4.3 Full side view of standing bison on polychrome ceiling, Altamira. This Magdalenian era image is of the extinct steppe bison (*Bison priscus*). Museo de Altamira y D. Rodríguez, CC BY-SA 3.0 via Wikimedia Commons

This vegetation pattern suited many of the herbivores, such as grazing bison and horse. Some, like deer and bear, were able to root beneath the snow for mosses or scrape lichen off the rocks, others were carnivorous. In the far south of the continent, grasslands dominated, as they do today in parts of Siberia south of the tundra zones. In the cave paintings there are no images of this vegetation or of the surrounding landscape. Some commentators have pointed to their absence in justifying claims of magic or ritual intent in the art. They have seen the paintings as integral to the hunt, a kind of imitative magical invocation, or perhaps celebration of great success. Lion and wolf were among the species that hunted other animals in the tundra, but humans were the most cunning and effective killers – their prey the large herds of horse, deer, bison and woolly mammoth which most readily yielded the flesh and skins humans needed for survival and comfort.

Some of the animals are shown on the European rock faces, with spears thrust into their bodies, pursued by spear-throwing hunters or in evident flight. Curiously, the hunters are stick-like, often crudely painted as in North American rock art and sometimes so stylised as to seem clumsy, motionless even – as they often are in African rock paintings and engravings. By contrast, the animals, and especially horses and bison, are painted in full, rounded forms. They are vigorous, possessed of a sinuous vitality while the hunters, in their more geometric and stick forms, express less in their whole bodies and more in their spear-thrusting arms. The horses, necks outstretched – and often only necks and heads or even heads alone – are racing. Just the heads are enough to suggest motion, energy, vitality – a kind of exuberance of spirit. Or is it, perhaps, the fear and desperation of the pursuit? Or are they, as some anthropologists have argued, images of spiritual forces, figures of dreams and visions? We return to this interpretation below.

The painted Altamira bison in its full, rounded muscularity undoubtedly expresses animal agility, power and beauty of form, whatever magical or mystical meanings may also be read into the images. The rich red, orange and violet tonings of the earth pigments, the black outlines of the body, the absence of "landscape" – of plants, stones, water, ice, snow, mountain slope – create a sense that these animals utterly dominate, delivering no space to their surroundings. Only they matter, and even when pursued and brought down by hunters, or resting and sleeping, they are still the dominant force in the virtual space created on the rock by the artists.

Great technical skill is shown in the artistic renderings: protruberances of the rock face and hollows are used to create depth and perspective; movements are often hinted, or suggested; and colouring, with the patina of time, can seem like the rich and unexpected tones of a potter's glaze. Shadows, cast by flares or in those places where daylight seeps in, serve to bring out three-dimensional forms, like hatchings in an etching. The artists have rightly been called master craftsmen and it is reasonable to suppose that when young they had been tutored or schooled. They could not have come to their projected assignment unprepared for a task which, it must be remembered, when not in the dim light of a foyer was mostly accomplished in the total darkness of the cave, often far from the entrance and lit by fires, lamps or torches which cast flickering light on the uneven rock faces.

Due very largely to the impact of studies by the nineteenth century pioneers of modern anthropology and by the inquiries and reports of antiquarians, historians and travellers investigating so-called "primitive" or "savage" societies, the earliest efforts to provide a rational explanation for the mystery of the cave paintings were focused on the magic and ritual mentioned above. The caves might thus be seen not so much as *galleries* but *temples* of art, with controlled access, perhaps governed by taboo. Most obviously, it has been suggested that a kind of sympathetic or imitative magic was in play, the painter either identifying with the hunt and imagining success in locating and killing the animals, or effectively bringing them down as a witch doctor might by casting spells. In the spirit of the Scottish social anthropologist, James Frazer's *The Golden Bough* (1890), popular at the time, the paintings could seem to early commentators as a way of multiplying the herds – an invocation which, incidentally, they came to be much in need of in later historical times. Multiplication would mean life-giving food, horns, bone and skins for the clan, and survival for the species.

Philosopher, Martin Heidegger, treats the concept of *techne*, often translated "technique", as knowledge of a kind that discloses or makes real "being." Hence art conceived as skill or technique – which the bison paintings certainly display – is a "work not primarily because it is wrought, made, but because it brings about being in an *essent*; it brings about a phenomenon in which the emerging power, *physis*, comes to shine – confirming, making accessible, explicable and understandable something as being or not being – stabilizing, manifesting being – realizing, opening, keeping open – which is knowledge." In this interpretation, the work of art both confirms the being, existence of the animal and provides accessible knowledge to artist and viewer.[7]

Magic, the attempt to multiply numbers, to bring or celebrate success in the hunt was, for early researchers, an obvious painterly motivation. It was thought that rituals may have taken place, the caves decorated as special places for the enactment of some kind of ceremony, on the analogy of mediaeval and nineteenth century gothic churches. Ritualistic or ceremonial claims such as these are speculative and no records, in the form of suggestive objects such as scratched or incised bones bearing ancient messages, to this effect have been found. Nor is there any oral tradition to substantiate these interpretations. Nevertheless, bold efforts in recent years have been made to strengthen claims of magic and ritual.

Cave art has been intensively studied and theorised especially since the middle of the twentieth century, with some intriguing conclusions. In *The Mind in the Cave*, South African anthropologist and African rock art authority, David Lewis-Williams, for example, has produced a closely-argued explanation of the origin and significance of cave art in South-western Europe that challenges commonly held views about its meaning and purpose. On a lengthy scholarly journey, he has aimed to demonstrate that, for the Upper Palaeolithic occupants of this region, the figurative and geometric subjects, their patterning and location in different parts of the caves, constitute a rich, complex *code* of spiritual messages and beliefs about the nature of the universe and life within it. So far from this being the work of a "primitive" people, the art signifies a proto-religious belief system, a spiritual "world view" or cosmology. Lewis-Williams has argued that cave painters (as distinct, he claimed, from the Neanderthals alongside whom they lived for millennia) were, in evolutionary terms, "modern" people. Their brain structures were identical with ours today and although hunter-gatherers, lacking written language and with a stone-age technology, they were "fully human". Could we speak to them, they would be recognisably human to time travellers from today's science-saturated culture. Their faith in a spiritual world, their superstitious beliefs about the natural environment and its inhabitants, and how to relate to it, differed only in degree, not in kind, from much of what is widespread in life today.

Lewis-Williams has sharply challenged the views of the abbé Breuil and other early scholars and researchers who treated the cave paintings, the scratchings and other marks as expressions of an innate aesthetic sensibility and their imagery and location as evidence of simple hunting-magic rituals. Such explanations are for him too vague and inadequate to account for such aspects as the complex patterns of the animal groupings and the location of many of them in dark, remote and inaccessible reaches of the caves.

Marshalling evidence and arguments for his appraisal of the cave artists and their fellows from a wide and imposing array of twentieth century research and theorising in the social and physical sciences, Lewis-Williams summons even poems by Coleridge and Wordsworth as supporting witnesses (dreams, visions and other altered states of consciousness). As well, he adduces anthropological field studies of African and North American rock art in addition to modern European studies, in support of his case. Evolutionary psychology and neurobiological research are called upon to testify to the identity of human brain functioning during the Upper Palaeolithic era with that of people today. From Marxist social theory, he draws a line to demonstrate unequal power relations and the material substructure of ideas in ancient culture. Linguistic and structuralist concepts, such as bi-polar opposites, dichotomies and the shaping and patterning of expression in communication systems are woven into his account to explain features in the figures and designs, as he builds toward an impressive, inclusive explanation of the mysteries and puzzles of the cave art.

That explanation, Lewis-Williams finds in the practices and beliefs of the spiritual system of Shamanism. Essentially, his argument is that the art work, integral to Upper-Palaeolithic culture in South-western Europe, is a system of meaning and communication which illuminates or affirms a rich and pervasive belief system, a cosmological order of spiritual beings (e.g. monsters and animal-friends), ways to manipulate natural forces (weather, illness), food supply etc., and the pattern of hierarchical social relations within communities. Shamans, in a state of exalted consciousness, in dreams and through hallucinations, access an alternative reality, a spiritual world. They manoeuvre their way through its hazards and dangers and bring back to their communities enhanced powers and beneficial accomplishments like healing. The trances in which the dreams and hallucinations occur may be experienced through bodily movements and vocalisation. They are remembered and find visual and material expression in the paintings, carvings, line drawings, sculptures and small art-craft objects in the caves.

The idea that the superbly naturalistic animal images were the product of remembered or imported visions and imitation by shamans rather than direct observation in the field merits further analysis. Lewis-Williams argues that the caves are to be understood not as *galleries* of art conceived as a distinct, stand-alone domain of aesthetic sensibility, but as *temples* of heightened consciousness, alternate realities and ecstatic or terrifying experience – a vital element in the spiritual well-being of the community. The bison, horse, auroch, deer, feline and other animal figures patterned in the caves, are not to be taken as literal representations of living creatures but as the medium of a visionary apprehension of a spirit world, the manifestation of a cosmic order made visible through the agency of the shamans and those to whom spirits communicate their messages from another world. The shamans themselves may paint or their followers may, but to do so they must have clear memories and a firm hand, since painting and carving of such fineness seem not to be possible in the intense ecstasy of a trance. Paintings might also be in the form of copies, just as the abbé Breuil copied the bison rock paintings when visiting Altamira and artists today copy the cave painting on the concrete walls of neo-caves for modern tourists to enjoy.

What the cave art expresses, according to Lewis-Williams, is thus a visionary apprehension of a spirit world, by those with special access, experience and communicative capabilities. To the people of the time, art was, he argues, integral to a cultural process of extending, reinforcing and sharing belief systems, a way of understanding and acting in the everyday world. The art had a purpose of fundamental importance and value to an intelligent but superstitious people searching for meaning and developing a distinctive cosmology or world-view. Interestingly, this interpretation has resonances with nineteenth century European ideas about the uniqueness and cultural value of visionary art, and the altered state of consciousness of the inspired artists.

Lewis-Williams has produced a persuasively argued case. Whether it becomes the standard or received scholarly explanation for our time is a matter for continuing debate, research and theorising. However, whatever may be concluded about the purposes and functions of the art in the minds of ancient people, the work has other meanings and values as well, not least for contemporary materialist viewers sceptical of "spiritual experience". Because he did not accept as adequate the "art for art's sake" hypothesis of such early judges as abbé Breuil, Lewis-Williams almost totally ignores the very features they drew attention to, and which are so striking to art lovers today. He says little about the supremely elegant and sinuous outlines of the animals, the visual impact and artistic effect of the rich colouring, the patinas, the beauty of individual movements and groupings. He makes no attempt to see the work as contemporary art lovers might – those who do indeed love art "for art's sake". Picasso, himself a superb draughtsman and painter, is said to have remarked that he and other contemporary artists could not match the artistry of the cave painters – their *artistry*.

In other words, fascinating and closely reasoned as the shamanist interpretation is, it still remains open to the generations since the modern recovery of the cave paintings and objects to see them as *masterpieces of art* not only as archaeological documentation of the development of human consciousness. Whatever may have been their meaning for their upper Palaeolithic creators and spectators and for modern anthropologists and ethnographers, the cave paintings are of unique importance and value as supreme *works of nature art*. Their study provides rich insights into what is meant by the love of nature.

Thus, as well as treating the cave art as cues to ancient magical beliefs and ritual practices, to spirit worlds and an ancient cosmology, we can engage with them as expressing awe, admiration, love, fear and understanding of phenomena of the natural world – their beauty, force, majesty and terror. Moreover, these sentiments need not be thought of as unique to the artists themselves, since they might be shared by family and others associated with the artists.

If the paintings served some social purpose – magical, shamanistic, artistic or otherwise – it would not be unreasonable to suppose that there was a division of labour, painters fed (and perhaps trained) in return for their service to family and clan. It would be expecting rather much that the community supported the artists to paint for the love of art, or of animals for that matter, independently of an important return from the material world or the domain of the spirits. Some social or shared purpose is indeed indicated, by the existence of paintings with many common features, in different and scattered locations across Southern Europe. Evidently there is a community purpose or value in the art, about which, however, we can now only speculate.

Just as the anatomist dissects the body in order better to understand the skeletal structure, sinews, muscles and vital organs, and thereby gain not only knowledge but appreciation, so in painting the bison and other animals the painter studies, learns about and develops a profound appreciation of them. The paintings not only inspire our awe and admiration for the skill, knowledge and talent of the artists; they can be seen as giving artistic expression to profound feelings for the beauty and majesty of the animals themselves. In doing so they can tell us something important about the artist as creator, the people then living and ourselves as viewers and appreciators. At the same time, they invite respect for the co-operating puny hunters, with their modest weapons, overcoming the majestic beasts.

Of the ancient art of the Pyrenees and Perigord, Roman Catholic priest, Pierre Teilhard de Chardin, wrote:

> We could make mistakes in interpreting in modern terms the prints of hands, the bewitched bison, and the fertility symbols which give expression to the preoccupation and religion of an Aurignacian or Magdalenian man. Where we could not be mistaken is in perceiving in the artists of those distant ages a power of observation, a love of fantasy, and a joy in creation (manifest as much in the perfection of movement and outline as in the spontaneous play of chiselled ornament) – these flowers of a consciousness not merely reflecting upon itself, but rejoicing in so doing.[8]

He continued that it is indeed and in the fullest sense present day humans at whom we are looking. (Fig. 4.4.)

In his essay, *The White Bird*, English artist, novelist and Marxist critic, John Berger, finds beauty in nature conceived as "energy and struggle".[9] He might have been (but was not) writing of the interaction between the spear-throwing, stick-like human hunters and the arching back, the sheer muscular force of the bison, and perhaps, as Lewis-Williams thinks, of the male-female and other opposites in the painted figures. Berger argues that alike for the artist and the observer of art, there is recognition and affirmation (and for the living, subject animal too?): "What has been seen is recognised and affirmed and, at the same time, the seer is affirmed by what he sees. For a brief moment one finds oneself … For an instant, the energy of one's perception becomes inseparable from the energy of the creation." In these elliptic remarks, Berger is searching for a way to show how through animal art (in *The White Bird*, wood sculptures of birds) artist, art work and viewer become as one. We might add, the living subject animal as well, as recognition and affirmation precipitate self-awareness and self-knowledge. This insight provides a cue for "reading" the bison paintings. We do not need to know who the artist was or what he or she intended.

To those ready to scrutinise and to enter into the spirit of what American philosopher, Susan Langer, termed the emotional space expressed by a painting[10] there are depths of meaning which enhance our awareness of ourselves, the art and the life of animals. Langer also found in what she called "savage art" insight, a discerning, active eye and the ability to articulate visual forms of significance. "His (the 'savage' artist) model is the human (or animal?) body … But as he imitates such models for practical ends, he sees more than the practical import of their shapes; he literally sees the reflection of human feeling, the 'dynamic' laws of life, power, and rhythm,

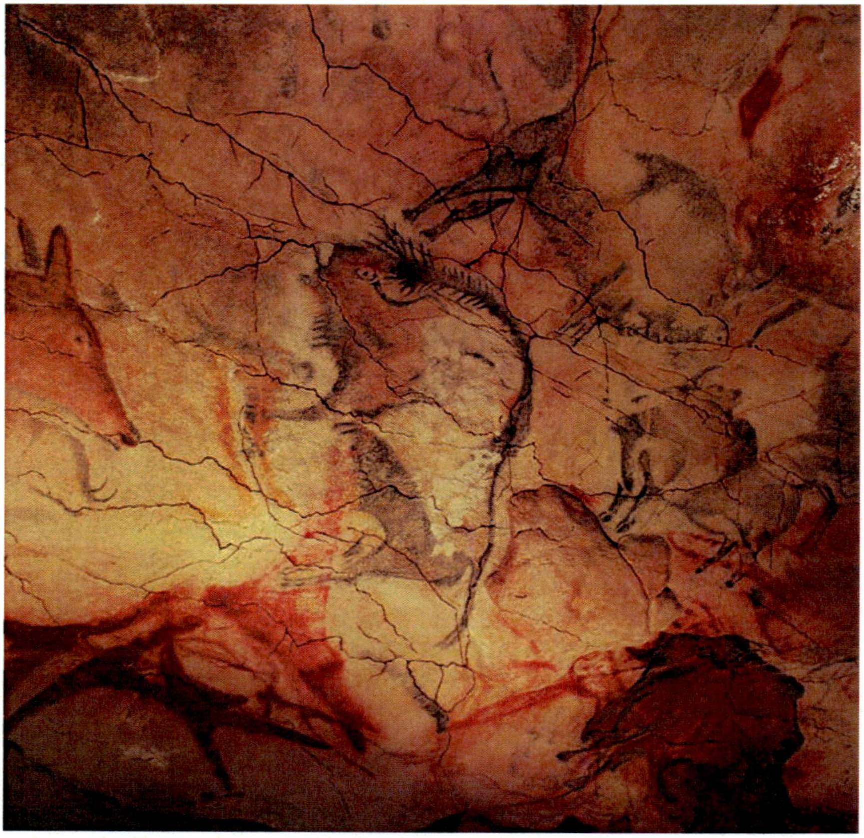

Fig. 4.4 Standing bison with other animal images on Altamira's polychrome ceiling. Yvon Fruneau, CC BY-SA 3.0 IGO, via Wikimedia Commons

in forms on which his attention is focused; he sees things he cannot name, magical imports, rightness of line and mass, his hands unwittingly express and even overdraw what he sees, and the product amazes and delights him and looks 'beautiful'. But he does not 'know' in discursive terms, what he is expressing, or why he deviates from the model to make the form more 'significant'. When he emerges from his savage state and takes discursive reasoning seriously he tries to copy more accurately, and the ambition for naturalistic, literal representation, for rational standards of art, moral interpretations, and so forth, confuse his intuition and endanger his visual apprehensions".[11]

Certainly formal, imitative and visually bland art succeeded that of the cave painters in the ancient world. Although in this ancient art there is clearly an imitative element, its effect is to enhance the "dynamic laws of life" and to give the most powerful expression to what Langer calls significant form.

Appreciating the immediacy, the directness and the intensity of the relationships that Berger and Langer have in mind depends on a willingness to enter fully into the emotional space of the cave paintings, not only transcending the physicality of the cave setting but also the cultural domains of palaeontologist, anthropologist, ethnographer and art historian. While they all provide vital cues, there is another realm of meaning to explore. "Bison", as an artistic creation occupying a significant part of the emotional space of ancient humans, can also provide a telling metaphor for modern humanity's relationship to fellow creatures. That may be conceived through images of superstition, magic and hunting, one "skin" as it were of the paintings in the dark fastness of their caves, or it can be embraced as a celebration of the beauty of nature and a determination to live in harmony and balance with the natural world. This is not the tension, the dialectic that Lewis-Williams identifies in the cave, but the reconciliation of potentially warring or oppositional forces.

We can imagine that the cave painters had many motives, as any painter might. In order to produce such vivid and lifelike representations they must have studied the forms and movements of animals closely, in depth and in many settings. Like George Stubbs, the great animal painter in eighteenth century England, they may even have made a close study of animal skeletons. Trances may be spiritually energising, but they do not give insight into bodily structure or control over line and composition. Capturing the colours and texture of the tufted hair and beard of the bison is, to a considerable degree, a function of the pigments that were available, the means of applying them and the nature of the rock face. But how the painter designed and disposed of pigments over the animals' bodies tells of how he or she saw the animal and its movements in the light of day and of feelings. Through the pleasure, delight, and sense of mastery as he or she transformed inert material into vivid, living form and on an unforgiving rock face, with fingers, hands, sticks and twigs, the painter seems to feel, to be, absolutely at one with the animal, even if his or her mind is simultaneously aware of supernatural forces and imagining an alternative reality (engagement of hand eye, paint, implement and rockface is, too, "alternative" to mundane life). Whether the purpose, at some other time, is to kill and eat, or to engage with spirits, now in the act of painting, the painter loves the animal form and ever so clearly expresses not only a deep feeling for, but a profound understanding of, a kindred form of life.

Although the abbé Breuil, in introducing the term *conceptual realism* as against "sensory realism", appears to have meant by it something like the "more general physicality of the animal" or "the animal as such" not an individual specimen or collection of individuals, the term might equally apply to the artists' expressive concept of animal power and beauty, which we can dwell upon, as we conceive it. But because we do not certainly know what motivated the artists and can have the merest glimmering of how they lived and what they believed and thought, some construction is necessary. Inevitably, this leads toward personal, subjective impressions. We can approach the cave paintings as a profound expression of feeling for the animal world and appreciation of its majestic and beautiful creatures.

We might compare this way of interpreting the cave paintings with that commonly followed by critics and historians commenting on historical as distinct from pre-historical periods of art. Swedish philosopher and artist, Göran Hermerén, has referred to the *communication situation* in analysing ways that art historians and critics have found meaning in paintings.[12] Essentially, this entails "placing" the art by searching for the motives and interests of artists whether explicit or hidden, examining cultural, social and psychological contexts, determining historical movements, styles and reactions to them, finding links and defining influences. For the cave artists, we lack most of these vital clues to the meaning of their work. These canons of art history and criticism seem unavailable to us, although the shamanistic interpretation of "the mind in the cave" and earlier interpretations can be extrapolated to inform aesthetic judgments. Our artistic interpretations, today, are inevitably subjective.

A former Slade Professor of Art at Oxford, Eric Newton, felt unable to recognize the cave painters as "artists"; they were "magicians", he said (as if it were not possible to be both). Paradoxically, a beautiful reproduction of an Altamira bison is the first of the 178 illustrations of what Newton has selected as his masterpieces of art.[13] Other commentators have reacted differently to the paintings, as when their acceptance as authentic in Cathaillac's *Mea Culpa* was likened to the explosive impact of the Fauves group of painters at the Paris Salon d'Automne in 1905[14]; and, as noted above, Picasso deeply admired them.

While we have bold and scholarly interpretations of what the cave paintings may have meant to their ancient creators and the communities in which they lived, as subjects of aesthetic interest, as works of art, we can approach the cave paintings in many richly rewarding personal ways. And we can do so without being bothered by our meagre knowledge of their origins and the enticement of anthropological explanations, as well as our inability, now, to see them other than in reproduction. Ironically, this latter point corresponds to our inability to see the bison themselves in the wild, but only as cosseted, fragmentary remnants in game reserves and zoos. Yet, by study and reflection and by whatever efforts we might make in support of their protection and conservation, the life of endangered animal species need not be extinguished. This hopeful call is now coming loud and clear from many sources. To see and appreciate the cave paintings as objects of beauty and arguably of deep love, expressed in artistic terms, is an invitation to value the life, now as well as then, of the creatures they depict. However, this is not without challenges as we contemplate subsequent events in the story of the bison, not only in Europe.

4.2 The Bison in History and Later Art

4.2.1 The Fate of the European Bison

Over the millennia, the massive glaciers, the vast ice sheets and blankets of perpetual snow and frost of the last great Ice Age gradually retreated northwards. By eight to ten thousand years ago, the warming climate had resulted in the replacement of the central and Southern European tundra, ice and snow by marshland, then huge forests and, towards the south of the continent, extensive grasslands and scattered forests. Animal life changed correspondingly. It seems that the great herds of bison headed to the north and northeast. The end of this Ice Age roughly correlated with the beginning of human agricultural societies and permanent settlements, the Neolithic revolution and what we refer to, if at times mockingly, as the beginnings of civilisation.

Revisiting the homeland of his forebears, English historian, Simon Schama, related that, in historic times, the forests of Lithuania and Poland were still home to huge herds of bison.[15] As wild animals, they were to be decimated and ultimately destroyed by their human competitors. In shocking detail, Schama recorded the destruction wrought over many centuries in and around the northern habitat of the bison – the great primeval forests of Bialowieza in his ancestral Lithuania. As the forests were violated, so the once huge herds of the European bison declined. Visiting the forest in the early 1990s, Schama recorded that, in total, fewer than 500 animals remained: 240 in the forest, 200 in a state park in the south of the country and a pair here and there in zoos. In fact, by the 1920s, bison were extinct in the wild and their slight recovery was the result of a scientific breeding program started in the 1930s, drawing breeding stock from zoos.

What or who caused the destruction? Climate may have played a part as has for example been claimed in the controversy over the extinction of Australian megafauna (Chap. 16). But for the European (and American) bison it was that great destroyer of other species – humans, our selfish ways and destructive tendencies, that almost certainly decimated the bison herds. Schama wrote that the slender stocks of just a few hundred animals surviving in the wild at the outset of the First World War were butchered by starving German conscripts. For many centuries previously, the combination of royal and aristocratic hunters and hungry woodsmen had depleted stocks and destroyed habitat with only occasional efforts, mainly during the Enlightenment period, to introduce sound resource management regimes. But it was also during that period that "management" of another kind was adopted. It followed in the footsteps of Western European and British land clearing, extended scientific agriculture, industrialisation and the ruthless exploitation of forest habitats for timber.

At the same time, and by contrast, the bison was often treated by poets and patriots as a symbol of national glory, its forest habitat an emblem of pride in the native landscape. In the sixteenth century, poet and hunter, Nicolas Hussovianas, had inscribed to Pope Leo X Polish-Lithuania's first and only full length ode to the bison, "Its eyes shining with a fearful red rage". Both in the ode and in many accounts in following centuries the bison, in Schama's words, "was depicted as a miraculous relic of a presocial, even prehistoric past – a tribal, arboreal world of hunters and gatherers, at the same time frightening and admirable. A talisman of survival". Ironically, it was not the bison but its hunter that was to be the recipient of this talisman. The hunter survived, the bison did not.

Among the several attempts to improve the overall conservation management of the rich resources of the Bialowieza forest, notable was that in the early nineteenth century by royal forester, Julius von den Brincken. Appalled by the accumulated effects of centuries of mindless exploitation, avarice and the cunning inroads of aristocrats and capitalists alike, he initiated a census as a step toward preservation of the dwindling bison herds – much as biologists and conservationists today use the census in studying threatened species (Fig. 4.5).

The tumultuous history of Poland throughout much of the nineteenth and twentieth centuries included political uprisings, expansionist ambitions of Germany to the west and Russia to the east and the precarious geopolitical position of the country in and following the two great wars of the twentieth century. These played havoc with the efforts of those who, like von den Brincken, searched for ways to maintain and develop appropriate habitats and management practices that would result not in species destruction but in conservation. Whether bison, a wild species, subject to other environmental pressures such as disease, climate change and "natural" competition, would have survived and even flourished had this work been sustained, is unknown. But it could be considered a definite possibility, under other circumstances.

As a footnote, Schama reminded us of the conservation role of the brutal Nazi, Hermann Göring, who during the Second World War arrogantly positioned himself as a kind of overlord and chief protector of the northern forest and its animal occupants. A dedicated hunter, admired even by the local hunting fraternity of the Nazi-occupied territory and visitors from other countries, Göring forbade the (unauthorised) killing of animals and birds in "his" domain. Whatever the grandiosity of his motives, it does seem that only the assertion of the highest political authority, even that of the repressive Nazis, has ever offered the prospect of a sustainable

Fig. 4.5 Lithuanian bison. Juljusz Brincken, baron, from his 1828 *Mémoire descriptif sur la forêt impériale de Bialowieza en Lithuanie*. Public domain

conservation regime. Underlining this grim thought is that, throughout recorded history, a strong survival prospect in the wild for the untamed European bison seldom surfaced and was eventually snuffed out.

4.2.2 The Fate of the American Bison

The European bison or Wisent (*Bison bonasus*) is one of two great species. The other is the American bison (*Bison bison*), commonly but incorrectly known as "buffalo," as in the lines from the popular ballad about the home on the range where "the deer and the buffalo roam"("bison" would not scan quite as well). The term buffalo correctly applies to certain related African and Asian mammals of the cattle family (*Bubalus* genus which includes the Asian water buffalo, an endangered species and *Syncerus caffe*, the African buffalo or cape buffalo, not listed as endangered). Despite common usage, I will avoid the term "buffalo" for the American animal and use instead "American bison".

The American bison is smaller, with a fuller coat and beard than its European cousin. It has shorter legs and a downward thrust of shoulder and upper body (see Figs. 1.2 and 4.5). Perhaps this suggests a less majestic beast or at least attributes that might indicate a less fearsome creature. Be that as it may, it is not so much contrasts in build and character as similarities in the life history of the two bison species that brings them together in a single, sorrowful account. Why sorrowful? Because the American bison suffered the same fate of virtual extinction at the hand of humans as did its European cousin.

Even J.J. Audubon, whose "naturalism" included shooting birds and pinning them into "natural" poses with wires, set against squared paper, was horrified by the mass slaughter of bison, likening their fate to that of the Great Auk.

Grazing on the once rich, flowering and seed-bearing grasslands of the Great Prairie of North America, the bison herds could be numbered in the tens of millions at the time of European settlement. Over many centuries they had been hunted for meat and hides by Indigenous North Americans who even when introduced to horses and European firearms did not exterminate them. It was floods of European settlers who, in the course very largely of the nineteenth century, achieved this disgraceful feat, through a combination of mass slaughter and almost total displacement of the bison's grassland habitat by agriculture. Bison tongue became a prized gastronomic delicacy in America, much as shark fin is in Asia today. As the railroads reached across the continent during the nineteenth century, tourist excursions were arranged so that bison could be shot for sport from the speeding – or halting – trains.

The vast grain fields now stretching across the mid-western states yield corn for the hogs that provide meat for North Americans, as well as grain exports to the world and corn syrup for the processed foodstuffs which have accelerated obesity rates in humans everywhere. In the tiny pockets of indigenous prairie that have been preserved, there is scarcely space for bison. The infinitely small numbers that have survived are sheltered in reserves. At the beginning of the nineteenth century, at the time when the much celebrated Jackson and Clark expedition (authorised by Enlightenment luminary, President Thomas Jefferson) "opened up the West", it has been estimated that in excess of sixty million bison roamed the plains. By the middle of the nineteenth century, such was the pace of westward expansion, agriculture and industrialisation, the animal was extinct in the eastern half of the country. By 1900, only two wild herds remained, one was in Yellowstone, which in 1872 had been declared the nation's first national park (Chap. 1), and the other in Canada. The passage of protection laws from late in the nineteenth century has resulted in a gradual recovery but on a miniscule scale by comparison with the grandeur of the Great Plains herds before the West was "won". It is ironic that Jefferson, by encouraging westward expansion which he hoped would be the dawn of a democratic agrarian Utopia, combining handwork and learning with care for nature, opened the way to massive species destruction and mono-cultural agriculture on a vast industrial scale that has destroyed habitat for many species in addition to bison.

The decline of the bison in Europe and America forms part of a global trend of extinction, endangerment and decline in animal numbers and species large and small alike. Some see this as the inexorable process of natural selection; others as "progress" in the era of consumer capitalism and others again as the destructive aggression of mindless and uncaring human population growth, settlement and industry.

It is cold comfort for the annihilation of the American bison that, as with the cave paintings in Europe, artistic representations give us today an opportunity to see and connect, if not to the living animals, then with vivid images of them. They live, at least in our imagination and feelings, and serve as a fierce challenge to modern humans to learn to live in partnership and to share the earth rather than attempt to dominate and, through thoughtless action that is masked by the banners of development and progress, to destroy so many fellow creatures. Despite the massive destruction, it is thanks to modern conservation and especially to Yellowstone that total annihilation of the bison, at least in North America, has been avoided. Through private as well as public effort, prairie land, especially in the far West, is being purchased from ranchers and returned to wilderness. The aim is to enlarge and connect with publicly protected land and to reintroduce diverse wild life. Herds of bison are among the animals being seen once more on parts of the prairies.

4.2.3 Representing the American Bison in Art

The Indigenous North Americans, as we saw above, hunted the bison but not to extinction, just as they themselves have managed to survive as a distinct culture but only in reserves designed to prevent further destructive inroads. Rock art, much of it vandalised and not all "authentic", together with more recent adornments of scattered rock faces, have been studied and photographed. An interesting depiction is of a pregnant animal, a signal that conditions favourable to reproduction of an individual are vital for the survival of the species. Habitat loss and ecological destruction which privileges just one species – human – is a violation of nature and of humanity's own educated sense of its place in the scheme of things.

American artists during the nineteenth century saw westward expansion and the revelation of stupendous geological formations, forests and vast skies over towering mountains as an opportunity to broaden their artistic horizons and connect with a fast growing urban middle class of prospective patrons. Painted scenes of richly coloured mountains and river valleys, dazzling peaks of stone and snow reaching improbably to the sky, great savannas, tumbling waterfalls, distant golden horizons, clear lakes shimmering beneath towering cliffs, and the Indigenous American posse on horseback, were to become enormously popular evocations of a glorious natural world (Fig. 4.6). They brought fame to Thomas Cole, Frederick Church, Albert Bierstadt and Thomas Moran among others. Romantic or noble images of Indigenous American encampments, chiefs and braves, were produced by Charles Bird King, John Mix Stanley (his splendid collection of some 150 Indigenous American paintings destroyed in a fire in the Smithsonian awaiting purchase by the government) and by the colourful adventurer, George Catlin.

It was the Pennsylvanian Catlin who, over six years during the 1830s, from a base in St Louis, traversed thousands of kilometres of remote western land in expeditions designed, as he said, to "rescue from oblivion" the visages, dress, customs and accomplishments of a vanishing Indigenous American culture. He likened their beauty and nobility to that of the ancient Greeks and Romans and struck a chord with painter, Delacroix, and poet, Baudelaire, in France, on the Rousseauist theme of the Noble Savage. Catlin, trained as a lawyer, was a painter, entrepreneur,

Fig. 4.6 Albert Bierstadt. Among the Sierra Nevada, California. 1868. Smithsonian American Art Museum. Albert Bierstadt, Public domain, via Wikimedia Commons

enthusiastic anthropologist and traveller who kept a journal, exhibiting, lecturing and promoting the Indigenous American cause. Living as a young man in Philadelphia, he imbibed the culture of a city renowned as an outpost of European natural history, science and progressive political and social thought (Chap. 15). Returning to the East from his westward journeys, in 1838, Catlin produced two major collections of Indigenous American items and paintings; from his journal he published books about his travels in North, Central and South America. He set up an Indian Gallery from which he launched lecture tours which saw his collection of art and artefacts move to London, Brussels and Paris. A large collection is preserved in the Smithsonian Institution in Washington; also, nearby, is the Paul Mellon collection of Catlin's portraits of Indigenous Americans, donated to the National Gallery in Washington in 1968.

Catlin's lifelong interest and mission centred on the close portrayal and valuing of what he feared was a dying Indigenous American culture. A number of his paintings also feature the American bison, mainly in Indigenous American hunting scenes. The collapse of Indigenous American culture was hastened by the deliberate efforts of the settlers. As late as the reconstruction period following the Civil War, under President Ulysses S. Grant, the United States Army slaughtered huge numbers of bison in order to deprive Indigenous Americans of a food source. As in Australia at that time, it was expected that the indigenous people would die out or at least become wholly assimilated as white settlers spread across the interior (Chaps. 15 and 16). Reducing the Indigenous American food source would hasten the process.

One of the most poignant of Catlin's Indigenous American animal paintings is *Dying buffalo shot with an Arrow* (1832–33). Gifted to the Smithsonian, it is regarded there as "an icon of animal painting".[16] The animal, its tail twitching and its mane, beard and coat adorning a superbly muscular body, sinks to the prairie, bare except for a few tufts of grass. In its one visible eye there is an unmistakable note of sadness, mingled with reproach. Its mouth is downturned, as if in proud and bitter contempt of its assailant. In Catlin's 1845 *Buffalo Bull, Grazing on the Prairie*, the bison's massive head, shoulders and forequarters are clad in a great shaggy coat while the tail is given a light-hearted upward flick. The animal completely dominates a landscape which was soon to witness its annihilation (Fig. 4.7). Another painting, John Mix Stanley's *Buffalo Hunt on the Southwestern Prairie*, (1845), shows Indigenous Americans on horseback, with bison dead or dying on the plain, against a sombre sky.

The painterly sympathy of both Catlin and Stanley appears to lie mainly with the Indigenous American hunters, one of whom, spear upraised, mane of hair flying, is shown as if in complete unison with his galloping steed. Together they bring the bison to earth (Fig. 4.8). Himself at times a hunter who shot a bison in order to sketch its dying moments, Catlin saw the animal through the eyes of the Indigenous American hunter, as a means to his sustenance and vitality. Although he captured the beauty, the energy and the immense force of the bison, and although pity for its fate shines through some of his pictures, Catlin did not appear to realise that the fate of the Indigenous American and of the bison was as if intertwined. The spread of the

Fig. 4.7 Buffalo Bull, Grazing on the Prairie by George Catlin, 1832–1833. Smithsonian American Art Museum. George Catlin, Public domain, via Wikimedia Commons

Fig. 4.8 John Mix Stanley. Buffalo Hunt on the Southwest Prairies. 1845. Smithsonian American Art Museum. John Mix Stanley, Public domain, via Wikimedia Commons

railroad, the ploughing of immense tracts of the prairie, the white man's diseases and unhealthy indulgences, in their "conquest of the West", destroyed both the noble wild animal species and the culture of the Noble Savage.

Catlin and fellow artists, when painting bison, worked in a long-established American tradition. The Plains Indigenous Americans painted the animals on cave walls, representing them in homes deep in the earth. The buffalo (bison) featured in legends, as for example in "Buffalo rock" (usually a fossil shell) which Blackfoot Indigenous Americans revered as a magical object capable of summoning buffalo to their tribes in times of hunger.

4.3 Some Other Animals in Art

The known beginnings of both naturalistic and symbolic animal art, and some of its finest expressions, are to be found in the ancient cave paintings in Europe. Other traditions of rock art, in Australia, Africa and North America, some of which may be older and more enduring (as with the continuous tradition of the Australian Aborigines, Chap. 16), contain the most vivid images and do indeed represent the power, beauty and vitality of animals. They have commonly been invested with spiritual, magical and ritualistic significance and are of immense importance in our understanding of human groups, their belief system, values and everyday life. Nevertheless, I have chosen in this chapter to focus on the European cave paintings and to single out the bison for several reasons which are as much to do with the animals' fate at the hands of man as with their representation and symbolisation in art.

Foremost among these reasons, the bison on the ceiling of the Altamira cave, as conceptualised, represented and described, conveys a deep, pure and vivid sense of oneness with the animals. It is as if the artist is not only seeking beauty in natural forms which is what English art connoisseur and social critic, John Ruskin, in the nineteenth century argued should be a primary mission of art, but directly bonding with them, expressing a feeling of unity in life and a common human-animal domicile in the natural world. I see in these paintings the beginnings of a grand tradition of nature-feeling expressed through art. Although trees, shrubs, flowers and non-vascular (so-called lower) forms were not portrayed then, as they have been by later masters (the Frontispiece of this volume, for example), in contemplating the prehistoric cave paintings of southern Europe, we are at one of the earliest stages of a great line of naturalistic animal representations and their domiciles.

Despite the time lapse between the most recent Ice Age cave paintings and the emergence of the agricultural and urban civilisations of Mesopotamia, Egypt and China, the artistic representation of animals has provided a link between them and the modern world. Whether as mythical creatures, objects of worship, symbols, or closely observed in movement and lovingly rendered, down to the last hair or twitch of nostril, animals have been a constant theme, investing art with rich cultural meaning and value. A few examples from different cultures and historical periods must suffice to indicate the range of meanings and the values expressed. They may at the same time underline the fragility of animal life, while stimulating feelings of love and unity.

Australian Aboriginal rock art, specimens of which may predate by many centuries that of the southern European caves, celebrates indigenous fauna. Protected by ledges and overhangs, and retouched from time to time, this art has profound cultural meaning, not only for the First Nations people. (Sect. 16.4)

In Egypt in the period of the New Kingdom (eighteenth to eleventh centuries BCE), there was an avenue of stone rams connecting the temple of Luxor with that of Karnak. Naturalistic by contrast with the imperious mythical Sphinx (half-human, half-animal and an expression of divinity), they seem to belong more with bison than with the half-human half-bird or animal stone figures of the Egyptian deities.

Religion of one kind or another has provided great scope for the artistic representation of animals, often the vehicles of powerful symbolic messages. An example is Giovanni Bellini's mid fifteenth century painting, *St Jerome in the Wilderness* or the *Desert*. Dürer, Leonardo, El Greco and Caravaggio are among the numerous artists who were captivated by the Saint's asceticism. Bellini's painting features the Lion, from whose paw according to legend the compassionate Jerome bravely plucked a thorn. Seated on a rock, with background depicting a pastoral scene – the life he had left – Jerome faces the Lion, one finger raised heavenward, the other hand outstretched toward the Lion holding a book which is presumably the Bible. The Lion sits patiently, a paw reaching out to the Saint – presumably in gratitude for the removal of the thorn but perhaps as well signalling acceptance of the Biblical message (Fig. 4.9). This beautiful, symbolically rich painting suggests unity and harmony between the man and the animal under the aegis of Christianity. As we shall see in the following chapters, for Christianity at least, attitudes toward animals have varied from the peace and harmony of the presumably vegetarian Eden of the book of *Genesis*, to a sharing of the hardship of expulsion, to a prideful assertion of superiority and right of exploitation.

The famous and deeply symbolic image of Agnus Dei, the Lamb of God, appeared in numerous Christian paintings in Mediaeval Europe; for example, the celebrated Ghent altarpiece by Hubert and Jan Van Eyck where the lamb is portrayed side-on (as were the Altamira and North American bison) on an altar in the middle of a field and surrounded by worshippers. Curiously, as recent restoration has revealed, the Lamb was painted with human, not sheep's eyes. Echoing the sacrificial rites of the ancient Hebrews, it leads viewers to contemplate sacrifice and redemption but also to admire the graceful naturalism of the portrayal. Many other religious paintings, for example of the ass – bearing Mary and infant Jesus on the flight to Egypt – carry a heavy symbolic load. Yet we can immediately feel for the physical endurance of the animals portrayed and perhaps too their animal servitude at the bidding of the human and the divine.

In Domenico Ghirlandaio's *The Adoration of the Shepherd* altarpiece in Florence, there is a distinct suggestion, in the clutches of the devotional Biblical animals, of an obligation falling on all forms of life to worship a remote divinity from whom,

Fig. 4.9 Bellini's image of St Jerome in the Desert. Barber Institute of Fine Arts. Giovanni Bellini, Public domain, via Wikimedia Commons

however, in the Old Testament especially, there can be no assurance of reciprocity. The religious intent and symbolism notwithstanding, we find in some of these remarkable paintings a deep feeling for animals just as we marvel at their celebration of physical as well as spiritual beauty, and of the earthly as well as the heavenly kingdom.

The inspiration of classical mythology and tales from ancient history equalled when it did not exceed that of religious devotion in generating splendid animal paintings, carvings, sculptures and decorative motifs. Superbly carved horses were to be found most energetically pulling chariots on the pediments of the Parthenon. Now stationary on the floor of the British Museum, they writhed and tumbled then in battle and in fatal encounters with lions and leopards in the hunt, and they served as seats for emperors and generals. Travellers to China today are likely to encounter images of the 2000 year old masterpiece from the Eastern Han Dynasty, the bronze *Galloping Horse Treading on a Flying Swallow* (34.5 cm. high and 41 cm. long) (Fig. 4.10). In the exuberance and freedom of this superbly self-assured creature, we feel the admiration of a master of the sculptor's craft. It comes as something of a

Fig. 4.10 Galloping horse treading on a flying swallow. Eastern Han dynasty, 25-220 AD. Gansu Provincial Museum, China. G41rn8, CC BY-SA 4.0 via Wikimedia Commons

shock to realise that a classic portrayal of what may be the legendary steed of heaven is now a symbol of the Chinese tourist industry.

Horses in particular were important partners with humans in the iconography of Western art, as in everyday life, from the Greek and Roman period until recent times. They shared the triumphs, victories and disasters of their human riders, often eliciting great admiration and pity alike for their role in human exploits and misadventures, notably in heroic battles of the *Iliad*. Among the brilliant anatomical drawings attributed to Leonardo da Vinci that are held in the Royal Library, Windsor, are acutely observed studies of galloping horses, horses trotting, horses recumbent and in battle with dragons, together with rock formations, clouds, plants and other natural phenomena. Unlike the mediaeval theologians, who sought to elevate humans and their reasoning powers above the animals, Leonardo "sought to establish the essential unity of structure of all living things, the earth an organism with veins and arteries, the body of a man, a type of that of the world."[17]

In the paintings of eighteenth century British artist, George Stubbs, horses are gloriously displayed in their own right. The son of a currier, Stubbs studied horses closely and was the author of a book of engravings which became a text of animal anatomy. His mastery is demonstrated in one of the greatest of all animal paintings, that of the champion racer, *Whistlejacket*. Held in London's National Gallery, this picture shows a majestic prancing animal, rising high on powerful hind legs, chestnut coat glistening over rippling muscles, full bushy tail and mane, and head arching elegantly toward the awe-struck viewer (Fig. 4.11). The artist has captured a bursting vitality and spirited pride as if the animal is no mere mortal creature but an

Fig. 4.11 George Stubbs. Whistlejacket. c1762. Close anatomical study and superb, affectionate artistry. National Gallery, London. George Stubbs, Public domain, via Wikimedia Commons

eternal force of nature. By contrast, his *Mares and Foals in a River Landscape* is a lovely evocation of an idyllic world where foals feed and their mothers calmly greet one another, living their own lives, comfortably together and neither dependent on nor under the saddle of man. In yet another contrast, Stubbs captured savagery in the force of nature in his shocking painting *Horse attacked by a Lion*.

Although of all the land animals represented in art over the centuries since classical times, horses appear to be the most numerous, reflecting their place in human affairs, there are many others. Among those that form an integral if not dominant part of great works of art are bulls – as in the cave paintings – lions, leopards, tigers, jaguars, elephants, wolves, foxes, hounds, domestic dogs and cats. Humanity's dependence on and active partnership with animals both wild and domestic have frequently been recounted with great love and affection in literature. For centuries, there have been many parallels in art to the animal tales about both living and imaginary creatures.

One of the most sensitive and beautifully observed of all animal portrayals is that of a resting hare, by the sixteenth century German master, Albrecht Dürer. Celebrated for his visionary religious work, Dürer is no less famous for his acute observation of the natural world, whether in a clump of meadow grasses, a forest lake or a living animal. His artistic rendition of the hare at rest, with its front paws neatly tucked beneath a firmly set head and great towering ears suggests a more ordered vitality than that that of the Altamira bison but it is a vivid affirmation of pulsing life and of the artist's oneness with the creature before him (Fig. 4.12).

Centuries later, painter Franz Marc, a leader of the Munich-based Der Blaue Reiter (Blue Rider) group of artists which included Wassily Kandinsky and August Macke, heralded "a spiritual revolution" in art, not of form but of content, and expressed symbolically in the use of colour. The Russian Kandinsky, the theorist of this movement as well as one of its leading artists (and following prototypes, a founder of the abstract movement in twentieth century art), in 1911 produced a manifesto on the spiritual meaning and significance of art. Translated soon afterwards, under the title The Spiritual in Art, this book was widely influential. Its message was that art should be a force of self-expression and self-development through a rich, expressive spiritual life. For his paintings of this period he used the term "soul-expression". Kandinsky was not alone in this. His influences included composer, Arthur Schoenberg, and the theosophist, biodynamicist and educator, Rudolf Steiner. It was a time when progressive education reformers in England, Continental Europe and the United States were growing restless over the uniformity and emotional barrenness of mass schooling, and seeking ways of fostering children's expressiveness, in art as well as in other ways (Chap. 16). For Der Blaue Reiter group, these broader ideas about the need to release and nourish the human spirit found expression in the exuberance of horse paintings and the symbolism of colour.

Together with Macke and Marc, Kandinsky found beauty and symbolic significance in animal vitality. His title page design for the group's *Der Blaue Reiter*

Fig. 4.12 Albrecht Dürer, Young Hare. 1502. Acute observation and admiration. Albertina Museum, Vienna. Albrecht Dürer, Public domain, via Wikimedia Commons

Almanac was of a leaping horse with rider. It may echo his interest in Russian fairy epics but in any case is a graphic rendition of the theme of unity of animal and human.

Amongst other utterances about his spiritual interests, Kandinsky wrote "each colour lives by its mysterious life". As a colourist, he was quite brilliant, not only in the vividness of his palette but in the extraordinary juxtapositions of conventionally discordant, clashing colours, yet in the picture they unify and show each other off to the best advantage. What has this to do with animals in art? Only that it was the beauty, vitality and vigour of form and movement of animals that he carried into his grand colour experiments. Significant form became significant emotional intensity.

The idea of spiritual regeneration was expressed in Marc's joyful study in 1911 of three wild, spirited young horses titled *Red Horses*. Their playful circling, like free spirits, on a valley floor is suggestive of the Three Graces, daughters of Zeus and representatives of beauty, charm and other sources of delight. Just as Kandinsky was fascinated by the combination of horse and rider, Marc had an adoration of horses as symbols of a purity and vitality in nature as distinct from the corruption and materialism of early twentieth century Europe. There is a curious resemblance of his 1911 woodcut of a horse resting to the bison couched on the ceiling of the Altamira cave – a spiritual bond of artists and art across the millennia. As with so many other young men, Marc saw the outbreak of the First World War as the forerunner of a spiritual renewal of Europe. It was not to be. The war cost him his life. Aged 36, he was killed in 1916, in the slaughter house of Verdun, one of the 330,000 Germans and 360,000 French killed or wounded. Marc's death followed on that of fellow Der Blaue Reiter August Macke, killed at the beginning of the Great War.

By bitter contrast, the great eighteenth century bronze horses bearing emperors and kings – such as those of Joseph II, Holy Roman Emperor (by Franz Anton Von Zaumen), Peter the Great (by Etienne-Maurice Falconet) and Elector Frederick William of Brandenburg (by Andreas Schluter) express an altogether different spirit. The horses are merely supports for the proud, imperious, stolid figures representing the power of state authority and military conquest.

The use of animal representations for purposes of ceremony and the affirmation of state power has continued from the earliest times until today, with horses commonly in the role of symbolic or literal support for the authority figures; for example, carrying England's Queen Elizabeth II to the ceremony of the Trooping of the Colours. However, as we have seen, they have also been widely represented in quite a different relationship to humans, as admired partners not mere props.

In the modern era, Jack Yeats superbly epitomises both love for and bonding with an animal in his paintings of horses and riders, many in the West of Ireland. I think of them as another kind of abbé Breuil's conceptual realism as mentioned earlier in this chapter – the realism of a vision of nature, human as well as animal but, by contrast with the cave painters, in equal proportions. In the National Gallery of Ireland are many beautiful paintings and drawings by Jack Yeats, brother of poet, William Butler Yeats, and member of a family whose creative genius in the nineteenth and twentieth centuries is handsomely recognised in the naming of a "Yeats Museum" within the Gallery.

Among many other items in the Dublin gallery is a lively watercolour of a three-horse rein pulling not a machine of war but a circus chariot. The circus was a favourite motif in the days when animals had as much a part in the performance as humans, if not more. In *The Double Jockey Act* a galloping horse, legs plunging forward, mane flying and eye flashing, expresses sheer mastery by the horse of an act that seems to cause the riders some anxiety. Yeats pictures other horses quite docilely with men hauling seaweed out of the waters of Connemara. In a short-lived monthly publication, a finely drawn horse leaps across mountains while a bearded countryman doffs his hat, cringing in awe.

As he moved from his early style as an illustrator, Yeats gave his horses and especially his racehorses something of the feeling of a Degas grouping (*Before the Race, Exercising Racehorses*), but is at his best in late expressionist and visionary pictures of solitary animals or horse and rider in some west of Ireland landscape touched with enchantment. The painting, *Morning*, coveted – and bought in instalments – by playwright friend, Samuel Beckett, first appears to the gaze as a simple scene, thinly painted. A boy is seated on a horse, gesturing westward toward the town of Sligo which he is leaving. The horse and the boy are clearly delineated, the town is insubstantial, dissolving into a yellow and blue haze. Of the two departing, the horse, four feet firmly on the ground, is the more substantial. The horse looks down to a cat sprawling on the ground; it is staying whereas the horse and boy are setting forth like Samuel Becket to France, or Yeats himself, on an adventure into the unknown.

Of Yeats' many other later paintings I will mention only one, through which shines alike the artist's deep affection for the horse and the joyous spirit of the traveller. *The Singing Horseman* is dominated by a yellow horse prancing through a visionary landscape, a golden leg firmly raised, head held high and a deep dark eye penetrating something known to itself but not us. The boy throws up his head singing, lost in some ecstatic state. The horse strides on, assured, confident, bearing a precious load (Fig. 4.13).

Before leaving the field of naturalistic, yet highly emotionally charged representation of living animals, I shall mention an animal sculpture movement which, reaching its apogee in nineteenth century France and continuing to the present day, signified both a public and a domestic feeling for animals in art which parallels the worldwide passion for household animal pets.

Some art historians have termed the nineteenth century "the age of bronze". It could as well be described as the age of animals in bronze, due to immense interest by public institutions and private art lovers alike in acquiring animal figures in bronze – and iron as well. The term *animaliers* was coined by critics who were, at first, disposed to ridicule what they felt was an obsession with an ignoble lower order – animals. The sculptors, Antoine-Louis Barye (1795–1875) and Christopher Fratin (1801–1864), to whom the abusive term *animaliers* was initially applied, were not deterred by the art-establishment view that sculpture was properly a noble art, its purpose to dignify and glorify human power and achievement or to arouse the passions over heroic defeats. It was appropriate to have an emperor or statesman on

Fig. 4.13 The Singing Horseman. Jack Yeats, brother of poet William Butler Yeats, has painted a noble steed in a visionary landscape. The yellow horse reaches forward while the rider, singing, throws up his head ecstatically. Horse and rider are as if one joyful spirit. Jack B. Yeats, Irish, 1871–1957 The Singing Horseman, 1949. Oil on canvas 61.4 × 92 cm. Photo © National Gallery of Ireland. © Jack B Yeats/DACS. Copyright Agency, 2021

a caparisoned charger, or a dying warrior sinking to earth but not to represent "lesser" creatures, wild or domestic, unless at the point of a threatening spear.

Nevertheless, public enthusiasm and official patronage proved too strong and in the galleries and shops appeared a great array of domestic-scale creatures, including monkeys at play, dogs, sheep, turtles, lizards, goats and numerous others. For more spacious settings there were lions, tigers, jaguars, crocodiles, elephants, cattle, horses, dogs, eagles, wild boar and many more. As scientific knowledge and public understanding of the animal kingdom increased throughout the nineteenth century, so did the variety not only of affectionate and respectful bronzes but also of paintings.

Animal representations in art were popular in nineteenth and early twentieth century England where painter and sculptor, Edwin Landseer (1802–1873) gained a wide following across the English-speaking world and beyond. His animal paintings were enormously popular, both those that seemed to have a comfortable domestic flavour (the dog, the horse) or others which seemed to reach out from their romantic landscape settings for human applause and appreciation. *The Monarch of the Glen* is a proud creature, its head held high and antlers stretching skywards, with the additional attraction of a romantic setting, capturing the creation of the Highland legend by novelist, Walter Scott – whose novels Landseer had earlier in his career been commissioned to illustrate. The handsome, powerful animal in combination with a background of the romantic, no longer barbaric, highlands of Scotland raised animal art to a new level in popular imagination.

It is easy to understand the official commissioning of large animal bronzes. The four lions grouped around Nelson's Column in London's Trafalgar Square constantly roar a patriotic message to the thronging crowds, most recently in the battle over Brexit. The growth in the number and size of cities during the nineteenth century, with their parks, public gardens, squares, ceremonial avenues and other prominent open spaces almost required the presence of monumental animals. There was a tradition of ornamentation on this scale to draw upon, dating to the earliest civilizations in Mesopotamia and Egypt, as mentioned earlier in this chapter.

Industrialisation and colonialism provided wealth to the great European cities and their increasingly prosperous middle classes avaricious of possession to display their taste and wealth. Both public authorities and citizens had the means to patronise the new age of bronze. Personal and family wealth grew enormously in the nineteenth century. An affluent middle class had the means to purchase not only prints and books but also original paintings, hand-made furniture and objets d'art. Small-scale sculpture was within the reach of an expanding market, no longer an aristocratic preserve or a function alone of government or church patronage. But why the burgeoning interest in animals as distinct from, say the interest of scholars in busts of Cicero or Socrates, of scientists in Archimedes and Newton, and of music connoisseurs in Beethoven and Schumann? One explanation is that for the aspiring middle class animals were both familiar and a source of pleasure. From late in the eighteenth century the pulse of scientific inquiry had quickened. Much work was being done throughout the nineteenth century in palaeontology, botany, zoology as well as physics, chemistry, astronomy and geology. The discovery of fossils resulted in a closer interest in animal anatomy, not only of extinct animals but that of animals familiar in everyday life. The researches and publications of such as Buffon and Cuvier in France, Owen and Huxley in England were widely known amongst the educated middle classes – and not only them, since due to popular lectures, and courses for workingmen, they radiated throughout society. Imperial expansion, colonisation of Asia, Africa, Australasia and Oceania, travellers' tales, reports by voyagers, explorers and missionaries generated a huge interest in natural history, in plants as well as animals (Chap. 15).

It was also in the nineteenth century that the late eighteenth century Romantic movement in poetry and literature intensified and took hold in philosophy, music and art. Early in the century, Géricault and Delacroix responded to the English Lake Poets, Byron and Shelley. This, combined with the Rousseauist elevation of emotional sensibility, found expression in animal painting, especially of the horse. Gericault's *Horse Frightened by Thunder and Lightning* shows a black-legged, pale yellow animal trembling as lightning streaks along a bank of dark, thundering cloud.

With both popular art and science feeding a wide public interest in the world of animals and of nature more widely, it is unsurprising that animal sculptures should come to rival if not supplant the austere themes of classical and neo-classical times reaching new and more diverse groups of art lovers. These sculptures were displayed not only in castles, palaces and palatial public buildings, but also on the domestic hearths, hallway entrances, mantelpieces and cabinets of new urban villas. A favourite spot for an eagle, wings outstretched and hooked beak about to pounce, was a raised shelf where visitors could readily catch its eye and admire the artistry. Such evocative bird poses also occurred in church pulpits.

It is impossible to know just how far the nineteenth century "age of bronze" took people interested in this art form toward an understanding of the animal kingdom and greater respect for the creatures represented. At least, a wider public became more familiar with the diversity of animal life, with the form, grace and movement of animals, their aesthetic sensibilities stimulated by new kinds of artistic representation. Many of these pieces survive, in galleries and auction houses as well as in their public and private settings. Vast numbers, probably most of them, were melted down to provide armaments and equipment for the succession of European wars. However, there is today a continuing lively business in the reproduction and sale of these small animal bronzes.

The tradition of noble and expressive animal art is not confined to naturalistic renderings of observed animals, allusions to ancient mythology and dramatic encounters between humans and wild animals. There is a large gathering of imagined animals: Pegasus, the winged horse, dragons, the half-animal half-human centaurs, satyrs, sphinxes, and the animals construed by artists not from direct observation but on hearing and reading travellers' tales and the records of voyagers, with accounts of sea monsters, the Eastern Fastitocalon, the Icelandic Kraken, Lewis Carroll's Jabberwocky and gentle animals with human characteristics as Beatrix Potter's Mrs. Tiggi-Winkle or the fearsome inhabitants of distant mountains and remote jungles who inhabit numerous children's books and now feature in virtual reality games.

These mythical and magical creatures have exercised a strong grip on the imagination. Their symbolism, and uses for purposes of moral edification, suggest rather more than mere light-hearted sallies of pen or brush. Whatever specific meanings or symbolic value may be attached to these mythical creatures, clearly art as well as literature, both naturalistic and fanciful, has for millennia provided a rich medium for incorporating animals, fanciful as well as natural, into human affairs.

The English visionary poet and artist, William Blake, in one of his magnificent engravings illustrating *The Old Testament Book of Job* has God, sided by a pair of angels pointing down to Behemoth, a rhinoceros-like animal (probably modelled on Dürer, of whose prints Blake had a set). Below is Leviathan, a fearsome, writhing sea-monster with gaping jaws. Job and his children cower beneath God who repeats a creation message, "Behold now Behemoth which I made with thee." Thus religion joined with art and science in proclaiming the unity of man and (symbolic) animals in nature, for Blake a "fearful symmetry".

As the ideas of naturalistic and figurative art yielded to the various forms and movements of modernism in the twentieth century, animals appear very largely to have vanished as a source of inspiration, as a form of creative artistic expression and as a way to enliven public understanding of art and nature alike. It will be claimed that they continue to hold a place, and a very prominent one, in photography and documentary film and in popular films such as *The Lion King*. These tend to focus on wild animals and exotic creatures – animal life at a safe distance. Just as in botanical art, it is through painting and drawing that more is expressed than a physically accurate representation. In drawing, painting and sculpting animals, the artist finds meaning and value that eludes the camera. As the claims for veracity and authenticity of modernism in music are challenged by a revival of melody and harmony, so we may find the need for a revival of animal painting as a way of reconnecting art with a media-saturated public. Picasso's brilliant drawing of a cockerel bursting with pride of life has no less artistic authenticity and veracity than his experiment in creating a rectangle for a human head with eyes attached at diagonally opposite corners.[18] If less in evidence in contemporary art, animals have nevertheless continued to occupy a vital place in our awareness and understanding both of who we are and our place in the natural world.

4.4 Modern Paradoxes

Science, especially since the time of Darwin and the popularisation of the theory of evolution as espoused by his followers, has been responsible for a vast increase in systematic knowledge of animals, their biological attributes and functions, life conditions, habits, behaviour and ecological value. Through popularisation of science, including literature, film and television, animals of all kinds have increasingly become the subject not only of study and understanding but entertainment. At the same time, study of animal behaviour, diseases, nutrition and so on and the economic and social value of animal products have resulted in a new understanding of the nature of the ancient partnership between animals and humankind.

The progressive development of zoos and zoological practice over the nineteenth and twentieth centuries, better ways of managing animals in enclosures, closer understanding of animal reproduction, co-habitation and habits whether in captivity or the wild, legislation designed to protect animals from abuse and exploitation, widening public interest in conservation and preservation of threatened species – and the growing popularity of vegetarian and vegan diets – all these are indicators that consciousness of animals as an integral part of human life is taking new and positive forms. There are what appear to be successful efforts to maintain a balance between human hunting cultures and the survival of wild animal herds. In Arctic Canada, for example, hunting by the Inuit of the Arctic musk-ox (*Ovibus moschatus*) which had survived from the Ice Age, had by the mid-twentieth century led to the near extinction of the species. That has changed thanks to the adoption of conservation and enlightened management practices in partnership between scientists and the Inuit communities. Hunting is now "sustainable", meaning that for the present at least, herds will be maintained and the Inuit will continue this aspect of a traditional way of life. Similar efforts are being made in Alaska, Russia, Greenland and Scandinavia. But, with the resumption of commercial whaling by Japan in 2019, a major setback has occurred.

Other social, cultural and economic forces are such that the overall message of the "Bison story" remains highly pertinent. As humans, we seem to be quite successful in promulgating and extending the reach of ways of life which are enormously threatening to many other species than ourselves and to their habitats. At the same time we seek to enjoy their presence whether live or in artistic representation, adopting enlightened attitudes toward them and recognising their right to survive and thrive. In this respect, the modern temperament is deeply conflicted.

What has been termed a conservation ethic, together with movements such as animal welfare and animal rights have combined with vastly increased knowledge and understanding of the natural world to provide us with a most powerful apparatus of values, knowledge and a potential will to act. As mentioned above, the philosophers Peter Singer and Tom Regan, respectively from Utilitarian and Kantian ethical standpoints, have argued persuasively for acceptance that animals as well as humans have rights which must be acknowledged and respected. It seems, however, that from these rights can be bracketed, in public policy and public sentiment, all kinds of uses and treatment of animals that entail a severe qualification of any rights that might be assigned them. Legislation aimed at protecting animals from cruelty has been an important step forward but varies considerably in scope and effect. Public disclosure of cruel and barbaric practices, for example in slaughterhouses, transportation of animals, intensive chicken farming and animal training testifies to shortcomings in legislation and monitoring, reflecting a degree of public passivity. Beyond the much needed changes led by animal rights activists there is need for a wider change in community consciousness. An appropriate growth point for this is the universal primary school, where all children can learn about and come to understand animal life as sacred, like that of humans (Chap. 16).

Of particular concern is a community-wide reluctance to address the dilemma of species decline and extinction in the face of urban spread, industrialisation and substantial habitat pollution and loss, as forests are stripped out and farmlands turned into vast tracts of monoculture or housing estates. Addressing this issue requires attention to these matters at all levels of education including adult education or so-called lifelong education. It is paradoxical that this destruction is accelerating while great affection is today shown for animals as pets, and as historically represented in art over the millennia.

Predators, poaching, fishing and whaling, climate change, disease and the persistence of the cruel and degrading handling of animals as a human food source are among the chronic challenges. Too often, over-simplified options are presented, as for example when in Australia in 2015 the destruction of the habitat of the koala in southeast Queensland was justified by public planning authorities. They claimed that the encroaching urbanisation provides employment as well as housing – people's needs versus the life of animals. Human needs, however, can be met in ways that are not detrimental to koalas. Some challenges are amenable to carefully considered political action, others to scientific investigation and others again to concerted, voluntary public effort. All have an important part to play together with significant popular publications such as Douglas Adams' global round-up of endangered species in *Last Chance to See,*[19] the follow-up film on threats to nature by

Mark Cawardine and Stephen Fry, and the efforts of television celebrities like Sir David Attenborough and popular science journals, such as *National Geographic* and *New Scientist* among others.

It may seem ironical that the "cave man's" bison art preserved on the rock face at Altamira cave should be treated in this chapter as an epitaph to "civilised man's" destruction of the creatures that inspired these ancient masterworks. The once living animals, those that have followed them over millennia, and their artistic representations combine in an affirmation of unique beauty and vitality in nature and culture. They call for human reverence, concern and practical action not only by public authorities, conservationists, scientists, scholars and innumerable voluntary groups, but above all in our everyday lives, thoughts and feelings.

Perhaps what is lacking is the understanding that action, if it is to be effective, must concentrate much more than hitherto on our total and everyday way of life. We need to focus efforts on the changes called for in our habits and daily practices. Only by working actively to make the changes needed for the rich diversity of animal, plant and human life have we any prospect of attaining a new balance and harmony with the natural world.

Notes

[1] J.R.R. Tolkein's poem "Fastitocalon," set in Middle-earth, is included in *The Adventures of Tom Bombadil*; Jorge Luis Borges (2002) "Fastitocalon" in *The Book of Imaginary Beings*. London: Vintage Books p.61. Translated by Norman Thomas di Giovanni.

[2] Lewis-Williams, D. (2002) *The Mind in the Cave. Consciousness and the Origins of Art*. London: Thames and Hudson; Bataille. G. (2005) *The Cradle of Humanity: Prehistoric Art and Culture*. New York: Zone Books; Curtis, G. (2006) *The Cave Painters. Probing the Mysteries of the World's First Artists*. New York: Knopf; Powell, T.G.E. (1966) *Prehistoric Art*. London: Thames and Hudson.

[3] Huyghe, R. (ed) (1966) *Larousse Encyclopedia of Prehistoric and Ancient Art*. London: Hamlyn; Moreno, A.M. (2014) *Altamira Museum*. Spain: Ediciones Palacios y Museos; Powell, T.G.E. (1966) *Prehistoric Art*. London. Thames and Hudson. p.46.

[4] Grigson, G. (1957) *The Painted Caves* London: Phoenix House. pp. 67 and 70.

[5] Read, H. (1945) *Art and Society*. London: Faber and Faber. 2nd edition p.11.

[6] French scholar Joseph Déchelette termed one of the chambers at Altamira "the Sistine Chapel of Quaternary Art" *Manuel d"Archéologie préhistorique* Vol. 1 1908 Paris. Alphonse Picard. p.241.

[7] Heidegger, M. (1959) "The Ode on Man" in "Sophocles' Antigone" in *An Introduction to Metaphysics*. New Haven. Yale University Press. pp.86–100.

[8] Teilhard de Chardin, P. (2002–1955) *The Phenomenon of Man*. New York: Harper Perennial. p.203

[9] Berger, J. (2009) "The White Bird" in *Why Look at Animals*, London: Penguin Books.

[10] Langer, S. (1957) *The Problems of Art*. London: Routledge and Kegan Paul.

[11] Langer, S. (1976/1941) *Philosophy in a New Key. A Study in the Symbolism of Reason, Rite and Art*. Cambridge, Mass.: Harvard University Press. 3rd edition. pp.251–2.

[12] Hermerén, G. (1969) *Interpretation and Meaning in the Visual Arts*. Stockholm: Scandinavian University Books.

[13] Newton, E. (1960) *The Arts of Man* London: Thames and Hudson.

[14] Myers, B. and Copplestone, T. (eds.) (1985) *The History of Art*. London: Hamlyn. Introduction.

[15] Schama, S. (1995) *Landscape and Memory* London: Harper Collins.

[16] Kloss, W. (1985) *Treasures from the National Museum of American Art*. Washington and London: Smithsonian Institution Press.

[17] MacCurdy, E. (ed) (1938) *The Notebooks of Leonardo da Vinci Vol.1* Preface, p.16. London: The Reprint Society.

[18] Gombrich, E.H. (1972) *The Story of Art*. London: Phaidon. Figs. 12 and 457.

[19] Adams, D. with Carwardine, M. (1990) *Last Chance to See*. London: Pan Books. Also *Last Chance to See – in the footsteps of Douglas Adams* – BBC film with Stephen Fry and Mark Carwardine.

Complete Listings for Wikimedia Commons Images

4.1 Maria Sanz de Sautuola. https://commons.wikimedia.org/wiki/File:Mar%C3%ADa_Sanz_de_Sautuola.jpg Attribution: Santander Prehistory Museum, Public domain, via Wikimedia Commons.

4.2 Drawing of polychrome ceiling published by Marcelino Sanz de Sautuola (*Dibujo del techo de Polícromos publicado por Marcelino Sanz de Sautuola*). https://commons.wikimedia.org/wiki/File:7_Dibujo_del_techo_de_Pol%C3%ADcromos_publicado_por_Marcelino_Sanz_de_Sautuola.jpg Photo: Museum of Altamira. Attribution: Marcelino Sanz de Sautuola. Fotografía por el Museo de Altamira, CC BY-SA 3.0 <https://creativecommons.org/licenses/by-sa/3.0>, via Wikimedia Commons.

4.3 Magdalenian era polychrome image of bison. (*Bisonte Magdaleniense policromo.*) https://commons.wikimedia.org/wiki/File:9_Bisonte_Magdaleniense_pol%C3%ADcromo.jpg Attribution: Museo de Altamira y D. Rodríguez, CC BY-SA 3.0 <https://creativecommons.org/licenses/by-sa/3.0>, via Wikimedia Commons.

4.4 Cave of Altamira and Paleolithic Cave Art of Northern Spain –110,113. https://commons.wikimedia.org/wiki/File:Cave_of_Altamira_and_Paleolithic_Cave_Art_of_Northern_Spain-110113.jpg Attribution: Yvon Fruneau, CC BY-SA 3.0 IGO <https://creativecommons.org/licenses/by-sa/3.0/igo/deed.en>, via Wikimedia Commons.

4.6 Albert Bierstadt – Among the Sierra Nevada, California – Google Art Project. IQE1CY9y_Rfy5A at Google Cultural Institute Attribution: Albert Bierstadt, Public domain, via Wikimedia Commons.

4.7 George Catlin – Buffalo Bull, Grazing on the Prairie – 1985.66.404 – Smithsonian American Art Museum. https://commons.wikimedia.org/wiki/File:George_Catlin_-_Buffalo_Bull,_Grazing_on_the_Prairie_-_1985.66.404_-_Smithsonian_American_Art_Museum.jpg Attribution: George Catlin, Public domain, via Wikimedia Commons.

4.8 'Buffalo Hunt on the Southwestern Prairies', oil on canvas painting by John Mix Stanley, 1845, Smithsonian American Art Museum (Washington D.C.) https://commons.wikimedia.org/wiki/File:%27Buffalo_Hunt_on_the_Southwestern_Prairies%27,_oil_on_canvas_painting_by_John_Mix_Stanley,_1845,_Smithsonian_American_Art_Museum_(Washington_D._C.).jpg Attribution: John Mix Stanley, Public domain, via Wikimedia Commons.

4.9 Giovanni Bellini – San Girolamo nel deserto. https://commons.wikimedia.org/wiki/Category:Paintings_of_Saint_Jerome_in_the_desert#/media/File:Giovanni_Bellini_-_San_Girolamo_nel_deserto.jpg The Yorck Project (2002) *10.000 Meisterwerke der Malerei* (DVD-ROM), distributed by DIRECTMEDIA Publishing GmbH. ISBN: 3936122202. Attribution: Giovanni Bellini, Public domain, via Wikimedia Commons.

4.10 Gansu Museum 2007 257. https://commons.wikimedia.org/wiki/File:Gansu_Museum_2007_257.jpg Attribution: G41rn8, CC BY-SA 4.0 <https://creativecommons.org/licenses/by-sa/4.0>, via Wikimedia Commons.
4.11 Whistlejacket by George Stubbs edit. https://commons.wikimedia.org/wiki/File:Whistlejacket_by_George_Stubbs_edit.jpg Attribution: George Stubbs, Public domain, via Wikimedia Commons. Source: http://www.nationalgallery.org.uk/paintings/george-stubbs-whistlejacket.
4.12 Albrecht Dürer – Hare, 1502 – Google Art Project. https://commons.wikimedia.org/wiki/File:Albrecht_Dürer_-_Hare,_1502_-_Google_Art_Project.jpg Attribution: Albrecht Dürer, Public domain, via Wikimedia Commons.

Chapter 5
To Be Religious

... For Man is everything, And more: he is a tree, yet bears more fruit; A beast, yet is, or should be, more; Reason and speech we only bring ...

George Herbert: Man

For we see that, in matters of Faith and Religion, we raise our imagination above our reason, which is the cause why Religion sought ever access to the mind by similitudes, types, parables, visions, dreams.

Francis Bacon (1893) *The Advancement of Learning.*

To uproot by argument habits long embedded in the character is a difficult, perhaps an impossible task. We may, I take it, regard ourselves as fortunate if we can get some portion of goodness by acquiring for ourselves all the recognised means of becoming good.

Aristotle, Ethics (Bk.10, Ch.9)

I, for one, am very grateful to nature, not just when I view it in that aspect which is obvious to everybody, but when I have penetrated its mysteries; when I learn what the stuff of the universe is, who its author or custodian is, what god is; whether he keeps entirely to himself or whether he sometimes considers us; whether he creates something each day or has created it only once; whether he is part of the universe or is the universe; whether it is possible for him to make decisions today and to repeal in part any sort of universal law of fate; whether it is a diminution of his majesty and an admission of his error that he has done things which had to be changed.

Seneca the Younger, Naturales Quaestiones. Book 1, Pref.3.

5.1 Diversity of Religious Beliefs and Practices

In this, the first of a cluster of chapters on the origin and development of religious ideas concerning "man's place in nature", we consider what appears to be the most powerful force that historically has shaped attitudes towards nature in the Western world. From time immemorial, numerous religions and cults of quite diverse

M. Skilbeck, *Loving and Studying Nature*, International Explorations in Outdoor and Environmental Education 10, https://doi.org/10.1007/978-3-030-80751-1_5

character have risen, fallen and declined, some to rise again. Most are theistic, with either one or many gods as the source of life and objects of awe and worship. Through communal rituals, expressive dance, art, music, laws, doctrines, creeds and ceremonies, relations between humans and the deity are affirmed and rehearsed by religiously-minded people, in both cosmic and moral terms.

From this great diversity of religious experience and expression, I have chosen to focus on aspects of Judaism, Christianity and Islam. The evolution of these global religions has substantially shaped Western and Middle Eastern civilisations. Attitudes towards nature arising and cemented in them form the subject of my overall enquiry. The expression of those attitudes and the rationale for them in Judaism, early Christianity and Islam call for particular attention, providing as they do durable foundations of a variety of beliefs and practices relating to nature that have persisted in some form to the present day. While there will be passing reference to major Far Eastern religions, it would extend the range of the enquiry too far and too superficially to do these religions the justice they deserve.

For Islam, the period of the Mogul Empire in East Asia and Northern India is the subject of a later chapter (Chap. 14). It illustrates a dimension of Islamic faith and practice that is relatively little appreciated, except by scholars, in the West today. The Mogul period has some parallels with the mystic Sufism of mediaeval Persia, another dimension of Islamic faith and practice that is little appreciated in the West today, except by historians and other scholars.

In discussing Judaeo-Christianity, I have used the *King James Bible*. For simplicity, and respecting widespread literary and historical usage, as noted in the Preface, I have frequently adopted the traditional religious usage of the terms "man" and "mankind"in reference to human beings and humanity and "he" or "He" for God and the deity. This usage carries no sexist or gender implications but I apologise should it offend any reader's sensibilities.

Doctrines and creeds and the organisations that promote them are key sources of any discussion of religion. No enquiry into Judaeo-Christianity could proceed without making the *Old* and *New Testaments* of the *Bible* the primary source. Likewise, for the Islamic faith, the *Q'ran* is the foundational holy text. These are the "religions of the book" – the ancient texts and their interpretation over time.

The rise to dominance in Western culture of the Catholic Church and the Protestant challenges from the sixteenth century onwards have been the principal carriages of Western Christianity. Through penetration of social and political affairs, these religions have immensely strengthened their impact on the everyday life of their individual adherents.

Beliefs and doctrines expressed in and engendered by the *Bible* and the *Q'ran*, together with participation in associated communal rites and ethical practices, are widely and understandably thought to be the essence of a Western and Middle Eastern religious life. However, there are religious temperaments, states of mind and notions that are not, or are no longer, anchored in such institutional observance and theological doctrines. Quakerism is a good example in Christianity and perhaps some mediaeval Sufi poets were religious more in their poems than at the mosque.

Some years ago, an English Anglican Bishop perceptively – and controversially - wrote that while God might be dead, as Nietzsche had proclaimed, religion was not.

So what might religion be, cast loose of its ancient theistic moorings? Are the beliefs and attitudes they engendered in the distant past necessarily binding on or appropriate to present life, even for the most devout? Is the God of the twenty first century the same as the Biblical and Q'ranic God? To some, such questions are blasphemous, but no discussion of ancient roots of modern religious attitudes should ignore them.

5.2 Religious Sensibility

Among the varieties of religious experience are moods and sentiments of awe and wonder, images of the overwhelming grandeur and beauty of nature, dazzling intuitions of truth (such as the biblical merchant, Saul's illumination when on the road to Damascus), inexplicable mystic trances and calm reflections on the unknown universe. German theologian and philosopher, Rudolf Otto, in an influential book, *The Idea of the Holy* (1917), analysed what he termed "numinous experiences", a form of spirituality whereby individual persons directly apprehend some spiritual force greater than themselves, a depth of feeling for the sacred, free of rational and ethical considerations. Such feelings may be stimulated by places with long-established religious practice of diverse doctrinal significance, or indeed none at all. Not all religions are those "of the book" and not all religious feeling is encapsulated in the doctrines and creeds of the churches.

William James, psychologist and philosopher of pragmatism, in his *Varieties of Religious Experience* (being the Gifford Lectures on Natural Religion, 1901–1902), wrote about a "sense of presence, reflections on being, meditation, quiet cosmic consciousness", and other feeling states (Fig. 5.1). He regarded them as more to the point in understanding religious *experience* than analysing the intellectual doctrines of the theologians. Feeling and conduct, not doctrines and creeds are the universal, constant elements: the end of religion is not God but "a larger, richer, more

Fig. 5.1 William James (1842–1910), pragmatist philosopher, psychologist, brother of novelist Henry James. Houghton Library, Public domain, via Wikimedia Commons

satisfying life". For James, theologians' efforts to prove the existence of God are irrelevant. Similarly, the contemporary meta-narrative of "Biblical theism", with its roots in seventeenth century Calvinistic Puritanism and the much more recent Charismatic movement, draw inspiration less from theological doctrine and more from peoples' direct, felt experience of the Holy Spirit at work.[1]

Outlooks and tendencies, analogous to those that William James identified as religious, are found in the lives of even avowedly anti-religious people. Thus, in the experience of even such a renowned atheist and severe critic of Christianity as Bertrand Russell, feelings and apprehension that William James would term "religious" were expressed in an episode Russell recalled from his first visit to Greece, as an elderly man, in 1952. In company with his wife, Edith Finch, and after visiting the classical monuments, he records: "I found myself in a little church belonging to the days when Greece was part of the Byzantine Empire. To my astonishment, I felt more at home in this little church than I did in the Parthenon or in any of the other Greek buildings of pagan times. I realised then that the Christian outlook had a firmer hold upon me than I had imagined. The hold was not upon my beliefs, but upon my feelings."[2] The difference Russell felt between the Parthenon and the small church is that the latter, he wrote, evoked a sense of sin, which powerfully affected his feelings, whereas the former did not. It is the eternal punishment in hell, inflicted on unrepentant sinners, that Russell lists as one of the reasons he gave for rejecting Christianity, "as a disease and a source of untold misery to the human race".[3] The title he gave to a famous book was *Why I am Not a Christian*, not *Why I am Not Religious*. Russell's feelings on this occasion were something of an echo of his early upbringing in the religious atmosphere of Victorian England (Fig. 5.2).

Fig. 5.2 Bertrand Russell (1872–1970), mathematician and philosopher; author of *Why I am not a Christian*, but touched by religious sentiment. Photo 1957. Anefo, CC0, via Wikimedia Commons

There are other places in Russell's writings where he expressed feelings and a sense of spiritual presence in nature similar to those William James characterised as religious. From prison in 1918, to which he was sentenced for his active support for conscientious objection to serving the war effort, he wrote to Ottoline Morrell that in the freedom of his mind he entered, "the souls of animals and even of the lowest animals".[4]

Russell's religious sensibilities are an interesting contrast with those of his one-time mathematics teacher at Cambridge and collaborator (*Principia Mathematica* 1910–1913), Alfred North Whitehead (Fig. 5.3). In philosophy, Russell turned to a materialistic logical atomism, whereas Whitehead, a neo-Platonist, developed a highly abstract metaphysical system whereby creative process, not material substance, is the most fundamental constituent of the universe. In his major metaphysical work, *Process and Reality*, Whitehead imaged God as the humility of Christ and the figure of Christ the person as "concretization" of the abstract idea of God. Thus, God is not, as in early Christian theology, an omnipotent entity, the unmoved mover of Aristotle and Aquinas, or the stern moralist of the *Old Testament*. In his companion volume, *Science and the Modern World*, Whitehead wrote of, "A steady fall in religious tone" and that, "Religion will not regain its old power until it can face change in the same spirit as does science. Its principles may be eternal, but the expression of those principles requires continual development."[5] Examples of such development feature in several of the following chapters.

In an essay *The rhythmic claims of freedom and discipline*, Whitehead wrote of "the present (that) holds within itself the complete sum of existence, backwards and forwards, that whole amplitude of time which is eternity." And it is in the present that the values of religion (as in science, art and morality) are to be realised in what Whitehead calls "the great art of life".[6] The idea of God as eternally unrealised possibility requires the world to actualise him. The creatures give present actuality to God. It is one of them, man, that created him as the ultimate, potential source of actual events in the world. Man creates God and God is actualized in man.

Fig. 5.3 Alfred North Whitehead (1861–1947), mathematical tutor and collaborator of Russell, and religious metaphysician. Photograph. Wellcome Trust. V0027330. CC BY-SA 4.0 via Wikimedia Commons

Whitehead's philosophy has been influential among modern theologians; for example, John Cobb, Charles Birch and Charles Hartshorne and the University of Chicago School of Divinity. His metaphysics is far removed from the beliefs and practices of Evangelical Fundamentalists.

In *Religion In The Making*, Whitehead famously defined religion as "what the individual does with his own solitariness". More specifically, religion is those purposes and emotions, "due to the personal measures of intuition into the ultimate mystery of the universe". Elsewhere, he refers to "Divine immanence" and the religious motive as, "the response to the divine persuasion".[7]

Russell experienced feelings of mystery and awe which he regarded as religious in a purely personal sense. In a letter in 1941, to his classicist friend, Gilbert Murray, he also declared a liking for the Church of England, "because it is the most purely Platonic form of Christianity. Catholicism is too Roman, Puritanism too Judaic."[8] In another letter, to Ottoline Morrell, "God … if I thought there was a God."[9] Whitehead's, by contrast, was a metaphysical quest to find connections between an abstract divinity and its working in the human mind and the world. The striking differences in the efforts of these two leading twentieth century philosophers to explain religious experience may be contrasted with what seems to us now as the common body of thought within which the thirteenth century Scholastic philosophers and Catholic theologians operated, their differences notwithstanding. God is conceived as a supernatural entity whose commandments are knowable through son and witness, Jesus Christ.

Recent developments in Protestant thought, drawing on the Continental Protestant reformers of the sixteenth century and the seventeenth century English Puritans, have invested all living things with spirituality that can be directly experienced, an echo of pagan people's investment of nature with spirits. Many dispersed groups, like the Lutheran Germans who migrated to Australia in the 1840s to escape political upheavals, were deeply conscious of their dependence on land which they were able to work as free settlers instead of serfdom in their homelands. Land was reverenced for its spirituality, not only cultivated for crops. This may be compared with the Australian Aboriginals' relation to the land – an enduring spiritual bond which co-existed with use of the land for sustenance (Chap. 16).

In Christianity, the image of a knowing and loving God is intact in everyday relations with the earth and can be apprehended by man, despite being damaged by the Fall and prone to disaster. Through redemption, and Jesus's ultimate return, damaged man and nature will be restored. The powers of understanding man lost at the Fall can be restored – through enquiry, experimentation, criticism and exchange. Essentially, God is to be known not only through the *Bible* and other texts but by direct activity, emotional experience and knowledge of his creation.

The literal meaning of the terms "religion" and "religious" is "to be attached" regardless of creeds, doctrines, rituals, and not necessarily to some divinity or divine force. However, in common usage these attachments entail beliefs in other than physical forces, in a spiritual realm. They evoke senses of mystery, awe and – usually – worship, no matter how vague or intangible the object of worship may be.

Over time, the attachments have changed. Beliefs and firm doctrinal positions have been modified or abandoned, new rituals and creeds adopted, yet there are attitudes and values that persist. As we shall see, attitudes arising from or supported by the ancient doctrine of man's superiority to the rest of the natural world have been forceful and persistent over millennia, not only in a religious context.

5.3 God, Man, Nature

In Judaeo-Christianity and Islam, the sacred texts have been presented as the word of God ("God breathed"). Their subject, however, is man. Some Christians today regard the God of the *Scriptures* as one of the supreme creations of the human imagination – humanity's perception of a cosmic order within which man and his needs and interests are of overwhelming importance. The *Old Testament* elevation of man over the earthly kingdoms of both animate and inanimate nature, hence his self-importance, has been and remains a dominant motif in Western culture.

By contrast, there are also in the monotheistic religious traditions of Judaism, Christianity and Islam striking episodes of direct, loving human communion with animals and plants, notably the sharing of the Earth as in the Garden of Eden, in Sufi and Christian nature poetry and in the love of nature expressed by Mogul Emperors and their courts. Together with the mediaeval figures, Hildegard of Bingen and Meister Eckhart, they challenge the orthodoxy of man's dominion, suggesting a partnership or a sharing of the beauty and fruits of the Earth.

These positive attitudes to nature and the living world are the subject of later chapters. First, we need to examine the religious doctrine of a hierarchical order of nature whereby animal and plant life have been treated as subsidiary not only to God and the angels but also to man. Their purpose, according to this doctrine, is to satisfy man's wants as stated most incisively in Chap. 1 of *Genesis*, the first book of the *Old Testament*. The development of the modern movement of eco-Christianity among both Protestants and Catholics draws upon ecological science both to refute the notion of a hierarchical order of nature as a metaphysical principle and to develop an alternative and better grounded vision of man's responsibility for the future of life.

5.4 Persistence of the Religious Impulse

Since, as we have seen, religious feelings and outlooks are not necessarily theistic and monotheistic religions are not the only systems of belief, why bother with ancient texts? Even for active believers, understanding and belief have moved on, greatly affected by modern scholarship, science and the experience over millennia of lived religion. Sceptics and atheists dismiss the ancient religious doctrines as the products of diseased imagination, demonstrating the distortions and misunderstanding of nature that were inevitable prior to the development of scientific knowledge and methods of enquiry. Atheists, from the French *philosophes* in the eighteenth century to Christopher Hitchens and Richard Dawkins in the twentieth have aimed to undermine any grounds for faith and belief, whether ancient or modern.

Whatever position one may adopt regarding the continuing controversies between science and religion, there can be no doubt that religion has been and still is a major factor, not only in the West and Middle East but worldwide, in shaping humanity's view of itself and its place in the world. Religious ideas have lodged in human minds and cannot be ignored in the course of modern efforts to understand human life. The ancient religious texts and commentaries are their most authoritative and still living source.

German philosopher, Martin Heidegger, in an essay on the magnificent chorus from Sophocles' *Antigone*, "Ode to Man", expressed an insightful opinion about the origins of ideas and beliefs.[10] He wrote that it is fallacious to believe, "history begins with the primitive and backward, the weak and helpless. The opposite is true. The beginning is the strongest and mightiest. What comes afterwards is not development but the flattening that results from mere spreading out; it is inability to retain the beginning; the beginning is emasculated and exaggerated into a caricature of greatness taken as purely numerical and quantitative size and extent." We should not dismiss ancient religions in the belief that their impact has passed, through "spreading out", but early beliefs and doctrines (such as man's dominion) may be supplanted over time.

Since I have contended that one of the problems humanity has created for itself in its early religious quest is the exaggerated importance attached to our own species as a factor in living nature, we may pause for a moment to consider what Sophocles intended in his "Ode to Man". The Ode addresses man's formidable power, his surmounting of the huge waves of the ocean, his strides across the land, his cunning and his dominion. A wonderful being, as some translators have it; strange and monstrous as others do. "He wearies even the noblest of the gods, the earth indestructible and untiring, overturning her from year to year, dominating beasts, birds, the seas, earth". But "rising high above his place" in death "he comes to nothingness".

For the Greek tragedians hubris, overweening pride and self absorption, commonly preceded nemesis, or collapse into misery and death. Our creation myths and religious doctrines have placed us, *Homo sapiens*, at the centre of the universe. The gods we worship are our creation, images of ourselves in our best and worst aspects, what we aspire to and, conversely, what we despair of. Nothing like us has yet been discovered anywhere in the universe. Religion has grown into a grand narrative of life in which man is constantly at centre stage as if both ringmaster at a circus and magician practicing black arts. Giving such close attention to our own being, elevating it above nature, we have prided ourselves on our God-like magnificence, self-important even when sinning. A religious impulse from the time of its early manifestations has led us to demean the natural world or at best to place it in bondage to our interests.

Fig. 5.4 E.O.Wilson (b. 1929), biologist, leading ecologist and environmentalist. Jim Harrison, CC BY 2.5 via Wikimedia Commons

The American ecologist, Edward O. Wilson has been among those scientists and thinkers who for decades have warned of the likely consequences of our failure to see that it is the whole of nature that is our shared home and that we must turn from a mentality of dominance and exploitation to one of care and restoration, that is, to intelligent and ethically strong stewardship.[11] (Fig. 5.4) Religious belief is not inevitably inimical to this and can be turned towards the love of nature and restoration of the damage we have done.

Just as in the seventeenth century we experienced the Copernican revolution that shocked our ignorant belief that the solar system revolved around the earth, we need a modern revolution to shock us into the understanding that nature does not revolve around man. Humans, however, possess the intellect, imagination and capacity for moral wisdom that will enable us to become the responsible regent that the early worshippers and creators of God once expected us to be.

Notes

1 James, W. (2002) *Varieties of Religious Experience. A Study of Human Nature.* Centenary Edition. London and New York: Routledge. p.391.; Baker, S. "Faith-Based Schools and the Creation Controversy – The Importance of the Meta-Narrative" an essay in a survey of the interpretation of religion in schooling in different countries: Chapman, J. et al., eds. (2011) *International Handbook of Learning, Teaching and Leading in Faith-Based Schools.* Dordrecht: Springer.

2 Russell, B. (1971) *The Autobiography of Bertrand Russell. Vol.III.* London: George Allen & Unwin Ltd. p.68.

3 Russell, B. (1927) *Why I am not a Christian.* Pamphlet of talk to National Secular Society at Battersea Town Hall.

4 Russell, B. (1968) *The Autobiography of Bertrand Russell. Vol II.* London: George Allen & Unwin Ltd. p.94.

5 Whitehead, A.N. (1926) *Science and the Modern World.* Cambridge. The Syndics of the Cambridge University Press. pp.233 and 234.

[6] Whitehead, A.N. (1978) *Process and Reality.* New York: Free Press. p.207; Chapter 3 "The rhythmic claims of freedom and discipline" in Whitehead, A.N. (1950) *The Aims of Education and Other Essays.* London. Ernest Benn Ltd., p.45.

[7] Whitehead, A.N. (1926) *Religion in the Making*. pp.189–91; William James had also defined religion with reference to the solitude of individuals: "the feelings, acts and experiences of individual men in their solitude, so far as they apprehend themselves to stand in relation to whatever they may consider divine." (*op.cit.* pp.28–9).

[8] Russell, B. (1968) *The Autobiography of Bertrand Russell. Vol. II.* London: George Allen & Unwin Ltd. p.250.

[9] Russell, B. *op.cit.* p.90.

[10] Heidegger, M. (1959) "The Ode on Man" in "Sophocles' Antigone" in *An Introduction to Metaphysics*. New Haven: Yale University Press. pp.86–100.

[11] Wilson, E.O. (2003) *The Future of Life.* New York: Random House.

Complete Listing for Wikimedia Commons Images

5.1 William James b1842c. https://commons.wikimedia.org/wiki/File:William_James_b1842c.jpg Attribution: Houghton Library, Public domain, via Wikimedia Commons.

5.2 Bertrand Russell 1957. https://commons.wikimedia.org/wiki/File:Bertrand_Russell_1957.jpg Attribution: Anefo, CC0, via Wikimedia Commons.

5.3 Alfred North Whitehead – cropped. https://commons.wikimedia.org/wiki/File:Alfred_North_Whitehead_-_cropped.jpg Attribution: Wellcome Trust, CC BY-SA 4.0 <https://creativecommons.org/licenses/by-sa/4.0>, via Wikimedia Commons.

5.4 Plos Wilson https://commons.wikimedia.org/wiki/File:Plos_wilson.jpg Attribution: Jim Harrison, CC BY 2.5 <https://creativecommons.org/licenses/by/2.5>, via Wikimedia Commons.

Chapter 6
Christianity Spreads from Ancient Roots

The Christian religion, a dominant force in Europe and parts of the Middle East for 2000 years since its inception with Jesus Christ and his disciples, and now a global force, grew out of the beliefs and practices of the Israelites, as expressed in the Old Testament stories. Greek philosophical thought also had a considerable impact, as we shall see in Chap. 7 when discussing views on the nature of the human soul. Jesus Christ was a Jew, as were his immediate disciples and many of his earlier followers. Paul, however, as a Greek was in touch with Hellenic thought. Christian beliefs and doctrines did not develop independently of other belief systems in the pagan world. Moreover, they were formulated and recorded over succeeding centuries by church people forming a social superstructure whose very existence rested on the foundations of labour provided by serfs. When they expressed views about nature (the earth, animals, plants), they did so from a standpoint socially and cognitively different from that of the masses. Paradoxically, Christianity, a religion for the masses, came to be articulated and administered by elites.[1]

6.1 *Old Testament* and Other Ancient Religions

Despite the unity of the *New and Old Testaments*, which was declared by the early Church Fathers and shown in the doctrine of redemption revealed in the *Old* and fulfilled in the *New Testament*, there are very significant differences between the two. It is instructive, for example, to compare the Sermon on the Mount with the Mosaic Law. God ultimately forgives but is stern and punishes most severely in the *Old Testament*; in the *New*, although at times angry, Jesus is by contrast gentle and forgiving.

We will, however, better understand the part played by Christianity in sustaining *Old Testament* beliefs about humanity's superior place in nature by reflecting on

M. Skilbeck, *Loving and Studying Nature*, International Explorations in Outdoor and Environmental Education 10, https://doi.org/10.1007/978-3-030-80751-1_6

their continuity with ancient roots and sources, pagan included. The religion of the Jewish people, the Tribes of Israel, was but one among many ancient religions. Its guardians, the nation of the Jews, fought against the other religions to achieve and retain political power and religious dominance. Occupancy of the Promised Land was vital for the Jews not only to enable their grip on political power but to confirm their religious ascendancy, just as recording the stories and the history of the Jewish people in the *Old Testament* was to give them intellectual and moral authority. This was the stern, proud religion of a historical nation with its unique holy book. That authority, however, was challenged from within as well as from without.

Idolatrous worship of the Sun, possibly personified in Baal (the Canaanite peoples' deity), by dissidents within the Jewish flock, was extremely threatening to the unity of the Jewish nation. So it was severely punished, as God inflicted disaster after disaster. For a lengthy period, following the flight from serfdom in Egypt, the Israelites succumbed to the temptation of idol worship. They mingled with pagans, whose gods of nature – such as Baal, a fertility god, a deity of rain and thunder – had to be brought down in the wrath of the super-natural God. The tolerance shown by pagan sects for one another was not reciprocated by the Jews. There was no respect for other religious systems. Thus the *Old Testament* provided a precedent for the later intolerance by Christianity of other religions and the determined repression of dissension within its own ranks.

The threat of succumbing to nature worship (to be distinguished from respect for nature) was not wholly overcome by the Israelites. Pantheism from time to time threatened orthodoxy. God had created nature and permeated nature, but for the most orthodox believers, God's presence was above and beyond nature, just as man was part of nature yet was above and had dominion over it. Such fine distinctions within Orthodoxy were at times lost in the vagaries of ordinary life. But a wrathful *Old Testament* God was ever-present to punish the wayward.

Blood-letting and the taking of life deriving from magical rituals were an integral part of ancient religions. Propitiation of the gods – and of the Hebrew God – through sacrifice and worship was all but universal in the ancient world. Sacrifice of living creatures became embedded in human consciousness as a fit and proper form of worship, thanksgiving and repentance for sin. The life of Jesus was sacrificed for the redemption of humanity and his blood, according to some interpreters, spilled on the Cross. This image of sacrifice is perpetuated in the drinking of wine and eating of the wafer in the present day Catholic mass and Anglican communion.

The loss of life in war is universally commemorated by religious ceremonies whose motif is noble sacrifice. Bloodletting, either physically or symbolically, remains a fundamental element of what we understand as the normal condition of human life, responding to our choices such as, for example, in the slaughter of animals for human consumption. We return to the importance of sacrifice and bloodletting below.

Worship and sacrificial rites require holy places and they involve sentiments and actions in the name of religion. These holy places, from the earliest times, have included sacred groves, springs, wells, trees, mountains, water courses and rivers, some celebrated over many centuries, such as Green Men and Green Women in

Celtic churches (Fig. 6.1). Natural features and elements, they have served as abodes of the spirits and sites of worship and religious festivities. On coming to land, as the floodwaters receded, Noah immediately erected an altar for sacrificial worship.

Ever since Helen, mother of Emperor Constantine, popularised pilgrimage, Christians have flocked to holy sites, some, such as Lourdes, renowned for claims of miraculous healing powers. A devout Christian travel entrepreneur, Thomas Cook, popularised pilgrimages in the nineteenth century to holy sites in Palestine and Syria, and thereby prospered. For Muslims, a pilgrimage to Medina or Mecca is a lifelong goal, enjoined on the faithful. Temples, churches, mosques and so forth succeeded the natural holy sites but they have not replaced them. The pagan belief in spiritual presences in natural sites lives on.

6.2 Carryover into Christianity

The close relationship of a spiritual life with living nature and natural motifs carried over from ancient cults into Christian art and architecture. Natural motifs play a vital part in the decoration and ornamentation of religious buildings, from the time when Solomon decorated his house with "carved palm trees and open flowers" (*Old Testament I Kings 6.35*) and in the pictorial embellishments of mediaeval religious texts. Together with structural features, trees, flowers, leaves and vines maintain the image of religion in a close association with natural forms as if itself an outgrowth

Fig. 6.1 Green Man. The underside of a misericord seat in the choir of the Abbaye de la Trinité, Vendôme, France. Jacques CHAZARD, CC BY-SA 4.0 via Wikimedia Commons

of the richness and diversity of the natural world. Even so, the early nature religions were attacked as Christianity strove to establish itself.

Stone Gothic vaulting in Christian cathedral naves and cloisters echoes the living canopies of forest trees. Twining leaves on the capitals of Corinthian columns in the naves of cathedrals are those of the acanthus plant, inspired by those of ancient buildings (Fig. 6.2). Carved fruits and flowers, wheat sheaves and innumerable images of agriculture, viticulture and horticulture, in the painted glass windows of the great cathedrals of Europe, link the heavenly with the earthly city. Nevertheless, as pointed out by American historian, Lynn White, the Christian attack on the pagan guardian spirits of sacred springs, graves and so on and the substitution of heavenly abodes for the angels, "made it possible to exploit nature in a mood of indifference to the feelings of actual objects".[2] It has fallen to twentieth century Eco-Christians to reverse this exploitation as an element of religious belief and practice.

Religious belief and virtuous everyday life were barely if at all distinguished as separate domains in the *Old Testament*, nor were they in the Homeric poems, in Hesiod's Theogony, or in the Greek myths. Gods were integral to everyday life, consorting with humans in affairs of war and love and either hindering or advancing everyday human life. Christianity emerged into an already rich, diverse and pervasive religious culture. It is in the context of both the *Old Testament* and other earlier forms of religious expression and observance that the injunctions and admonitions of the Sermon on the Mount need be seen. There was both discontinuity and continuity, but over and above the antecedents was a formidable new energetic charge of love, devotion, charity, fellowship and care for others than oneself. These and related elements distinguished Christianity from its forerunners. Lessons on

Fig. 6.2 Use of natural form (acanthus leaf) as decoration in architecture. Corinthian capitals of Roman era Maison Carrée, Nîmes, completed about 2AD. © José Luiz Bernardes Ribeiro/CC BY-SA 3.0

forgiveness and redemption were further strengthened by Paul's *Epistles* and his missions to the Greeks and Romans. Over several centuries, through the writings of theologians, missionary endeavours, preaching by priests and the witness of saints, all of this was gathered together into a new design, a spiritual and material reality which gave meaning to the existence of all people everywhere and directions for a good life in the eyes of God. But this new design included many elements of its forerunners.

The Crucifixion and Resurrection had proved to his followers that Jesus was the incarnate God and Saviour, "Death hath no more Dominion." The *Old Testament* expectation of redemption for a chosen people through an inspired Messiah had been realised in Christ (the Greek name for the Messiah). Paul (killed in 69 A.D.) spread the new message across the Hellenistic and Roman worlds. But, new as this message was, it served rather to reinforce man's sense of his elevated perch in the hierarchy of nature, than to align him with all creatures, great and small.

6.3 Early Spread of Christianity

Over the centuries, from its modest and almost fatal beginnings, Christianity spread across the Middle East, into Western Europe, Northern Africa and, ultimately, around the world. It became a profoundly transformative force in the social and political spheres. It saturated culture and endeavoured to fight off heresies and alternative systems, notably that of Islam while doing its best to root out surviving pagan cults and their deities.

The first Christians were the immediate disciples and followers of Christ in his short life on earth. Conversion of Jews and adherents of other religions and cults followed across the Roman Empire. Initial persecution by the Roman state and the contempt of its official pagan religions did not succeed in quelling the Christian zealots. In the course of its first three centuries, Christianity secured a strong foothold in the Roman civic order. As Roman political power weakened, Christianity moved to fill a vacuum. Of the utmost importance was the consolidation of the faithful into the structures of the Church and their acceptance of its doctrinal affirmations and authority, its claims to universalism - hence exclusionism. From a loose association of largely independent and small gatherings of believers, scattered across Southern Europe, Northern Africa and the Middle East, the Catholic Church grew into a highly organised hierarchical institution. Priesthood emerged from the apostles; bishops, who initially were only first among equals and elected, became both instruments of ecclesiastical authority and, following the fourth century Council of Nicaea where doctrines were debated and over which Emperor Constantine presided, hierarchical functionaries of Empire. At Constantinople, the Emperor gave Christianity a new centre of power which eventually separated from Rome as Byzantine Christianity. It is not evident that Jesus of Nazareth anticipated such accession of power, although the *New Testament* contained warnings about temptations of the earth as well as flesh.

Numerous debates occurred and many disagreements arose over the meaning in Biblical passages of such mysteries as a trinity of God the Father, Son and Holy Ghost, and the flesh of the divine Christ. Synods assembled and Councils were held to root out heresy and establish orthodoxy. At the end of the second century, the Church fathers set up a canon or "rule of faith" as a criterion of orthodoxy. Emperor Constantine (272 – 337), himself a Christian, by the Decree of Milan in 313 AD, required tolerance of Christianity within the Empire, and by Christianity of other sects. His acceptance of Christianity was a major triumph for the Church. He was, as at the Council of Nicaea (325 AD), active in its affairs although he also retained the title of Pontifex Maximus, head of the Roman pagan priesthood.

6.4 The Catholic Church

By the fourth century, the Church had ceased to be simply a self-contained religious institution bursting with enthusiasm. It had become a most earnest sociopolitical movement that, over the following centuries, grew to rival secular authority. In the course of time, what had started as a frugal, subsistence way of life for humble believers became, for many of the elevated leaders, sumptuous living in the grandeur of the Church complicit in exploitation of the land and the masses. Emperor Constantine had set in motion what became one of the great conundrums for the Church, its relation with secular power. Early in the fourth century, Ambrose, as Bishop of Milan, claimed equality with the Roman emperor, if not superiority. He established the principle that there were matters on which State must yield to Church. The impact of his consolidation of the growing power and influence of the Church and of its leading dignitaries was to be felt over many centuries.

Other Church fathers of this period contributed pillars to strengthen the burgeoning power of the Church. Origen of Alexandria (c.185–c.254), never canonized because of his unorthodoxy, provided early Christians with "truths" as a basis for faith, interpreting Scripture both literally and allegorically. Jerome (347 – 420), of mythical fame for extracting a thorn from the paw of a lion, conferred with Hebrew and Greek scholars in producing the *Vulgate*, a standard Latin version of the *Bible*, which, with further revisions, was in the sixteenth century pronounced by Pope Clement VIII the authoritative text for the Catholic church (the *Clementine Vulgate*) (Fig. 6.3). Twentieth century editions of the *Vulgate* maintain a continuous textual tradition dating to the days of Jerome and the Roman Empire.

The Church was also a fortress of hope as barbarian invasions and internal strife weakened the Empire, ultimately to the point of collapse in the West. It was Augustine, Bishop of Hippo, (354 – 430), another of the early Church Fathers, who, in his monumental *City of God Against the Pagans*, gave the most forceful and compelling expression of this hope. Earth having failed, a Catholic-defined Heaven beckoned, as we see in the following chapters.

The power of the Church was not merely a function of the collapse of the Empire. In order for individuals to aspire to eternal life with God, they had to be baptized,

Fig. 6.3 Harmony between early Christianity and the natural world. Albrecht Dürer's 1514 etching of St Jerome in his study. St Jerome in his Study. Kupferstich-Kabinett Dresden. Albrecht Dürer, Public domain, via Wikimedia Commons

thereby becoming members of the Church. The doctrines concerning salvation were the property of the Church as were the rituals and ceremonies. Most of all, the Church was the mediator of the *Bible*. It was in control of the Word of God and his earthly interpreters. As Edward Gibbon put it in his *Decline and Fall of the Roman Empire* (1776 – 1788): "The loose wanderings of the imagination were gradually confined by creeds and confessions; the freedom of private judgement submitted to the public wisdom of the synods; the authority of a theologian was determined by his ecclesiastical rank; and the episcopal successors of the Apostles inflicted the censure of the church on those who deviated from orthodox belief."[3] (Fig. 6.4).

Fig. 6.4 Edward Gibbon (1737–1794), Enlightenment historian who upset many Christian readers by his criticisms of the role of Christianity in the decline of the Roman Empire. Engraved by E. Scriven after Joshua Reynolds portrait of Edward Gibbon. British Library/ Mechanical Curator collection. Public domain, via Wikimedia Commons

The Church had at its disposal a considerable array of instruments, some highly emotive. These included the asceticism, fanaticism and martyrdom of early Christians who served as heroic models of pious conviction. The early monasteries, which grew out of hermit cells and camps in the Egyptian and Syrian deserts, were both a practical demonstration of the sacrifices attendant on extreme piety, and centres of relative calm, ordinary life as the Empire succumbed to intrigue, attacks and invasions. This at least was the model. It cannot be said that the various instruments at the disposal of the Church always performed to the expected standard, set too high by the saintly leaders for many of the followers, so scandals often marred the image of the holy self sacrifice. Nevertheless, a watchful Church continued to assert its ultimate authority.

In the sixth and seventh centuries, the Roman pagan religions were eclipsed by the Catholic Church and its doctrines, not without a protracted struggle. Thus the unifying and organising sociopolitical force of the Church was cemented in the West. Constantine had extended the Empire's boundaries to Istanbul and moved the capital from endangered Rome to the renamed city, Constantinople. The Eastern Catholic Church was formally separated ecclesiastically from Rome, the Pope no longer its head. As in the West, it aspired to increasing secular as well as religious authority. In secular matters, fealty was accorded the Emperor but the Church often secured authority over lesser kings and princes.

As a consequence of the engagement of the Church with civic life over the first few centuries after Christ, to be a Christian came to mean more than subscribing to particular religious doctrines and performing certain rituals. To identify oneself as a

Christian was to a degree to proclaim oneself as Roman, and vice versa. Yet there was an incipient contradiction in this identification since doctrinal and credal emphasis was on the human soul and its journey towards heaven, not on a person's place in the civil domain. A question which resurfaced over the centuries was to whom should a Christian soul be attached? It was answered in the twelfth century by Thomas à Becket, Archbishop of Canterbury, who paid with his life for his answer. Again, during the Reformation, the souls divided into Catholic versus Protestant and the secular princes made the decisions as to which territories were to be home to one or the other.

6.5 Man – God's Regent on Earth

Christians, as the Israelites before them, were at pains to demonstrate the uniqueness and superiority of their religion, the transcendence of its beliefs over all others. Unlike their pagan competitors Jews, and Christians after them, believed in one God, creator of the universe. The creator was uniquely omnipotent, omniscient, eternal. Creation was not a mere chance occurrence, it was purposive, and all of Creation existed to fulfil a divine mission. The mission was God's; man was his regent on earth, called upon to choose God as his lord and to abide by his commandments.

Humans were to live on earth, piously and devoutly, obediently and worshipfully, as prelude to the possibility of eternal bliss in the lap of God. This, however, was only a possibility for the few, chosen by God out of his grace. Yet all must live the good life in the hope of receiving that grace. Why people adopted Christianity, aware that the great bulk of them faced eternal damnation, calls for explanation. Was it a case of hope prevailing over common sense and reason? Were conditions of everyday life so appalling that even the faint prospect of salvation gave hope? In spite of the doctrines and the authority of the church, did the faithful in their hearts secretly reject the theological doctrine of salvation and damnation? Did they, living in the present, put aside all thoughts of the uncertain future? Whatever answers may be given to these questions, the notions of salvation through the resurrection of Jesus and eternal life of the spirit greatly strengthened the controlling hand of the Church.

In his *Decline and Fall of the Roman Empire*, Edward Gibbon accounted for the success of Christianity in a naturalistic way that aggravated many of his eighteenth century contemporaries. He identified: the Hebraic, intolerant zeal of the early Christians; the miraculous powers ascribed to the Church; the "pure and austere models" of people leading a devout Christian life; and the union and discipline of the Christian Republic.[4] Bertrand Russell, in his *History of Western Philosophy*, added to Gibbon's factors the terrifying threat of damnation and the possession of the sacred book, whose legends and history, people were encouraged to believe were a true and providential history dating from Creation.

Man, made in the image of God, was mortal yet in possession of a potentially immortal soul; his duty was to attend to the soul, to do everything possible to nurture its progress towards goodness, to exercise his will, implanted by God, to ward off temptation and evil. Man was born in Adam's sin but could be redeemed through a life of virtue, the life that was demonstrated in the *New Testament* by the living God, Jesus. To achieve merit, man had to cast his eyes heavenward, rejecting close attachment to earthly existence. In his *Confessions*, Augustine proclaimed that, "the love of this world is fornication against God."[5] Augustine himself, a powerful bishop, nevertheless played a most active part in the church's earthly affairs. As for loving the world being a fornication against God, when a young man, Augustine fathered a son out of marriage, and left the mother.

6.6 An Intellectual Boost to Faith: The Theological Impulse

Although it has been said, notably by followers of Karl Marx, that ideas alone do not change cultures and ways of life, as they become organised into systems of thought with implications and consequences for human social organisation and behaviour, they can become powerful drivers of cultural, social and political change. There can be no doubt that the closely reasoned ideas and behavioural prescriptions of its theological expositors and interpreters gave to Christianity a tremendous boost throughout its first 1500 years. They provided the intellectual drive and emotional conviction needed to carry the vagaries, doubts and complexities of faith over hurdles that stood in the way of unshaken belief and the triumph of the Church in its earthly mission. In this way, the ideas gave to correct, that is to say authorised belief, an importance equal to virtual action.

Theology was absolutely essential in systematising the diverse elements of religious belief into orderly human conduct and socio-political organization. The passionate affirmations, the stories, miraculous events, prophecies, insights, commentaries, and the assorted reports whether by Jews or Christians had to be verified and articulated. Further, for the very survival of these religions, rules were needed, structures had to be created, limits set and authority firmly established to beat off challenges as much from within as from other religions. Not all challenges could be lightly brushed aside; some had to be carefully studied and reasoned against. A superimposition of rational thought was necessary if religion was to have more than a uniquely personal and emotional appeal.

In the thirteenth century, in Paris, the master theologian and Dominican friar, Thomas Aquinas, and his great Franciscan rival, Bonaventura, reasoned that God's existence could be proved. Faith was the bedrock but it was not sufficient in a sceptical or indifferent world where other interests were pursued at the expense of concern for the human soul and its relationships with the Almighty. The task of theology was to build an impregnable fort of human reason in defence of Christian belief. One philosophy – that of the Scholastics, those devout thinkers and teachers in the cathedral schools and colleges – was to bind together one church, one ruler, one language, one civilization and one order of nature, combined in a broadly harmonious and impregnable unity.

It is fascinating to observe how that Western European Scholastic fortress, assembled in the course of the thirteenth century into a most imposing structure, has been reconstructed, even to the point where, today, its primary categories and philosophical foundations have come into question even in some colleges of divinity. Led by the Jesuits, there was a seventeenth century revival following the onslaught of the Protestant Reformation, and another one in the twentieth century under Pope Leo XIII and Neo-Thomist philosophers in Europe and America. The coherence, order, logical tightness and authority of the Thomistic system is known these days as the official philosophy of the Roman Catholic Church to priests in training, theologians and philosophers of religion. Vestiges survive in the wider Christian world but as an intact dominant theological system, its impact waned even in parts of the Catholic Church with the passing of the Middle Ages, as did all the other elements in the grand mediaeval synthesis. We return to Thomas Aquinas and other Scholastic views below, when discussing the order of nature.

6.7 Augustine, Bishop of Hippo

It was because he understood and felt threatened by the political as well as the emotional force of the Roman Pagan religions that, 1000 years before Aquinas, Augustine, Bishop of Hippo (354 – 430) wrote his polemical and theological masterpiece, *City of God against the Pagans* (413 – 426) (Fig. 6.5). Edward Gibbon, who criticised the early Christians for contributing to Rome's decline by their attack on the pagan gods, summed up the purpose of *The City of God*, "Designed as a solid and immortal proof of the truth of Christianity."[6] At great length, Augustine explained and elaborated the *Bible* in order to convince sceptics of its historical truths and its logical coherence. He sought to demonstrate the superiority of Christianity to paganism, both as a system of thought and belief and an ethically grounded, personally fulfilling way of life.

Produced at a time of schisms and heresies, barbarian attacks on the Roman Empire and frequent disruptions to normal life, *The City of God* projected a beguiling picture of an eternal, unified church rising far above the fray. It also visualised the ideal Christian, spurning temporal pleasures and earthly temptations while preparing for eternal life.

One of the four Doctors of the early Church (the others are St Ambrose, St Jerome and Pope Gregory the Great), Augustine is best known today for one of the most searching and influential of all autobiographies, *The Confessions*. Rousseau, Tolstoy and H.G. Wells are among those who, from quite different standpoints, followed his autobiographical lead in psychological insight and frankness, not to say self-flagellation.

The religious significance of *The Confessions* lies in its account of how, only after a long and devious journey, Augustine came to an emotionally overwhelming

Fig. 6.5 St Augustine, master theologian of early Christian church and of Protestant reformers. St Augustine arriving in Carthage. Fresco. Ottaviano Nelli. early fifteenth century. Sant'Agostino, Gubbio. Public domain, via Wikimedia Commons

realisation of what Christianity meant to his whole life. It is also, in its later sections, a remarkable dialogue between Augustine as the enquiring servant of God and his conscience, represented as God within.

Mischief and mistakes in childhood, for which Augustine severely castigated himself, were followed by the study of the classical liberal arts (which he later rejected for their pagan associations). He also studied the Greek sceptics and neo-Platonists in his youth and early adulthood, later attributing wisdom to them, but necessarily that of a lower order than the wisdom of God. During his years of early and middle adulthood, Augustine fell under the spell of the Persian Manichaean religion, an alternative to or a variant of Christianity led by a prophet, Mani, in the eastern frontier of the Roman Empire and spreading westwards. His conversion to Christianity came in a famous episode when he rushed in tears into a garden, throwing himself under a leafy fig tree and declaring his acceptance of what had always been his mother's religion. This was a great emotional release after years of self-doubt and gradual loosening of the Manichaean ties.

Much like the *Old Testament* prophets, Augustine felt himself in direct communion with God, no less as interlocutor than as medium taking messages from on high. As noted above, the text of the latter parts of *The Confessions* can be considered as a closely argued dialogue between the outer man, the Bishop, and his emotionally pitched conscience.

6.8 *The City of God*

While *The Confessions* is the story of a personal spiritual quest which can be followed by others - and indeed has been - the monumental scale of *The City of God* had a purpose relating more directly to the future of the Church. In the first place, Augustine had to refute the pagans and heretics, both the Manichaeans and the Pelagians, followers of the British monk, Pelagias, who, against Augustine, urged Christians to work for their own salvation and not depend only on God's grace, "We have free will and are not crippled by original sin." For Augustine, the ground was to be cleared of such alternative and dissenting viewpoints before the work of God could be revealed in its glory. Then the history of the world from *Genesis* in the past to the coming Day of Judgement was to be recounted from the standpoint of a Christian theologian and sacred historian. Whatever else has been claimed for it, *The City of God* is a masterwork of the literary imagination and that may be the best way to read it today.

The *Old Testament* history, which Augustine reviewed, was that of the unfolding of God's heavenly will and man's sinful impulses on earth. In this history, God's purpose was paramount but sinful mankind had to be urged and hectored to fulfilment through virtuous living, free of all pagan and heretical taint. Finally, Augustine showed how hope and perseverance in mankind's historic quest were to be inspired and sustained by the vision of a heavenly city awaiting the good and faithful. God's city was sharply contrasted in truth and beauty with the dullness and confusions of daily life on earth. Even so, Augustine invited readers to recognise elements of the heavenly city above in the existence and workings of the Church below. He was as much bishop as theologian.

While Augustine generally wrote in demeaning terms about man's absorption with earthly pleasures, there are passages in *The City of God* that show his reluctance to escape entirely from enjoyment of the natural world, "Behold the manifold and varied beauty of heaven and earth and sea, the abundance and marvellous quality of light, sun and moon and stars, shady woods, the colours and odours of flowers, the many kinds of birds who sing and display their bright-coloured plumage." Continuing, in terms that would delight Edward O. Wilson, distinguished modern day entomologist and ecologist, Augustine embraces, "the diverse kinds of so many animals of all sizes – and the smallest are most admired, for we are more amazed at the work of tiny ants and bees than at the huge body of whales."[7] However, these are the delights of unhappy and condemned men who are to be deprived of the vastly greater heavenly delights of eternal life. It seemed, too, that Augustine did not much care for the wilder aspects of nature. In discussing Varro's writings on agriculture, whereby Sylvanus, god of wild nature, is repulsed, he expresses his dislike for the fierce, horrid and uncultivated forest god.[8]

In order to encourage an acceptance of the vision of eternal life that he had set out in such detail, Augustine used fear and terror of the living, burning bodies of the sinners, eternally tormented in hell where not just burning humans but immaterial spirits – demons and angels who misbehave – were liable to never-ending burning and torture. Of all the weapons chosen by the early theologians, fear of eternal punishment on the one side and hope of salvation on the other were the two most often evoked.

Augustine made a theological point that was to be debated for well over a thousand years, resurfacing in the Reformation sects of Lutheranism and Calvinism even more than among Catholics. It was by the grace of God, not through virtuous acts and good works, that people entered heaven, and only a small minority at that. God, omnipotent, was not to be constrained by mere human virtue. Yet virtue and good works were still required and without faith none could reach the City of God. This conundrum was still being worked on by the Puritan divines on the frontier of Christianity in the wilderness, in early seventeenth century Massachusetts.

Relatively few Christians could have read or understood the voluminous *The City of God* or any other of the numerous books, tracts and letters that Augustine wrote. Readers would have been in the main other bishops, priests and some well-to-do educated lay people. The majority were expected to take their doctrines from priest-led communities of the faithful, within the ambit of the Church.

While the City of God could never be achieved on earth, the Church in Augustine's mind should embrace as many as possible of the Heavenly City's attributes. Aided by the Church, every believer able to reason could reach toward God. In his letters and tireless exercise of the duties of a bishop, Augustine constantly laid down the epistemic principle that God illuminates the mind, empowering humans to see the truth even when they failed to pursue it.

Humans have been endowed with reason, a point the great mediaeval theologians like Thomas Aquinas and John Duns Scotus were to seize on in their rationalist arguments for Christian belief and faith. Through reason, suitably guided, the faithful could grasp the essential truth of the *Bible*, even interpreting it metaphorically where a literal interpretation might confound reason. That, too, was a thought Aquinas shared with Augustine. It was to prove of great use when, in the eighteenth and nineteenth century, critics unpicked some of the fundamental historical and doctrinal claims of Christianity.

6.9 Being a Christian; the Biblical Hierarchy of Being

The faithful, through understanding the *Bible*, would learn to respect, admire and worship God's creation but not with an undue enjoyment of earthly things and not with any misguided opinions challenging man's proper higher place in the order of things. For the theologians over the centuries, of particular importance was respect for the Biblical hierarchy of being. This aspect of the creation story was to be taken quite literally. For Augustine, the hierarchy rose from: (1) things that merely exist, through (2) those that live, to (3) those that have reason (e.g. humans). Then, further up, were (4) angels, and (5) at the very peak, God. *Respect for and worship of God's creation meant acceptance of this hierarchical order of nature, a doctrine that has had profound consequences in human attitudes towards nature.* (Chaps. 7 and 9).

The City of God and *The Confessions* are among the most powerful and influential of all Christian theological works, together with Aquinas' *Summa Theologica* and *Summa against the Gentiles* and *The Commentaries* of John Duns Scotus. *The City of God* and *The Confessions* are great, reflective poetic works, masterpieces of Western literature and an inspiration to many philosophers including Schopenhauer, Kierkegaard and Nietzsche. Bertrand Russell, in his *History of Western Philosophy*, found considerable interest in Scholastic philosophy and wrote admiringly of Augustine's thought, including his speculations on the nature of time. Despite their majesty, to many Christians today, these classic works must seem like dust settling in rooms of an old, deserted mansion.

His strictures on the pagan philosophy that he had studied in his Manichaean period notwithstanding, Augustine drew quite a lot from it in expounding Christian doctrine. His lyrical, imaginative and spiritually elevated passages; for example, on the nature of the soul and the prospect of a wholly good eternal life for the virtuous, reflect the influence of Greek philosophy. The Platonic teachings on eternal, unchanging abstract ideas, with God (the highest good) at their pinnacle and the mundane world of nature scarcely worth attention, clearly informed Augustine's thinking. In Book VIII of *The City of God* he remarked that, "No school has come closer to us than the Platonists."[9] Humankind's ascent in Plato's *Republic* from blind ignorance to truth and reality was mirrored in Augustine's image of the ascent of the human soul through knowledge and virtue towards God.

The Greek philosophers did not disdain earthly life to the extent that Augustine did. Socrates, for example, was known for his enjoyment of drinking parties. Not so Augustine, who wrote in an admonitory letter to his friend, the Roman general and governor, Boniface, calling on him to abandon love of the world and enjoyment of a sinful life. This letter is typical of his rejection of the objects, the pleasures and temptations of sensory existence.[10]

To a wealthy widow, he prays her children will grow up in Christ and that, "young as they are, they perceive how dangerous and harmful is the love of the present world"[11]. Again, in a letter to the lords of pagan Mandaura, he deplores that man is "blinded by the desire for things visible". Yet in the same letter he refers to Christ "adorned with an earthly kingdom".[12]

As for the Greek philosophers, their main defect in Augustine's eyes was failure to understand Christian wisdom in pursuit of their own notions of wisdom. Some of them might even have been elevated as Christian saints, had they been born when Christian wisdom was available.

Such conceits aside, it is obvious that what it meant to be a Christian and lead a Christian life was understood in a number of ways, even in the earliest Christian centuries, before deviance or at any rate diversity became the norm rather than the exception. There were from the start the schismatics and heretics who offered different explanations, for example, of the mystery of the Trinity, but the unity and power of the Church were to be preserved against them. Not all unorthodoxy was, however, overcome by orthodoxy. Elements of pagan beliefs and rituals persisted. A person could still retain traces of paganism while proclaiming Christ the King. Philosophical thought gave depth and richness to some of the less well constructed

Christian entities, such as "soul" and the relation of knowledge to virtue. And, as the Roman Empire declined and Christianity absorbed its civic functions, civic affairs and politics gave a different colour to religious passion than was conceived and expressed by those first century fanatics who beat their fists against the walls of Rome.

6.10 Educating Christians

Augustine and other leaders provided Christian doctrine, belief and approved practice with an intellectual and moral foundation that has endured, if fragmented, to the present day. What it provided for the numerous and diverse religious insights, messages and values expressed in the Old and New Testaments did not fundamentally change for almost two millennia. The centuries of the decline of the Roman Empire led to the gradual development of new forms of civil society (ethnically-based principalities and kingdoms, aristocratic rural strongholds, feudalism, education systems). These were important social changes with implications for church doctrine and Christian practice. They also led to further exploitation of (lowly) nature and its resources. The spread of Christianity through monasteries, missionary activity, schools, colleges and universities brought with it a more studied understanding of the original message and the testing of its meaning for both secluded and public life.

For later theological development, of great significance was the establishment of schools and the pursuit of learning under the reign in the eighth century of the semi-literate Frankish King Charlemagne, the first "Roman" emperor for three centuries, following the collapse of the old Empire. In 800, the Pope in Rome crowned Charlemagne the Holy Roman Emperor. Said to be the beginning of the Middle Ages, Charlemagne's conquests and reign set fresh directions in scholarship, learning, education and religion (Chap. 10). These were to achieve lofty heights in theology with the flowering of Scholastic theology and philosophy, which as we have seen, in the thirteenth century, under Bonaventura, Thomas Aquinas and John Duns Scotus among others, marshalled human reason in explication and justification of faith.

Much had been written and debated since Carolingian times. Yet Thomas in his *Summa Theologia* and *Summa contra Gentiles* set out on a vast thought project to consolidate and systematise, so for theology there was to be a logical framework including inductive proof of the existence of God. Correspondingly, there was to be a clear ethical system providing specific directions for Christian life, and credal and worshipful practices within the framework of Church organisation. Although there have been many developments and modifications since, the Scholastic system of Aquinas and others, on the foundation of Augustine and other early Church Fathers, provided orthodoxy of belief, organisation and practical life broadly accepted by the Catholic Church to this day. By an 1879 rescript, Pope Leo the XIII required that the Thomistic theological system be required study in Catholic education institutions.

Notes

[1]On the growth and spread of the organized religion of Christianity I have drawn mainly on MacCulloch, D. (2009) *Christianity, the First Three Thousand Years*. New York: Viking Penguin.

[2]White, L. (1967) "The historical roots of our ecological crises" *Science* Vol. 155 No.3767.

[3]Gibbon, E. (1985) *The Decline and Fall of the Roman Empire. Vol III.* London: The Folio Society. p. 28.

[4]Gibbon, E. (1985) *The Decline and Fall of the Roman Empire. Vol. II.* London: The Folio Society. p. 94.

[5]Augustine (1977) *St Augustine's Confessions Vol.I* Cambridge, Mass.: Harvard University Press; London: William Heinemann Ltd. p. 41.

[6]Gibbon, E. (1985) *The History of the Decline and Fall of the Roman Empire. Vol III The Revival and Collapse of Paganism.* London: The Folio Society. p. 352.

[7]Augustine (1972) *The City of God Against the Pagans. Vol. VII.* London: William Heinemann Ltd:, Cambridge Mass.: Harvard Univ. Press. p. 335;

[8]Augustine (1968) *The City of God Against the Pagans. Vol.I.* London: William Heinemann Ltd.; Cambridge Mass.: Harvard Univ. Press. p. 343.

[9]Augustine (1968) *The City of God Against the Pagans Vol.VIII.* London: William Heinemann Ltd; Cambridge, Mass.: Harvard Univ. Press. p. 23.

[10]Augustine (1965) *St Augustine Select Letters.* London: William Heinemann Ltd.; Cambridge, Mass. Harvard Univ. Press. pp. 415–437.

[11]Augustine (1965) *St Augustine Select Letters.* London: William Heinemann Ltd., Cambridge, Mass. Harvard Univ Press. p. 183.

[12]Augustine (1965) *St Augustine Select Letters.* London: William Heinemann Ltd., Cambridge, Mass. Harvard Univ. press. pp. 463–477.

Complete Listing for Wikimedia Commons Images

6.1 Vendôme (41) église de la Sainte-Trinité 14. https://commons.wikimedia.org/wiki/File:Vend%C3%B4me_(41)_%C3%A9glise_de_la_Sainte-Trinit%C3%A9_14.jpg Attribution: Jacques CHAZARD, CC BY-SA 4.0 https://creativecommons.org/licenses/by-sa/4.0, via Wikimedia Commons.
6.2 Columns – Maison Carrée – Nîmes 2014. Photo: José Luiz. https://commons.wikimedia.org/wiki/File:Columns_-_Maison_Carr%C3%A8e_-_N%C3%AEmes_2014.JPG Attribution: © José Luiz Bernardes Ribeiro.
6.3 Dürer – Hieronymus im Gehäus. https://commons.wikimedia.org/wiki/File:D%C3%BCrer-Hieronymus-im-Geh%C3%A4us.jpg Attribution: Albrecht Dürer, Public domain, via Wikimedia Commons.
6.4 GIBBON (1862) p1.008 EDWARD GIBBON. https://commons.wikimedia.org/wiki/File:GIBBON(1862)_p1.008_EDWARD_GIBBON.jpg
Attribution: See page for author, Public domain, via Wikimedia Commons. Image extracted from page 008 of volume 1 of *The History of the Decline and Fall of the Roman Empire ...*, 1862 edition, by Gibbon, E. Original held and digitised by the British Library. Copied from Flickr. File is from the Mechanical Curator collection. from Flickr.
6.5 Ottaviano Nelli – St Augustine Arriving in Carthage – WGA16508. https://commons.wikimedia.org/wiki/File:Ottaviano_Nelli_-_St_Augustine_Arriving_in_Carthage_-_WGA16508.jpg Web Gallery of Art. Attribution: Ottaviano Nelli, Public domain, via Wikimedia Commons.

Chapter 7
Souls: Pagan, Christian, Animal

The notion of "soul" has a long and convoluted history from ancient pagan times to the present day. For many pagans it was a presence, a ghostly shadow which hovered over the body after death. In the Book of Genesis it represents the inner life force of the body (Gen:2.7), a vital aspect of the whole person. Throughout the Old Testament the idea of "soul" recurs, as in references to desires, hunger, thirst, appetites, suffering, sympathy, breath. In the New Testament, as in other religious doctrines, the body can die while the soul lives on. Like Plato and other Greek philosophers, some early Christian theologians believed in the pre-existence of the soul, and the doctrines of final judgement and resurrection required life of the soul after bodily death. For the purpose of this chapter, of particular interest are the claims respectively of Thomas Aquinas, of hierarchy in which human souls occupy a superior place, and of René Descartes, French Enlightenment philosopher, that soul and body are distinct and separate entities, and that only humans have souls.

7.1 What Is a Soul?

The Greek term "Psyché" and the Hebrew "Nepeth" (living being) in the Biblical and theological texts have been translated into English as "soul". This concept has acquired a multiplicity of meanings and associations in Christianity and other religions as well as in everyday life. Souls are generally thought of as belonging to individuals, the inner life, feelings, or sensations of a living, usually human, being. We also talk of the collective soul or spirit of the people, or a nation. To be soulful is to have calm, peaceful feelings or emotional qualities including devotion (the soulful, expressive eyes of a dog). To "soul search" is to look into oneself either to disclose guilt or to strengthen the will. "Soul music" is highly expressive of feelings of deprivation and hope.

For Christians, "soul" has a wide range of meanings and associations according to the nature and conditions of people's belief about their relationship to God. "Soul" is the breath with which God flooded Adam's nostrils, bringing him to life.

M. Skilbeck, *Loving and Studying Nature*, International Explorations in Outdoor and Environmental Education 10, https://doi.org/10.1007/978-3-030-80751-1_7

It is the dynamic working of God in his contention with Lucifer in humans. It is the God-given inner validity of the life of the individual person. It is a sustaining principle for the devout, or it is the troubled afflicted spirit in erring humanity; hence the religious emphasis on care for the souls of humans.

"Soul-mates" are those bonded by common feelings and beliefs, whereas someone or something that is "soulless" lacks depth of feeling, is emotionally and morally barren, devoid of integrity and value. In all such usages there is a strong element of feeling, commitment, intensity, devotion. "Soul" refers to positive qualities of value and importance, which call for care and love. The lack or weakness of "soul" has negative implications.

In the concept of "soul" as immortal, there is a suggested hypostatisation: "soul" is a metaphysical, eternal entity, distinguishable from the material entity, the body, but intimately related to it on earth and in time.

7.2 Capabilities of Christian Souls

The existence of soul as possessing immortal qualities was not part of the belief system of the first books of the *Old Testament* but it is a central tenet of historical Christianity as it is also of Islam. The nature of the soul and especially its relationship to God have been the subject of endless discussion from very early times to the present day. There are literally hundreds of references to soul in the *Bible* and this figure would run to thousands if we add theological and devotional works since Biblical times. The Christian soul contains and expresses an individual's emotions and attitudes, but it has also often been equated with mind or, simply, personality. It may be united with the body in a tense relationship, yet capable of life independent of the body, either mortal or immortal.

In *A Dialogue between the Soul and Body* (1681), the English metaphysical poet, Andrew Marvell, brilliantly played out this double tension:

Oh, who shall from this dungeon raise.
A soul enslaved in so many ways?
With bolts of bones that fettered stands.
In feet, and manacled in hands;
Here blinded with an eye, and there.
Deaf with the drumming of an ear;
A soul hung up, as 'twere, in chains.
Of nerves and arteries and veins;
Tortured, besides each other part,
In a vain head and double heart.

The flight of the soul beyond all earthly concerns and limitations is the subject of a magnificent sculptural setting by Bellini, *The Ecstasy of St Teresa* in the Basilica of Santa Maria della Vittoria in Rome. A Seraphim is shown driving the fiery point of a lance towards the heart of a swooning woman. The scene is flooded with golden light rays from heaven. Teresa, the sixteenth century Spanish Carmelite, was named a Doctor of the Church in 1970 by Pope Paul VI (Fig. 7.1).

Fig. 7.1 The ecstatic devotion of St Teresa captured by Bellini in the Basilica of Santa Maria della Vittoria in Rome. Gian Lorenzo Bernini, CC BY-SA 4.0, via Wikimedia Commons

In her *Autobiography*, Teresa wrote that she had in her youth practised mediaeval ways of experiencing spiritual ecstasy whereby her soul might achieve union with God and avoid sin. She wrote at length about the nature of religious visionary experiences; prayer was the first stage towards ecstatic devotion as the soul ascended towards God. Vision, she said, yields "a harvest of ineffable spiritual riches, and an admirable renewal of bodily strength".[1] In an ecstatic state, the soul not only transcends the world, it spurns it, "What Empire," she wrote, "is comparable to that of the soul who, from the summit to which God has raised her, sees all the things of earth beneath her feet and is captivated by none of them?"[2]

God and Lucifer, heaven and earth, contended to occupy the Christian soul. Like good and evil in the Manichaean religion, they are fighting a battle, not just for some individual but the whole of humanity. Those who, like St Augustine, saw soul as a sacred, immortal element, envisaged it ultimately at rest in the heavenly city. In some Eastern religious beliefs, its eternal life is experienced in a succession of residences whether in the bodies of humans or other creatures.

7.3 Soul in Greek Thought

Orphism was among the early Greek cults which introduced new concepts of the elevated self and a moral feeling for human life in which soul was to be kept pure throughout life. For the Platonic Greek philosophers, seeking to establish a philosophical foundation for the rule of moral law, soul is divinity in man, enabling him to free himself from ignorance of the good through study and knowledge. To no-one more than Socrates was the idea of *paideia* or the full cultivation of soul more important.[3] That feeling of the vital significance of the soul was taken from Greek philosophy into Christian theology.

In the two extraordinary Platonic dialogues, *Phaedrus* and *Symposium*, soul conceived as divine love achieves ultimate poetic expression. In his famous tribute to Eros in *Phaedrus*, Socrates speaks of a divine madness or inspiration that lifts the soul and sets it on a journey to truth and the highest beauty. In this turbulent process, soul is conceived as mind that must ultimately harness the very emotions that started it on its journey. Socrates regarded falling in love with visible physical beauty (and in some circles of fifth century BCE Athens, this meant men with boys) as the first impulse in the soul's quest for philosophical truth. Physical or earthly love was, however, only the start and must lead to acquired judgement, aiming at what the mind judges to be best in life. There is an echo of this in Augustine's discourse on the transition from earthly to heavenly life.

In a lengthy discourse in *Phaedrus*, which is perhaps unparalleled in Western literature for its exalted imaginative flights, Socrates finds in the nature of the soul both divine and human aspects, immortality and trans migration, together with a mix of wildness and temperance and a capacity for reason and truth. The soul, therefore, epitomises humanity in its highest aspirations as well as its fancies and follies. Nevertheless, it is, or is destined to be, most at home with philosophical beauty and truth.

Phaedrus was set outside Athens, on a grassy knoll in the countryside. It is a bucolic Virgilian scene, rare as a setting for a Platonic dialogue. Later in the day, as Phaedrus and Socrates head back to the city, Socrates praises Pan. This is an ironic touch, since at the beginning of the conversation with Phaedrus, he had remarked, "I'm a lover of learning, and trees and open country won't teach me anything whereas men in the town do".[4] Socrates does, however, find "Divine presence" in that delightful spot, close to a cooling stream and shaded by a plane tree.

In this charming, natural setting, Socrates conjures the mystical flight of the soul. He has it soaring above earth and drawn through the Empyrean, like a chariot pulled this way and that by two horses in conflict (Fig. 7.2). It is a magnificent parable, refreshed by the beauty of the earthly landscape outside Athens, and finished with a prayer to nature. However, nature and landscape are not to be taken literally. They are tokens of the beauty of the spiritual world into which Socrates has led the soul. In his *A Hymn to Intellectual Beauty*, Shelley, the early nineteenth century English Romantic poet, expressed a similar feeling. The truth and beauty of nature had calmed his restless youth. Now, wrote Shelley in adulthood, it is the beauty of the intellect that "gives grace and truth to life's unquiet dream".

In the *Symposium*, Plato's other grand myth of the soul, Socrates is found in more familiar haunts, a drinking party in the city. Again he addresses the theme of love. He makes the disarming logical point that the desire for love is neither beautiful nor good since its pursuit shows that it lacks these qualities. Logic notwithstanding, Socrates has to qualify this by allowing love a little of these qualities so that seeking more than it already has is not unreasonable. Love, he goes on, cannot be the God that some of his drinking companions had insisted it is. Instead, love is the driving force of the soul, a powerful spirit between God and mortal. Like a messenger, it flies up to heaven with our worship and praise, and brings to earth heavenly

Fig. 7.2 The ideas of divine love and the human soul in the philosophy of Plato and the neo-Platonist, Plotinus, were to impact on Christianity as it evolved. "Whither are you Going? And Whence have you Come? A Phaedrus Drawing." Graphite drawing of the charioteer from Plato's "Phaedrus." The charioteer drives two winged horses (or Pegasus) through the heavens beholding the Platonic Forms, before descending back to earth. Drawn by Keith Buhler in 2004, and gifted to philosopher John Mark Reynolds. Keith Buhler, CC BY-SA 4.0, via Wikimedia Commons

responses and commandments which, Socrates tells the party, "form the medium of the prophetic arts, of the priestly rites of sacrifice, initiation and incantation, of divination and of sorcery, for the Divine does not mingle directly with the human."[5] The Divine did, though, mingle with the human in the Greek myths and, in Christianity, through the incarnate Jesus and at the biblical Judgement Day.

Socrates, in his fantastic tales of the cosmic flights of the soul, anticipated but with infinitely richer imagery, one of Christianity's greatest concerns: the health of the soul or spirit and its proper destiny. The Greek philosophers also anticipated the importance of soul to Christianity as it developed into a socio-political movement. In the greatest of his dialogues, ***The Republic***, Plato set out a most elaborate, closely reasoned educational and political theory wherein the individual soul achieves its highest earthly level in the just society. The good society is a unified and elevated gathering of souls, each occupying a social role according to its distinctive character. Thus there are souls that rule and souls that are ruled. Just as with St Paul's Perfect Church, the Platonic Republic was no mere collection. It was a moral entity, unified and hierarchical. Similarly, like his teacher Plato, Aristotle conceived goodness as a "condition of the soul", to be striven for not only in the individual but in the rational society, on earth and in every-day life.

7.4 Animal Elements. Do Animals Have Souls?

But do all souls have capacity for goodness and heavenly destiny? To both the Greek philosophers and Christian theologians, there was in soul an "animal" element, a deep-seated tendency to act impulsively, greedily, lustfully and with disregard of both one's own best interests and those of the community. This animal element prevailed over the spiritual in many people and so had to be controlled and directed by the higher souls, as by the higher elements in soul as such.

Just as in the individual, there is a hierarchy of qualities, so in society there are those who are fit to rule and those whose destiny is to be ruled. For the philosophers, the hierarchy led upwards from those who expressed only ignorant opinions to those whose knowledge reached towards understanding of ultimate reality, the so–called forms. The theologians' corresponding ascent was from wickedness through repentance and piety to a reaching towards the wisdom of God. In the Church, hierarchy was shown in the deeper religious understanding of priests and monks, then bishops and, ultimately in the authority of the Pope.

It was the task of the priesthood to teach the faithful, thereby preparing their souls for eternity. According to the *Biblical Book of Revelation*, at Judgement Day the souls of the wicked, that is those who had failed to learn or to apply their learning in everyday lives, were destined for damnation. There were, too, those souls that might be redeemed by a lengthy period of purgation. Their learning had not stretched far enough or was not adequately applied in everyday life. Finally, there were a few truly virtuous souls that could be admitted immediately to eternal life. The centuries' long interweaving of religious and political ideas, their intertwining and mutual reinforcement gives to the religious notion of soul an earthiness that must have dismayed saints like Teresa.

So much for humans. The question of souls in animals gave rise to difficulties. Animality, or the satisfaction of material wants and appetites, together with existence solely on a sensory level, qualified the souls of the human masses to a place at the bottom of Plato's hierarchy. Christian theology, likewise, used animality and sense experience to contrast with spiritual striving towards God and goodness. Again, we are caught in the dichotomy of the two cities, the heavenly and spiritual at one extreme and the earthly and physical at the other. Between the two is a path that, according to the Greek philosophers and Christian theologians, humans can and should endeavour to tread upward whereas animals, plants rocks, minerals etc. cannot. Aristotle granted dignity, courage and fortitude to the animal soul but it fatally lacked reason. In his *Politics*, he allowed that some animals, together with slaves and children, could take pleasure in "beautiful tunes and rhythms".[6] Low creatures, such as frogs (which he thought were spontaneously generated from the earth) were, in their animality, necessarily earthbound.

In the *Kabbalah*, the collection of esoteric and mystical writings of the Jewish sages from the beginning of the Christian era onwards, the human soul is that which gives rise to the physical body. While it is the source of animalistic desires, it also generates kindness and compassion and, by the divine soul, can be turned from sensual desires towards spiritual pleasures. This contrasts with the demeaning portrayal of animality by Greek philosophers and *Old Testament* scribes alike. The *Kabbalah* hints at a unified view of spiritual and animal life.

The moral status of animals remained confused in Christianity over the millennia. Not until quite recently, and then with the aid of the secular moral philosophers, Peter Singer and Tom Regan, have animals been granted the rights of – in Regan's words – "subjects of life".[7] Almost universally, from the first book of *Genesis* onwards, animals have been regarded as objects, not subjects, their "fitness" or purpose in the scheme of things is that of man's disposal and use.

Genesis tells us that man, animals and plants came from the same "weeklong" birth period – a thoroughly modern scientific view – but there the unity of organic nature ended. Man was fashioned by God in his own image, animals were not. Nevertheless, when speaking in the *Old Testament* to Noah, God said his covenant was with "every living creature" (*Genesis 9.11*). From this perspective, as from their interpretation of "dominion" as meaning care, stewardship and responsibility, some modern Biblical commentators have inferred an intended partnership between man and all other creatures. It is difficult to agree with this interpretation when in so many places in the *Old Testament* the Israelite priests, following earlier pagan practice, sacrificed animals. For example, in *Leviticus*, the second book of *Moses*, the priest gives precise instructions on the performance of the first sacrificial ritual killing of animals, which was an integral part of the religious worship of the time. But does one really slaughter and then, cannibal-like, eat one's partner? It is quite clear that the core Biblical message regarding man's relationship to the living world grew out of the assertion of his superior status and the specific function of animals, which was to serve his needs (Chap. 9).

Thomas Aquinas adopted the Aristotelian notion of different levels and kinds of souls. The highly regarded and influential Persian Muslim, Avicenna, held similar views to Aquinas and it seems that he was familiar with them (Figs. 7.3 and 7.4). Soul was conceived as intellectual or emotional energy. Plant souls lacked sensation so were lowest on the scale of quality and virtue. Animals have sensations, so are higher than plants. But only humans have intellect and this places them at the top of the scale next to the angels and God. In the Mediaeval period, many images of human power and piety were created, from jewelled reliquaries to mighty cathedrals (Fig. 7.5).

Augustine twisted and turned as he grappled with the concept of animal soul in ways which suggests that he may have had some doubts on the matter of its character and status vis-a-vis that of the human soul. Although they cannot reason, through the senses animals have "some semblance of knowledge". In a most curious way, they play a part in humans' use of the "inner eye" to discover forms in nature: "their nourishment and generation is analogous to sensation. Yet, though they and all

Fig. 7.3 St. Thomas Aquinas, as pictured by Carlo Crivelli on the altarpiece for San Domenico, Ascoli Piceno, fifteenth century. Now part of the so-called Demidoff Altarpiece, held at The National Gallery, London. Carlo Crivelli, Public domain, via Wikimedia Commons

Fig. 7.4 The Persian Muslim, Avicenna, believed in different levels and kinds of souls as did the Christian theologian Aquinas. Portrait on silver vase. Museum at BuAli Sina (Avicenna) Mausoleum, Hamadan, Western Iran (7423560860). Image: Adam Jones from Kelowna, BC, Canada, CC BY-SA, via Wikimedia Commons

Fig. 7.5 Chartres cathedral, sublime Mediaeval expression of the Christian idea of heaven on earth. PMRMaeyaert, CC BY-SA 3.0, via Wikimedia Commons

physical things have causes that lie hid in their nature, they do display for observation by our senses their shapes, which gives shapeliness to the visible structure of our world; so that, it would seem, to make up for their inability to know, they have will, as it were, to gain notice and be known."[8] It would doubtless be gratifying to the animals, were their intellects sufficiently developed, to know that by their shapes they assist man mount his ladder of intellectual knowledge of shapeliness and other forms. Attributing to animals this "service" to man is yet further evidence of an overriding belief in an inferior rank for animals and plants in the order of things. If humans, Augustine wrote, "were beasts, we should live a carnal life and whatever accords with its sensuality."

In attacking heresy in *The City of God*, Augustine gave full vent to his conviction that the metaphysical superiority of man by virtue of his rational, spiritual soul, entitles him to moral superiority as well. Taking as his text Jesus' commandment, "thou shalt not kill", he condemned suicide and then took up the application that some heretics had made to animals: "… some try to extend this commandment even to wild and domestic animals and maintain that it is wrong to kill any of them. Why not then extend it also to plants and to anything fixed and fed by roots in the earth? … Are we so completely deranged that we assent to the Manichaean error? (As Augustine himself presumably did, in his Manichaean period). He went on: "Hence, putting aside these ravings, if when we read, "Thou shalt not kill", we do not understand this phrase to apply to bushes, because they have no sensations, nor to the unreasoning animals that fly, swim, walk or crawl, because they are not partners with us in the faculty of reason ... by the altogether righteous ordinance of the Creator both their life and death are a matter subordinate to our needs."[9]

There could be no more unequivocal assertion of man's rights to impose his will on organic nature to meet his felt needs, at the expense of other creatures. It has remained a central assertion or assumption throughout Christian history as it had been in Jewish history and it remains the standard acceptable principle of a supposedly natural relationship. Man's great misfortune has been to ground his pre-scientific attitudes in a hierarchical misunderstanding of this relationship and thereby to take a short-term and potentially self-destructive view of his own best interest.

Augustine struggled in reaching for metaphysical arguments that demeaned nature. In places, he did enlarge in positive terms on the qualities of animals, "They may be keener in their senses, more agile, swifter, stronger". They care for one another and their young in a peaceable manner, "For even the most savage beasts from whom (man) derived a part of his savagery (for he was in fact called half-wild), preserve their species by a sort of peace: by cohabitation, by begetting, bearing, suckling and rearing their young."[10] Nothing, however, deflected Augustine from his uncompromising position which is that God "gave man dominion over irrational creatures." It is the metaphysical misfortune of animals, in the cosmic order of things, to be among the perishable and changeable – like mankind's animal body. Lacking souls, or at any rate fully formed souls, animals are denied the happiness of "clinging to God". However, animals, irrational and lifeless things, are not out of touch with the harmony of the universe. Augustine did leave a door ajar which, millennia later, Christian ecologists would push open. The perishable things (cattle, trees etc) have been endowed by God with, "a pattern whereby, coming and going one after another, they produce a beauty of the lowest order, namely, that of the seasons such as harmonizes on its own level with the parts of this world."[11]

This is the natural order as understood by Augustine, the theologian, pontificating in his study. However, he continued that there is also an order of value whereby some non-sentient entities might be ranked above sentient beings. Had he the power, Augustine would have liked, "to banish the latter from nature altogether, either because we do not know where they naturally fit or because in spite of our knowledge we still put our own interests first."[12] An amusing illustration of significant variability in the operation of the value hierarchy is provided by that immemorial staff of life, bread, "For who would not rather have bread in his house than mice?"

Appealing in this way to a scale of values over the overarching natural hierarchy, why did Augustine not allow the logical possibility that the rational soul may have no inherent value greater than the animal soul? If bread can be more valuable than the life of the mouse there may be circumstances where man's submission may serve his and nature's interests better than dominion. Are there perhaps circumstances when, in the interests of the natural order overall, God would do better to listen to the souls of animals – plants as well – than to the self-justifying arguments of the theologically-attuned rational soul of man? At any rate, Augustine has provided a loophole in his own argument of which contemporary Eco-Christians might take advantage.

On this note, we draw the ancient story of the soul to a conclusion, or rather to three conclusions, or at least implications. First, the Judaeo-Christian religion from its inception adopted the principle that man's soul – his rationality – gives him a superior and dominant place in nature together with his right to use nature to satisfy his wants. Second, in satisfying his wants, man has the right freely to slaughter animals and to take from the earth its vegetable and mineral resources without restraint. Third, that the earth, for all its riches and beauty, is a place of transient existence for the human soul and, by inference any other souls. In the spiritual journey towards heaven, the earth itself is a temporal and inferior place.

These conclusions may not be held by many Christians today in their early formulations, but for practical purposes they remain strong if not dominant in Western culture. There are, fortunately, some signs of a breakthrough among them; for example, the 2015 Encyclical Letter Laudato Si′ of the Holy Father Francis on Care for our Common Home which recognised the damaging impact of human activity and called for restraints. At a public audience in 1990, Pope John Paul II said that animals have souls and called for human solidarity with them. He linked environmental issues with those of social justice, thereby making them interdependent. This had implications for Catholic schools which had previously emphasized social justice.[13]

Notes

[1] James, W. (2002) *Varieties of Religious Experience. A Study of Human Nature.* Centenary Edition. London and New York: Routledge. p.22

[2] James, W. *op.cit.* p.321

[3] Jaeger, W. (1944) *Paideia: The Ideals of Greek Culture. Vol.II. In Search of the Divine Centre.* Oxford: Basil Blackwell.

[4] Hamilton, E. and Cairns, H. (eds) *The Collected Dialogues of Plato.* Princeton, New Jersey: Princeton Univ. Press p.479

[5] Hamilton, E. and Cairns, H. (eds) *op.cit.* p.555.

[6] Aristotle.(1959) *Politics* trans. H. Rackham. London: William Heinemann Ltd./ Cambridge Mass.: Harvard Univ. Press. p.665.

[7] Regan, T. (1983) *The Case for Animal Rights* Berkeley: University of California Press; Singer, P. (1975) *Animal Liberation: A New Ethics for our Treatment of Animals.* New York: Random House.

[8] Augustine (1968) *The City of God Against the Pagans. Vol.III* London: William Heinemann Ltd.; Cambridge, Mass.: Harvard Univ. Press. pp539 and 543.

[9] Augustine (1957) *The City of God Against the Pagans Vol. 1* London: William Heinemann Ltd.; Cambridge, Mass.: Harvard Univ. Press. p.93.

[10] Augustine (1960) *The City of God Against the Pagans. Vol VI.* London: William Heinemann Ltd.; Cambridge, Mass.: Harvard Univ. Press. p.169.

[11] Augustine (1968) *The City of God Against the Pagans. Vol. IV.* London.: William Heinemann Ltd.; Cambridge, Mass.:Harvard Univ. Press. p.17.

[12] Augustine (1968) *The City of God Against the Pagans Vol. III.* London: William Heinemann Ltd.; Cambridge, Mass.: Harvard Univ. Press. p.491.

[13] Personal communication. Professor Peter Fensham. Sept 2019.

Complete Listing for Wikimedia Commons Images

7.1 Ecstasy of Saint Teresa September 2015-2a. https://commons.wikimedia.org/wiki/File:Ecstasy_of_Saint_Teresa_September_2015-2a.jpg. Photo: Alvesgaspar. Attribution: Gian Lorenzo Bernini, CC BY-SA 4.0 <https://creativecommons.org/licenses/by-sa/4.0>, via Wikimedia Commons.

7.2 Whither are you Going? And Whence have you Come? A Phaedrus Drawing. https://commons.wikimedia.org/wiki/File:Whither_are_you_Going%3F_And_Whence_have_you_Come%3F_A_Phaedrus_Drawing.JPG. Attribution: Keith Buhler, CC BY-SA 4.0 <https://creativecommons.org/licenses/by-sa/4.0>, via Wikimedia Commons.

7.3 Carlo Crivelli 007. https://commons.wikimedia.org/wiki/File:Carlo_Crivelli_007.jpg. Attribution: Carlo Crivelli, Public domain, via Wikimedia Commons. The Yorck Project (2002) *10.000 Meisterwerke der Malerei* (DVD-ROM), distributed by DIRECTMEDIA Publishing GmbH. ISBN: 3936122202.

7.4 Avicenna Portrait on Silver Vase – Museum at BuAli Sina (Avicenna) Mausoleum – Hamadan – Western Iran (7423560860). https://commons.wikimedia.org/wiki/File:Avicenna_Portrait_on_Silver_Vase_-_Museum_at_BuAli_Sina_(Avicenna)_Mausoleum_-_Hamadan_-_Western_Iran_(7423560860).jpg. Photo: Adam Jones. Attribution: Adam Jones from Kelowna, BC, Canada, CC BY-SA 2.0 <https://creativecommons.org/licenses/by-sa/2.0>, via Wikimedia Commons.

7.5 Chartres, Cathédrale Notre-Dame-F 149. https://commons.wikimedia.org/wiki/File:Chartres,_Cath%C3%A9drale_Notre-Dame-F_149.jpg. Photo: PMRMaeyaert. Attribution: PMRMaeyaert, CC BY-SA 3.0 <https://creativecommons.org/licenses/by-sa/3.0>, via Wikimedia Commons.

Chapter 8
Imagining Nature in Myths and Creation Stories

Whenever humans have settled and formed communities they have told stories about origins of themselves and the places they inhabit. Many stories extend to the cosmos and explain the existence of animals, plants, the Earth, stars – and the spirit world, elements of a unified whole. They have, as well, projected destinies for humankind which are inextricably linked to origins. Colonists have spread Judaeo-Christian origin stories far and wide. For example, when Europeans colonised America and Australia, they brought with them the creation myths of the Bible. The indigenous people had their own creation stories which have either been held onto by a slender thread, replaced as and when they were converted to Christianity, or to some degree blended with the stories of the colonisers (Chapter 16.4).

8.1 The Continuing Grip of Creation Stories

Post-Darwinian creation stories, such as those of Roman Catholic theologians, Teilhard de Chardin and Thomas Berry (Figs. 8.1, 8.2), drawing on the French philosopher, Henri Bergson's *Creative Evolution* (Fig. 8.3), lacking the arcane poetry of the ancient creation myths, tell of the interacting physical and chemical forces of the universe. Out of their patterns, a spark of life and a design emerged, eventually developing into that remarkable form of existence, *Homo sapiens*. According to Darwinists, not design and designer, as in the old creation stories, but randomness and evolution over immense periods of time account for the processes and outcomes, for which a designer would be superfluous. Although the old stories are now properly seen as a combination of the limitations of scientific knowledge at the time and the creative genius of the human imagination, they express a powerful symbolic truth about pattern and unity in nature. Their concreteness and emotional content make them more accessible to most people than the abstractions and puzzlements of modern mathematical physics.

Creation in these stories is not merely an event, it is a vital process starting in motion the whole of nature; God required more than a single day in creating the

M. Skilbeck, *Loving and Studying Nature*, International Explorations in Outdoor and Environmental Education 10, https://doi.org/10.1007/978-3-030-80751-1_8

Fig. 8.1 Pierre Teilhard de Chardin (1881–1955), twentieth century Roman Catholic intellectual who speculated on the origins of life and human spirituality. Unknown author, CC BY-SA 3.0, via Wikimedia Commons

cosmos. Creation stories poetically stir human understanding and generate the active life of humanity. How humans stood in relation to God or the gods above, to their power and goodness and to the earth and its creatures was of the utmost importance. For their very survival, let alone their well-being and happiness, humans needed to try to know where they stood in this remarkable order of things. What were their opportunities, rights, responsibilities; how could needs and wants be satisfied? What were the extent and limits of human power and authority and what consequences might follow should humankind overstep the mark?

Students of ancient creation stories and their role in community life have observed how the stories foster belief systems not only about origins and interrelationships in nature, but also about identity and ways of life for the whole community to take up. Indeed, the stories are not only religious in a narrow sense, they are clan and tribe-bonding and nation-creating and nation-unifying. Communities are bound together in common beliefs concerning who their members are, whence they have come, their place in the order of things, and the values, attitudes and everyday ways of life they should be pursuing. While literal belief in these stories fades (but has certainly not vanished) with the progress of scientific and social knowledge, metaphorically they retain a strong hold, while the values and attitudes they generate are sustained and continue to be acted on.

We cling to the myths and stories - or at least to fragments of them - and we may enjoy the imagery and language. They reside deep down in the treasury of the imagination, serving as a rich resource for poetry, literature, art and social justice. Scholars have argued that so far from being relics or fragmentary vestiges, these

Fig. 8.2 Thomas Berry (1914–2009), prominent Catholic theologian in twentieth century USA. Photo by Lou Niznik, Courtesy of the Thomas Berry Foundation

creation stories continue in the civilisations of which they are a part to structure pervasive, if not always articulate, patterns of thought and emotional states that we don't readily abandon to science and common sense. Even when not believed in literally, they impact in various ways on our actions as elements in the legacy of what we take for granted and don't think to challenge. Creationism, for example, with its doctrines harking back to literal, pre-scientific interpretations of the *Book of Genesis*, appears to be very difficult to root out of the human psyche by scholarly historical and scientific argument and data.

8.2 Functions of Creation Stories

The best-known and most widely diffused of Western creation stories is told in the opening chapter of *Genesis*, the first book of the (Jewish) *Old Testament*. Its subject is God's creation of the cosmos, and of the Earth within the cosmos as the earthly home of humanity. Little is said of the heavenly home, or of the human soul, on which, as we have seen, philosophers and theologians have dwelt. The metaphysical and ethical strands of *Genesis* are woven around notions of divine order, God's commands and the need for humankind to bind itself to them. The consequences of not doing so, of disobedience, were fateful and fatal. Creation itself is good, but in the *Genesis* story it is sullied by evil which is symbolised in the plucking and eating of forbidden fruit which in turn led to the expulsion of Adam and Eve from the

Fig. 8.3 Henri Bergson (1859–1941), whose *Creative Evolution* attracted a wide, sometimes hostile, audience. Henri Manuel, Public domain, via Wikimedia Commons

idyllic Garden of Eden. The function of *Genesis*, as the first book of the *Old Testament*, was to lay not only a metaphysical but also a political foundation for the Jewish nation. The biblical creation story is further discussed below.

It is above all in ancient creation stories and particularly that of the Old Testament, which is so well articulated, that we find the deepest and most far-reaching early religious expressions of man's relation to nature and his attitudes towards it. Nature is invested with life, which in the pagan myths is found also in the sun, moon, oceans, the abodes of spirits both benevolent and harmful. In Chap. 16, we consider the creation stories of indigenous Australians in which natural forces, animals and humans share a single imaginative domain which is sacred and must be cared for and protected. The land, no matter how barren and desolate it may appear to Western eyes, is no less treasured than the Promised Land of the *Old Testament* Israelites, owning the people as much as the people own it.

Stories of creation and origins bear further consideration here and are discussed further in Chap. 14. Whether *Genesis* and creation might be interpreted somewhat differently for modern purposes, compared with how they typically have been in the Judaeo-Christian and Islamic traditions, is one of the issues to consider. The *Old Testament* Garden of Eden, which is the setting for God's creation of man, has its origins and analogues in other ancient creation stories. Some of these still survive in religious pockets in the Middle East and in Asiatic and other religions. Ancient

Assyrian, Egyptian and Greek mythology had creation stories with tales of fabulous creatures, some part-human and part-divine, and heavenly forces which had an earth-shattering impact. Supernatural or divine forces in the cosmos not only permeated human life, structuring its course, they informed nature altogether with vitality and significant meaning, benign or terrifying. Polytheistic religions commonly invested the gods with specific natural forces. For example, in the Indian Vedic religion, Indra is associated with storms, Agni with fire and Vayu with wind. The human tendency to worship or to find ways to propitiate natural forces meant that the gods were frequently personalised and thereby made familiar.

It is evident that creation stories in the ancient world were not mere tales of mystery and imagination. They satisfied basic needs and they bonded tribes into nations. Presumably they still do for the approximately 40% of the American population said to believe in creationism while rejecting Darwinian evolutionary theory. What, then, are these needs? Assuming such needs to exist today, how are they best satisfied beyond literal adherence to ancient myths? In order to answer these questions, we should further consider what the creation stories have to offer, either to our reasoning or our imagination.

First, there is curiosity about ourselves in the world, famously explored by Gauguin in his Tahitian tableau painting, "D'où venons-nous? Que sommes-nous? Où allons-nous?"(Where do we come from? What are we? Where are we going?). A sense of mystery and supernatural force adds a spice of excitement that appeals to the imagination, even to a sense of adventure for the wayward human soul. Modern science and historical scholarship satisfy that curiosity in ways other than belief in myths, even if at times rather too prosaically.

In the second place, the subject matter of the stories is not confined to humans and animals. Much like science fiction, so popular in modern times, their scope extends to the cosmos: sun, moon, starry heaven, whatever may be imagined to lie in the furthest reaches of the universe, actually, though, conceived as close to earth in the understanding of the time. The seasons, day and night, light and darkness, wind, rain, thunder and lightning, the whole of nature exposed to the senses is accounted for, with depth and insight. But nowadays, through scientific inquiry and, for some, metaphysical reflection, we have achieved greater intellectual depth and richer insight.

Third, the elements and components of vital nature are not static in the origin myths. They are moved by forces, ultimately God-given; there is energy, growth, change which present opportunities for man. In *Genesis*, light from the sun meant daily activity, life for plants, visibility and warmth for people to enjoy. Rain from the sky started crop seeds into life in the dusty soil and set rivers flowing. The daily passage of the sun across the sky set rhythmic pattern of the seasons, hence of human activity. By comparison, how well does science yield personal engagement with these life-giving forces of nature? We can enjoy the ancient creation stories and value their imaginative genius while putting aside their claims to literal truth.

Such is the majesty and beauty of the *Book of Genesis*, the questions it raises and up to a point answers, continue to provoke thought. As Adam and Eve were chased out of Eden by the angel of God, they had a fair understanding of the answers and what lay ahead. Wicked as they were, they knew that on the one hand they were to till the soil and, on the other, to multiply the human race. Phrasing questions and articulating answers required language. Even prior to its development, however, humans could express intentions and hopes and enact rituals through dance and song. Flaking and chipping stone tools, hunting in teams, projecting images on cave walls and carving figurines were to be among many ways of expressing relations with the natural world.

We can hypothesise – as many scholars have done – that the need for support and assistance when danger loomed, invited assistance from others, animate or inanimate. If power greater than unaided man's could be invoked from elements or forces of nature – sun, moon, stars, magical creatures, mountains, in enchanted groves and so forth, and from their spirits – their invocation could not but help. The weaker supplicated the perceived stronger, and in return for obeisance and worship, expected care and assistance. To invest these strong entities with superhuman powers was a quite natural step. Pagan worship was not confined to remote gods, it was worship of nature and spirits in everyday natural objects. Furthermore, if it was not only support that was needed, but understanding of the phenomena and forces of nature, the superhuman forces might be invested with cosmic, not merely earthly, powers.

Gods or entities with superhuman powers had to be created by pagans – first imagined then reasoned - since mere sensory evidence of their existence as physical phenomena (sun, stars, mountains etc) conferred on them inadequate capacity to help mankind. The sun, with the figure of Apollo as its god, was much more powerful and approachable than the sun alone. A god was not necessarily a physical entity but a way of conceiving, of investing a natural phenomenon with a power sufficient for a constructive interest in human affairs.

The alternative, theistic view, as in the *Old Testament*, is that God existed prior to creation and that instead of being invented by man, he was the prior lord of all creation. It is striking that, apart from the visions, apparent miracles and everyday experiences of the Israelites, the *Old Testament* offers no proof for the existence of God. He is, from the beginning to the end of the Scriptures, just there: brooding, omnipresent and all powerful. His word was his proof and he was proof of the power of words. The powers and qualities attributed to God are those felt and expressed by the faithful – it seemed to suffice for the Israelites that the human mind was illuminated by the experience of God. Neither did the early theologians make serious attempts to prove God's existence by reason. As we saw in Chap. 7, that was the task Thomas Aquinas set himself in the thirteenth century, some two millennia after the first books of the Old Testament were produced.

8.3 Trading Creation Stories

Creation stories in the ancient world were exchanged amongst cultures, rather like trade in material goods. Mesopotamia was the locus of stories that appear to have had significant influence on *Genesis*. There are several parallels, if not direct antecedents, in the Babylonian myth, *Enuma Elis*. The universe is chaos before order is established, with a habitation (temple) for the gods (who are both male and female), a spring of freshwater (rising in the sea) and humans, created for service to the gods. There was a mother God, more than equal to her male counterpart although in the end it is the male, Marduk, chief god of Babylon, who rises, Jupiter-like, to command all the other gods.[1] (Fig. 8.4).

Dating from the eighteenth, or as some scholars suggest, from as late as the eleventh century BCE this Babylonian story is recorded in cuneiform on clay tablets which were recovered and translated only in the nineteenth century. While there is dispute over any direct influence on *Genesis*, the parallels are intriguing. One suggestion is that its significance lay merely in the determination of whoever authored *Genesis* to assert the supremacy of the Israelite Yaweh over Mesopotamian Marduk.

Fig. 8.4 Marduk mounted on the "fierce snake", stands in victory on the watery body of the vanquished Ti'amat on the occasion of the Babylonian New Year's festival. Statue of Marduk depicted on a cylinder seal of the ninth century BCE. Image from Weissbach, F.H. *Babylonischen Miscellen*. 1903. Franz Heinrich Weißbach, Public domain, via Wikimedia Commons

Creation stories fed one another. Since in retelling they would be modified over time, it is not surprising that as people moved about (nomadic pastoralists, traders, soldiers on military campaigns etc.) through the exchange of beliefs and cosmological stories, myths and creation stories acquired family resemblances. The tree of life is an example of religious exchange, probably appearing first in Mesopotamia in the fourth millennium BCE and, by the second millennium BCE surfacing in Egypt, Greece and the Indus civilisation. Its analogue, or perhaps another version, emerged early in the first millennium BCE as one of two notable trees in the Hebrew Garden of Eden. The other was the tree of knowledge.

The tree of life variously appeared in Buddhist art as well. Its wide dispersion, analogously in the form of a sacred bough was studied by Scottish archaeologist, James Frazer. The tree, in his *The Golden Bough* or parts thereof, became something of a universal myth of the succession of social power and the regeneration of life. In Syria, the tree of life was a decorative motif on royal garments, as if royalty commanded life. Animals, too, were figured, symbolising sexual potency and regeneration.

In all of these myths, the ideas of divine and earthly order and the exercise of power feature prominently. Life is not random, it is ordered and controlled; it is purposive. Thus in Assyrian mythology, a bird-headed King impersonates the divine world order which the king, as representative of the gods, maintains. This has been seen as a symbol of the perfection or of the perfectibility of man. Traces of this idea – which resurfaced in Europe in the eighteenth century – are to be found in the *Old Testament*. Before falling from grace man is as perfect – as near to God – as a human can be. In the *New Testament,* Jesus as the personhood of God represents the divine perfection and the perfectibility of mankind.[2]

A stylised version of the tree of life was a symbol of Judaism in the first century and is today a state symbol of Israel. Some scholars have claimed a definite link between the Assyrian and Jewish iconographies. The Menorah, a candlestick popular today amongst Jews and non-Jews alike, is in the form of a multi-branching tree, a visual image of the doctrinal structure of the *Kabbalah*, the mystical and prophetic discourses that, in the Christian era, complemented and interpreted the *Old Testament*. The tree, in this imagery, represents the creation of the world, charting cosmic harmony.

Scholars have found evidence of belief in successive emanations of God, a continuity of creation. As with the Assyrian tree, there are references to the ideal man, created in the image of God. Symbolically, superior, esoteric knowledge, proceeds like sap rising in the tree. At the roots, animal passion dominates. Flowing up the trunk, sap carries the soul upwards to spiritual awakening and, ultimately, superior knowledge. There is an allusion to this symbolic use of the tree in *Proverbs* 3:18, "She (that is the wisdom and understanding of God's way) is a tree of life to them that lay hold upon her; and happy is everyone that retaineth her."

It is, of course, a relatively simple matter to note superficial resemblances and echoes among myths and to attribute cross-cultural symbolic meanings to what may be coincidences or claims based on inadequate evidence. Caution is properly called for in our observations from present time back to the swirling and only partially disclosed and analysed array of ancient belief systems. Whatever interest there may be in tracing elements of ancient tales and their likely inter-relationships, we should not, however, ignore an association which is of utmost value in the legacy of the mythical tree of life. This is the importance of trees in the cycle of life. Trees matter for different reasons; for example, in Mediaeval times five acres of forest were consumed to produce charcoal for one ton of iron. Iron for military purposes became a key destroyer of forests in Europe. During the Renaissance, land-owners were encouraged to plant forests for the timber required by the English and French naval shipyards and for the construction of houses, public buildings, churches, bridges and other amenities of civilization. Living beings in their own right, trees in large, healthy, biodiverse forests spread around the globe are essential for all other living creatures, humans included. Trees of life indeed.

8.4 A Tremendous Story of Paradise on Earth: The Garden of Eden

> Thus was this place,
> A happy rural seat of various view;
> Groves whose rich Trees wept odorous Gumms and Balme,
> Others whose fruit burnisht with Golden Rinde
> Hung amiable, *Hesperian* Fables true,
> If true, here only, and of delicious taste:
> Betwixt them Lawns, or level Downs and Flocks
> Grasing the tender herb, were interpos'd,
> Or palmie hilloc, or the flourie lap
> Of som irriguous Valley spred her store,
> Flours of all hue, and without Thorn the Rose:
> Another side, umbrageous Grots and Caves
> Of coole recess, o're which the mantling vine
> Layes forth her purple Grape, and gently creeps
> Luxuriant.
> (John Milton *Paradise Lost*, Book IV 247-261. 1674 version.)

In the fourth book of *Paradise Lost*, Milton wrote rhapsodically of the beauties and pleasures of paradise on Earth. The first home of Adam and Eve was at the same time the site of the first sin, only redeemed through the resurrection of Christ whereby Adam and Eve, in Paradise regained, reached the heavenly Paradise. This magnificent epic poem not only gives rich dramatic depth to the battle between good and evil, God and Satan, symbolised in the eating of forbidden fruit in Eden, it filled the garden with a glorious abundance of verdure, fruit and flower. This enrichment of the biblical landscape was to inspire a romantic style of gardening, distinctively English and most appealing to nineteenth century country parsons (Chaps. 13 and 16). It is reflected too in the gardens of the Mogul emperors (Chap. 14). A far cry from the theologians' strictures on enjoyment of earthly delights, and something to maintain in creating paradise on earth.

The first five books of the *Old Testament*, the *Pentateuch* (i.e. "instruction"), in the most vivid, poetic language tell a tremendous story of the creation of the cosmos, the origins and beginning of life, and the struggle of the first people to survive and thrive. As English Jesuit poet, Gerard Manley Hopkins, wrote in his *Spring*:

> A strain of the earth's sweet being in the beginning
> In Eden garden.

The *Pentateuch* story is extraordinarily rich in imaginative insight, images and revealing episodes. While many religious people continue to treat these wonderful stories as historically accurate, replete with literal truths, I share the opinion that thanks to the theory of evolution, there are now better, even more dramatic understandings of the origins and development of physical life on earth and in the atmosphere than those of the *Old Testament*. The value of these old stories lies not in their historical verisimilitude or even metaphysical truth but in their inspirational symbolic and poetic qualities.

It is unfortunate that in the widely publicised squabbles between creationists and evolutionists, the imaginative and dramatic riches of ancient creation stories are generally overlooked, treated either as scientifically irrelevant fantasies or vehicles of God-given literal truth.

In the first chapter of *Genesis* we find God at the beginning of a massive, week-long act of cosmic creation, a metaphor for aeons of time and infinite space. Out of nothing, and at the beginning of time, he makes heaven and earth, the stars, sun and moon, the ocean, day and night, clouds and rain, herbs and grasses, fruiting trees, birds and animals, whales, fish, "cattle and creeping things". Seeing his creation richly furnished with diverse organic life God said that, "it was good". But something was lacking so God proceeded to create humankind, "in his own image, in the image of God created he him, male and female he created them" (*Genesis* 1:27). Animals and birds and sea creatures were all destined to be fearful of man so, "into your hand are they delivered … Every moving thing that liveth shall be meat for you; even as the green herb have I given you all things" (*Genesis* 9:2–3). God gave to humankind, "dominion over the fish of the sea, and over the fowl of the air, and over the cattle, and over all the earth, And over every creeping thing that creepeth upon the earth" (*Genesis* 1:26).

God blessed man and woman and said unto them "Be fruitful and multiply, and replenish the earth and subdue it" (*Genesis* 1:28). But would believers not agree that, in His omniscience and therefore knowledge of man's sinfulness and wickedness, God must have realised that powers of dominion and subduing can be – and would be – grossly abused?

Continuing his act of creation, God on the seventh day concluded that the heavens and earth were finished. The mighty act had been accomplished and the history of humanity on earth could commence.

What had been accomplished? First and foremost, other ancient creation stories and their associated cults and religions had been trumped by a work of the imagination on a magisterial scale, in a story capable of withstanding critical attack at the time of writing and for almost two millennia. Second, through *Genesis* and further elaboration, the world is established as a physical entity fit for man the pastoralist and cultivator of the earth, God's regent and master of the living world and the mineral treasures of the earth, and together with woman to populate it with their progeny. Third, the world is not only a physical entity, it is a sacred temple in which humanity, experiencing God's presence, is to worship and express obeisance to him. Fourth, the world beyond its physical reality and its holiness is a moral sphere wherein humankind is not only to worship and obey but, possessed of free will, is to choose the right or good and reject the wrong or evil. As to the substance of these moral categories, on the basis of the commandments laid down and the directives set forth in *Genesis*, there would be a progressive exegesis or revelation in the history of the Jewish people, as recorded in the *Old Testament*. Further, decisive revelation was to occur through the life and death of Jesus Christ. In *Genesis*, man is a vegetarian, as were the Manichaeans of Augustine's youth and early manhood. Only after the great Flood was permission given to eat flesh – a further step in the "service" animals would deliver to humans.

Something more needs to be added about Adam and Eve, the progenitors of humanity. They were created in God's image, that is at a level if not quite that of God and the angels, then not far below. What is more when, through the sin of disobedience, they destroyed that perfection, they remained perfectible. Life on earth, outside the garden was to be that of toil and pain but equally a quest to return to perfection in the eyes of God, a spiritual as much as a material quest.

The Genesis story progressed from the great cosmic drama of the creation in Chap. 1 to the first actions of humans on earth and their projected future. The second chapter starts with a brief recapitulation. In order for plants to grow, the dry dust of earth was rained on. Man had been created, and in his service to God he was to till the Earth, "so the Lord God formed man of the dust of the ground and breathed into his nostrils the breath of life, and man became a living soul".

Agriculture and horticulture, or gardening, were established simultaneously, in a plot of land "eastward of Eden". This, the famous "Garden of Eden", is often referred to as earthly paradise. Augustine, for example, used the term "material Paradise" to distinguish it from heaven's spiritual paradise. One of humanity's greatest images of beauty, desire and hope, Eden is the analogue of the heavenly paradise of eternal life in and with God, to which the devout, Muslims no less than Christians, aspire. Thus Eden drew its inspiration from both earthly and heavenly worlds (Figs. 8.5 and 8.6).

The topography of the Garden is not altogether clear since we do not know just how far it was being projected as a physical not a metaphorical space. There was a river, rising perhaps from a spring which, dividing into four as it flowed away, formed mighty streams to water the lands of Havilah, Ethiopia, Syria and to wherever the River Euphrates stretches. In other words, the drylands of North Africa, the near and Middle East were provided with life-giving water, Eden the source. Certainly there was a "topological" space, that is the subjective concept of space projected throughout history by artists and poets for whom Eden was a rich stimulus to the imagination. The theme of a fourfold division of water was taken up by Persians and the Mogul Emperors in their garden designs (Chap. 14).

Fig. 8.5 The Garden of Eden charged the imagination of painters and poets over many centuries and provided and continues to provide an image for garden lovers today. In 1612 Jan Brueghel the Elder painted this view of Adam and Eve in the Garden of Eden. Doria Pamphilj Gallery. Jan Brueghel the Elder, Public domain, via Wikimedia Commons

Fig. 8.6 Las Pozas, surrealist garden of Edward James in the tropical mountains near Xilitla, Mexico. Photo: Adam Jones, Kelowna, BC, Canada.(32553071348) CC BY-SA 2.0, via Wikimedia Commons

8.5 Uses of the Eden Story

Was Eden just another flight of fancy? In the *Q'ran*, Eden signifies heavenly bliss for the faithful. It is only in *Genesis* that Eden can be read as a physical space – a real garden. The *Q'ran* does, however, invest its account of heaven with quite physical details of the eternal life of the faithful. Throughout the *Bible* it is treated as a myth, but it is always symbolic. Hence the two trees, one of life, the other of knowledge. God intended humankind, as blessed beings, to choose life. Eve, however, tempted by the serpent (lowliest of creatures), chose the forbidden fruit of the tree of knowledge of good and evil, thus admitting sin into human affairs, but also knowledge. Augustine, Bishop of Hippo, ever in search of symbolic meanings and allegories, treated the Garden as a lesson in the lineaments of a virtuous life:

> ... as the life of blessed men, its four streams as the four virtues, namely, prudence, courage, temperance and justice, its trees as all the useful studies, the fruits of the trees as the conduct of the righteous, the tree of life as wisdom, the very mother of all good things, and the tree of knowledge of good and evil as experience that comes with the transgression of a commandment ...
>
> We may also interpret these objects in terms of the Church and thus regard them better as prophetic prior indications of what was to come. Specifically, paradise may point to the Church itself, as we read about it in the *Song of Songs*, the four streams of paradise to the four gospels, the fruit-bearing trees to the saints and the fruits thereof to their works; the tree of life as Holy of Holies is surely Christ himself and the tree of knowledge of good and evil is personal control of one's own will.[3]

Another suggestion is that the tree of life symbolizes nourishment, a special privileged spiritual nourishment that ensures eternal life of body and soul. According to *Proverbs* (3:18), "She [i.e. wisdom and understanding of God's way] is a tree of life to them that lay hold upon her; and happy is every one that retaineth her." There have been many other interpretations, as befits a rich symbol. For example, in *Revelation of St John the Divine*, "To him that overcometh will I give to eat of the tree of life" (2:8), whose leaves, "were for the healing of the nations" (22:2). A branching tree can suggest the descent of life – the progeny and descendants of Adam and Eve as recounted in succeeding books of *Genesis*. But it also suggests life as an energising force in the universe, the essence of being, a precious cosmic quality or, in religious terms, the spirit of God. Modern ecologists like Edward O. Wilson, in raising the issue of the future of life pose questions about humanity's role in jeopardising the vital quality of existence. The imagery of the tree of life calls for us to strengthen our understanding of the conditions needed if life is to continue to thrive, that is a kind of wisdom that robustly addresses the challenges to the future of life on earth.

As for the tree of knowledge, whose fruit was forbidden, surely humanity, endowed by God with reason and curiosity, would naturally seek knowledge, whatever dark secrets might be revealed on the way. Adam had been granted the right to name (to know the distinctive features of) animals and plants, that is to use language and to practise science (observation and sensory discrimination). Why, then, limit his further quest for knowledge? It has been suggested that this was to check his hubristic aspirations, to ensure obedience, that is the goodness of absolute obedience. Having made him in his own image, God nevertheless did not wish man's earthly eyes to be fully opened, that is, "to become as one of us" (godlike). There is an echo of this in the exercise of authority by the early and Mediaeval Catholic Church in its rooting out of dissent, its requirement of orthodoxy of belief as a proxy for the highest knowledge and then its hostility to Renaissance science. At the personal level, Jesuit poet, Gerard Manley Hopkins, bewailed his failure to reach to God's wisdom and mercy, just as St Augustine in his interrogation of God over some of the mysteries of religion was denied the understanding that could only come, if at all, in heaven. Man was God's regent, he was not to aspire to equality.

In the Garden of Eden, the human hand which stole knowledge was not therefore to reach forth, to, "take also of the tree of life, and eat, and live forever". Adam's regency of God was limited and after sinning, he and Eve were driven out of the Garden of earthly delights and away from proximity to heaven. To prevent their illicit return, to eat of the tree of life (immortality, which was to be God's gift to the elect), God placed, "Cherubims, and a flaming sword which turned every way". Adam and Eve set forth to experience the toil, pains, dreams and hopes of earthly life. In Chap. 2 of *Revelation*, where John is commanded by God to write to the church ministers: "To him that overcometh will I give to eat of the tree of life which is in the midst of the paradise of God" (2:7). It is a long wait from the expulsion from Paradise to re-entry (by the few) at the second coming. In the meanwhile, there have been many uses made of the image of the Garden of Eden as foreshadowing Paradise (Chap. 14).

The *Bible* has been the most widely distributed, expounded and read text in history, not only in the West but worldwide. It has had an immense influence on literature, art, music, philosophy, ethics, social conditions, political affairs, and beliefs about life. They, in turn, have played a part in reinforcing religious belief and values. The truth claims and moral precepts of the *Bible* are deeply embedded in human consciousness. Not least in its impact is the doctrine of a hierarchical natural order which theologians and church leaders in the Christian era extended from the natural to the socio-political world. We shall return to this notion of an order of nature since it is the source of a most damaging misunderstanding of life on earth and human responsibility for the future of life.

The image of the Garden of Eden, as we have seen, has inspired some remarkable feats of the imagination. We conclude the chapter with one more of them: the garden of Las Pozas (The Pools) in steep, mountainous, sub-tropical rainforest near the village of Xilitla in northern Mexico (Fig. 8.6). The garden, covering more than 32 hectares, comprises natural waterfalls and pools, trees, shrubs and a wide variety of indigenous and exotic plants including orchids, bromeliads, halconias, ferns, epiphytes and many others, set among surrealistic concrete structures – steps, pavilions, columns, pens for exotic birds and small animals, ramps, bridges, narrow walkways along valley walls, and a Staircase to Heaven. The structures in Las Pozas were created over some 35 years starting in 1949 by the wealthy English poet, sculptor and patron of the arts, Edward James, who wanted to create a Garden of Eden, and his friend and architectural collaborator, Plutarco Gastelum.[4]

Together with concrete serpents, birds of paradise, mushrooms and dream-like forms, some of them painted, there were also, in a zoo area, live crocodiles, snakes, deer, ocelets, flamingos, parrots and other creatures that James loved and talked to. In one of his poems he wrote of "the sound, the sound of green" as if the sounds of his jungle garden were as important as its colours and forms.

Notes

[1]Parpola, S. (1993) "The Assyrian Tree of Life. Tracing the origin of Jewish monotheism and Greek philosophy." *Journal of Near Eastern Studies.* Vol. 52, No. 3 July. pp.161–208. Published by University of Chicago Press. (Accessed 06/052010) https://www.jstor.org/stable/545436

[2]Different versions of ancient and Christian perfection and perfectibility are discussed in several chapters of Passmore, J. (1970) *The Perfectibility of Man.* London: Duckworth.

[3]Augustine (1966) *The City of God Against the Pagans. Vol.IV.* London: William Heinemann Ltd.; Cambridge, Mass.: Harvard Univ. Press. pp. 217, 219

[4]Hook M. (2006) *Surreal Eden. Edward James and Las Pozas.* New York. Princeton Architectural Press; Jeffrey Bale's blog: http://jeffreygardens.blogspot.com/2016/01/las-pozas-pools.html (accessed March 23, 2021).

Complete Listing for Wikimedia Commons Images

8.1 TeilhardP 1947. https://commons.wikimedia.org/wiki/File:TeilhardP_1947.jpg Aource: Archives des jésuites de France. Attribution: Unknown authorUnknown author, CC BY-SA 3.0 <https://creativecommons.org/licenses/by-sa/3.0>, via Wikimedia Commons.

8.3 Henri Bergson 02. https://commons.wikimedia.org/wiki/File:Henri_Bergson_02.jpg Photo: Henri Manuel (−1947). George Grantham Bain Collection (Library of Congress) Attribution: Henri Manuel, Public domain, via Wikimedia Commons.

8.4 Marduk and pet. https://commons.wikimedia.org/wiki/File:Marduk_and_pet.jpg Attribution: Franz Heinrich Weißbach, Public domain, via Wikimedia Commons. Source: Franz Heinrich Weißbach (1903), *Babylonischen Miscellen.* Wissen-schaftliche Veröffentlichungen der Deutschen Orient-Gesellschaft 4, p. 16

8.5 Jan Brueghel (I) – Garden of Eden – WGA03559. https://commons.wikimedia.org/wiki/File:Jan_Brueghel_(I)_-_Garden_of_Eden_-_WGA03559.jpg Web Gallery of Art. Attribution: Jan Brueghel the Elder, Public domain, via Wikimedia Commons.

8.6 Las Pozas – Edward James Surrealist Garden – San Luis Potosi – Mexico – 03 (32553071348). https://commons.wikimedia.org/wiki/File:Las_Pozas_-_Edward_James_Surrealist_Garden_-_San_Luis_Potosi_-_Mexico_-_03_(32553071348).jpg Photo: Adam Jones. Attribution: Adam Jones from Kelowna, BC, Canada, CC BY-SA 2.0 <https://creativecommons.org/licenses/by-sa/2.0>, via Wikimedia Commons.

Chapter 9
Hierarchical Nature. Did God Love His Creatures Equally?

The short answer to the question in this chapter title is that he did not or that, if he did, his earthly regent, man, did not act in the spirit of his God. The first episode in the Book of Genesis concerning man's responsibility to nature was the unnatural slaying of Abel, son of Adam and Eve, by his jealous brother Cain. This passage has been interpreted allegorically as a reference to the entry of violence into human life through the clash between hunter-gatherers and agriculturalists who need land for sustenance, to be safe – and then ever more land. Wrathfully, God called from on high, " The voice of thy brother's blood crieth unto me from the ground." Memorably, Cain responded: "Am I my brother's keeper?" We are to infer that he is, or should be. But is he also the keeper of the rest of Creation? Was it man's sinfulness that led him to exploit and destroy nature and, if so, from a religious perspective is that destruction to be seen as another act of defiance and disobedience? In terms of the historical injunctions and doctrines of Judaism, Christianity and Islam, the modern answer to these questions must be yes.

9.1 Preservation of Life by Noah's Ark

From Cain's murder of Abel and defiance of God, we move to Noah, singled out from the rest of humankind, so wicked on the whole as to cause God almost to regret Creation. Noah was sent on a survival mission to build an Ark, a large, sturdy wooden boat, and take aboard, across the floodwaters, a small remnant of humans and two of every other living thing, male and female (Fig. 9.1). The Ark was no mere wooden craft; it symbolised the survival of life in the face of life-threatening natural adversity. Noah was charged to feed and water the creatures, to keep them alive and, after 40 days, to bring them safely to dry land.

This story, so charming to recount to children with a model boat and small painted wooden animals scattered around, is yet another magnificent symbol for adults to ponder. Water, in the world of ancient physics, was one of the four elemental forces or elements of nature. Without water, plants die. With too much, they also die. Man did not control the initial source of water but could learn to make wise and productive use of it. This had been intimated in the spread of the four Rivers far into

M. Skilbeck, *Loving and Studying Nature*, International Explorations in Outdoor and Environmental Education 10, https://doi.org/10.1007/978-3-030-80751-1_9

Fig. 9.1 Noah's ark encounters stormy seas in this representation attributed to Mughal artist Miskin c1590. Illustration from *Divan* (Collected Works) by fourteenth century Persian poet Hafez. Attributed to Miskin/ Freer Gallery of Art, Smithsonian Institution, Washington, D.C.: Purchase – Charles Lang Freer Endowment, F1948.8

the drylands beyond Eden. Floods are part of nature; they can destroy, or humanity can adapt to them – and avoid their harmful impact by good land and forest management. In this instance, by building a boat and safeguarding life, Noah, his wife, family and a few other folk were to be preserved on the Ark, not merely for their own bodily sake but in recognition of their virtue and as stewards and conservators of other creatures.

Noah's own survival was conditional on his ensuring the survival and well-being of all the other creatures. Perhaps Noah should become the patron saint of modern ecology.

There is yet another aspect of the story of the Ark which has resonated over time. To test the extent and likely duration of the flood, Noah set a dove to fly forth. When it returned with the leaf of an olive tree in its beak, he concluded that the floodwaters were ebbing. After a terrifying, life-threatening ordeal, harmony and hope were restored. That life would continue was God's intention but its realisation lay in the good offices of a group of virtuous, practical and prudent people.

God in the Ark story once again called upon mankind to bring forth every living thing, "that they may live abundantly on the earth and be fruitful". While he resolved to punish mankind no more – a resolution broken numerous times throughout the *Old Testament* – God recognised that man's restless imagination would lead to further wicked ways and evil deeds. Nevertheless, as he savoured Noah's burnt offering laid on the altar he had constructed on the drying land, God promised that he would not again bring life on Earth to the verge of extinction. It is not clear whether Noah and his family realised that they had to be the agents delivering this promise.

The Noah story is more than a mythical episode in Jewish history. It raises several disturbing issues for life on earth today. There is no guarantee of the survival of life on earth since even the cosmos which envelops the earth will ultimately decay. But long before this distant eventuality, the ruthless destruction of the earth's living resources and the poisoning of the atmosphere pose a very real threat to the survival of life, or many forms of life.

9.2 Why Sacrifice Living Creatures?

The preservation of life having been ensured by the Ark, why was the offering, burnt on the altar of thanksgiving and worship, to be, "taken of every clean beast and every clean fowl"? Why would God wish for and enjoy the ritual slaughter of his own creation and presumably creatures that had been saved on the Ark? We recall that he neither enjoyed nor accepted Cain's murder of his brother. Unlike the humans excluded from the Ark and Cain, the "clean" animals had done no wrong. They were presumably fulfilling their purpose in the order of life by submitting to ritual sacrifice. At one and the same time, they were emblematic of the vitality and beauty of nature, yet treated sacrificially in the same way as in the other religions to which Judaism was claiming moral superiority (Fig. 9.2). Biblical burnt sacrifices of animals are not self-justifying, so what is the rationale? The answer of the Mosaic Law was that a covenant between God and man required a cleansing sacrifice whereby the blood of calves and goats was shed for purification. But what is purified by the shedding of blood? (Chap. 7).

The books of the *Old Testament* contain numerous sacrificial incidents, animals and grain replacing earlier human sacrifice. Twice daily there was a ritual burning of animal and bird pieces, atonement for sin and propitiation of divine wrath. There were many occasions for sacrifices: festivals, warfare, to ward off droughts, to attract a good harvest and so on. Analogously, redemptive sacrifice of a living being (Christ) is the essential theme of the New Testament. Although questioned by modern theologians, it remains a theme of Christian ritual and art to the present day. Is there not a contradiction between the Biblically affirmed divinity and beauty of created life on the one hand and, on the other, the sacrifice of life for thanksgiving,

Fig. 9.2 Sacrifice of boar depicted on an Attic red-figure cup. The religious sacrifice of animals was a common element of ancient pagan as well as Judaeo-Christian religion. Tondo from an Attic red-figure cup, ca.510 BC–500 BC, by the Epidromos Painter. Louvre Museum, Public domain, via Wikimedia Commons

for purposes of worship and as a symbol of redemption and the regeneration of life? If there were irrationality among the Israelites, continued by their Christian successors, in at least symbolic acts, surely this is evidence of it.

The images of blood splashed over altars, and flowing onto the earth as a sin offering (*Leviticus* 16–17) is deeply disturbing, as are the dead animal carcasses smoking and reeking and eventually feasted on by the priests. Their memory is kept alive today in the sacrament performed with wine and wafers – the blood and body of the sacramental Christ, subject to cannibalistic consumption.

In his vision of God's return to the sinful Israelites, when "the earth shined with glory", the prophet Ezekiel witnessed the building of a splendid altar. Priests were ordained "to offer burnt offerings thereon and to sprinkle blood thereon". Ezekiel is commanded to make a "sin offering" of a young bullock to "take part of the blood thereof" and splash it about. The bullock, slaughtered and bloodied, is then burnt elsewhere in the temple. The vision runs into a following day when a kid and then another bullock are slaughtered. The ritual extends over 8 days of slaughter, blood-letting and the sacrificial burning. That burning apart, it might be a modern abattoir in one of those countries to which exporting nations, such as Australia, shipload live animals.

We may well think of the graphic scene in Ezekiel's vision as a gruesome ritual belonging to primitive times. It was, however, one of the foundations of our modern relationship with the animal world. His nostrils filled with pleasing aromas as his creatures are sacrificed, God accepts the "peace offering". He takes the sinful Israelites back into his forgiving bosom. Ezekiel will have played his part as an intermediary and the animals theirs by submitting to man's dominion and his bloody knife.

9.3 The Recurring Dominion Theme

The dominion theme, affirmed many times and in different ways in the Scriptures, was no minor strand of the Creation story but one of its foundations and continuing themes. For example, it was sung in several lines of *Psalm* 8:

4. What is man that thou art mindful of him? and the son of man, that thou visitest him?
5. For thou hast made him a little lower than the angels, and hast crowned him with glory and honour;
6. Thou madest him to have dominion over the works of thy hands, thou has put all things under his feet;
7. All sheep and oxen, yea, and the beasts of the field;
8. The fowl of the air, and the fish of the sea, and whatsoever passeth through the seas.

As if this were not enough to puff man's vanity, in *Psalm* 91, God in warrior vein, and man's "refuge, fortress", will arm him against the wicked; his truth will serve as "shield and a buckle". Moreover, man will be empowered to defeat evil. He "shall tread upon the lion and adder; the young lion and the dragon shall be trampled under foot". This violence might be justified as vivid metaphor, but why is it that man's fellow creatures, wild animals, are cast as the epitome of evil and wickedness? One reason might be that when humans encroached on their habitats, lions and dragons (large reptiles) appeared fearsome and dangerous. Ironically, it is some men who are now the wild and dangerous ones, such as the wealthy twenty first century American dentist on safari, gun in hand, who wounded and then after days of pursuit killed a famous African lion. Metaphors are sadly abused when rendered literally by blood-thirsty humans.

Throughout the *Old* and into the *New Testament* references abound to the servitude of animals, to animals slaughtered, burnt, trampled underfoot. They draw carts and ploughs, carry humans, provide fleece, fibre, skins, bones, meat and milk. Man has come to depend on them for his very existence. I do not suggest that domestic animals were generally ill-treated by the Israelites and Christians, nor that their well-being was neglected. The issue is the imbalance, the complete one-sidedness of the partnership and the overriding bias towards man's self interest, and toward a hierarchical structure in church and society leading to the exploitation of nature.

When the Israelites, disobeying yet again their heavenly Lord, deserted Yaweh to worship the metal image of a calf, they were punished for idolatry. Animals were to serve man; they were not to be objects of reverence. We are told that, following the flight from serfdom in Egypt, in the fear and terror of the wilderness, the chosen people forgot God who had parted the Red Sea for them. They sacrificed their sons and daughters to devils and made a metal image of a calf to worship: "Thus they changed their glory into a similitude of an ox that eateth grass" (*Psalm* 106:20). Had Moses not stood upright before him, God would have destroyed his chosen people.

By contrast and as evidence of contrariness in the exercise of omnipotence, *Psalm* 104 is a glorious paean to the natural world, to God's creation. A sense of harmony and a shared life in the rich amplitude of nature pervades the text. God is apparently in a happy, equable mood. Similarly in *Psalm* 23 the psalmist basks in green pastures beside a quietly flowing stream, his soul restored and his head anointed by a good and kindly Lord.

God's kindly attitude toward his creatures is shown by Jesus when in the Sermon on the Mount, he calls on the doubtful to accept God's mercy: "Behold the fowls of the air: for they sow not nor do they reap… yet your heavenly Father feedeth them." Yes, God cares for his creatures but ranks them nonetheless and calls for their sacrificial death. Jesus continues: "Are ye not much better than they?" (*Matthew* 7:26). Again, while God would not allow a sparrow to fall, without his support, sparrows are of lesser account than men: "ye are of more value than many sparrows" (*Matthew* 10:31).

Similarly, on meeting two men possessed by devils, Jesus acceded to their request to cast the devils into a herd of swine. This he did and the whole herd of swine ran violently "down a steep place into the sea, and perished in the waters" (*Matthew* 8:52), or, in Mark's account, "were choked by the sea" (*Mark* 5:13). There could have been more humane ways of exorcising devils than the violent death of a herd of pigs. Other episodes in the *New Testament* suggest that Jesus, like his *Old Testament* forebears, thought man, with his spiritual soul, "better than a sheep" with its animal soul (*Matthew* 12:12). Sheep, nevertheless, played an important role in the *New Testament*, as in the image of the Lamb of God (*John* 1:29), Jesus the shepherd, and the parable of the lost sheep (repentant sinners) (*Luke* 15:4–7). Later, in Christian art, sheep and other animals performed symbolic services but neither in the *Bible* nor thereafter did "services" place animals in either the natural or spiritual world on an equal footing with man.

The animals, as we saw in Chap. 7, either had purely animalistic souls or no souls at all. This lack of soul in animals was a point the seventeenth century Jesuit-trained French mathematician and philosopher, René Descartes, was keen to demonstrate. He treated animals as mere automata, governed by the newly emerging laws of physics. Devoid of feeling and consciousness, as he believed, they therefore had no soul.

According to the Cartesian viewpoint, there is not a uniformity nor continuity of life but a sharp bifurcation. As in the Platonic order of knowledge and understanding, where reason and knowledge were qualitatively differentiated from mere apprehension and opinion, it was through theological and philosophical doctrines that man was placed with God and the angels, while animals, plants and the natural world generally were relegated to a distinctly different, lower order of being.

9.4 The Notion of Natural Hierarchies

It seems that the notion of natural hierarchies or orders of being has been deeply inscribed in the human mind. Originating in the philosophy of Plato, Aristotle, Plotinus and other Greeks in the ancient world, the idea of a so-called "Scala naturel", a stairway or ladder of nature was developed from the Middle Ages to modern times as a way of categorising both the natural and social worlds. This is the so-called Great Chain of Being, whereby all phenomena are linked, not randomly or in networks, but in descending scales of value and significance.[1] (Fig. 9.3).

As in the theological discourses building on *Genesis*, and noted above, at the top of the ladder are God, the angels and demons (fallen angels), the stars, the moon, kings, princes, nobles and, in descending order, commoners, wild animals, domesticated animals, trees, other plants, precious stones, valuable metals, other minerals. Humankind, differentiated by authority and status, acts rather like a fulcrum at the centre of the stairway or ladder. Above are pure spirit, permanent and unchangeable and, as Plato put it, the highest good. Below are animals, possessing motion and appetite, and below them we descend to earth whose only attribute is existence.

The logic of the hierarchy depended on a principle of accretion. Above mere existence, each rung of the ladder added something of value. Perched above the animals with their emotions and beastly appetites is man, an admixture of spirit and animal passion. The grosser distinctions in the chain had to be broken into finer gradations. Thus the father occupied a rung of the ladder above the mother and she above children. Animals were ranked from domestic to untameable animals, such as lions, then in descending order to snakes, who had not only disgraced themselves by the wicked serpent ancestor in the garden of Eden but according to the Augustinian notion of fitness, and together with the "smallest vermin" had to occupy "the inferior positions of thy creation". Much higher up the ladder, the angels were themselves ranked. As late as the thirteenth century, Thomas Aquinas developed a nine-fold order of angelic descent, but since the very existence of angels had only ever been affirmed and never proved, it is difficult to see what value this ranking had other than as a further affirmation of hierarchy.

The Great Chain of Being, initially seen as a God-given order of nature, became secularized. It attracted naturalists and philosophers as well as theologians. Aristotle had graded animals over plants, due to their mobility and animals themselves on structural-functional principles, their mode of reproduction, possession of blood and so on. Linnaeus, the great Swedish naturalist, in his *Systema Natura* (1737) devised the threefold scheme of kingdoms, those of minerals, plants and animals,

Fig. 9.3 The Great Chain of Being, initially a God-given order of nature ascending from mere being to God became a model for the political and social as well as the natural order. Diego de Valades "The Great Chain of Being". Copperplate engraving within, *Rhetorica Christiana*, Perugia. Petrus Jacobus Petrutius. 1579. Getty Research Institute. Didacus Valades (Diego Valades), Public domain, via Wikimedia Commons

together with the binomial principle of species identification (species: variety). These were part of the Chain.

The grip of the Great Chain of Being on Western thought was strengthened through its manifestation in political and social life and its impact on philosophical thought. Prior to the Darwinian revolution, which substituted continuity of nature for hierarchy and bifurcation, it facilitated identification and classification of observed phenomena, both natural and social. For example, at the end of the eighteenth century, an Irish judge declared, "Society consists of noblemen, baronets, knights, esquires, gentlemen, yeomen, traders and artificers."[2] Serfdom continued in

Russia even into the twentieth century and, in the United Kingdom, the ancient House of Lords continues as the upper chamber of parliament. Our interest in hierarchy is as a powerful reinforcement of two central religious themes, namely man's claims to rightful dominance over nature and man's self-importance as the centre of value and power around which the world revolves.

Christian belief and practice have, of course, remained strong in modern times, despite the inroads of science, anti-clericalism, agnosticism and atheism. Powerful attacks stretched from Hobbes, Spinoza, the Deists, Voltaire, Bayle and many others who, from the seventeenth to the nineteenth centuries and beyond, pitted themselves against not only religious belief and bigotry but the socio-political and ecclesiastical bulwarks that Christianity had erected over so many centuries. This sustained critique has succeeded, together with developments in theology and religious practice, in slowly bringing about a change in many of the ancient attitudes including those of human superiority and dominance over nature. We turn, now, to some examples of notable Christians who instead of lording it over the earth, found fellowship with it and its creatures.

Notes

[1] Lovejoy, A.O. (1964) [1936] *The Great Chain of Being: A Study of the History of an Idea.* Cambridge, Mass.: Harvard University Press.

[2] Young, G.M. (1963) *Victorian England: Portrait of an Age.* Oxford Paperbacks No.12. London: Oxford University Press. p.6.

Complete Listing for Wikimedia Commons Images

9.2 Sacrifice boar Louvre G112. Epidromos Painter. https://commons.wikimedia.org/wiki/File:Sacrifice_boar_Louvre_G112.jpg Photo: Jastrow. Attribution: Louvre Museum, Public domain, via Wikimedia Commons.

9.3 Great Chain of Being 2. Artist: Diego Valades. https://commons.wikimedia.org/wiki/File:Great_Chain_of_Being_2.png Source: Rhetorica Christiana, via Getty Research. Attribution: Didacus Valades (Diego Valades), Public domain, via Wikimedia Commons.

Chapter 10
Walafrid Strabo. Benedictine Monk, Virgilian Poet, Mediaeval Gardener

We move in this and succeeding chapters from religious attitudes toward nature as expressed mainly in early texts and doctrines and by their exponents, toward practical expression in a variety of ways of life and activities which embody loving feelings for organic nature. Instances, some familiar, others less so, have been selected from Christianity, Judaism and Islam which together indicate that within each of these religious systems, there is no less an appreciation of the pleasures of earthly life than of the spiritual virtues seen as essential for a heavenly hereafter. It is from this earthly strand of love, understanding and care for organic nature that, in modern times, has emerged such movements as Deep Ecology and Eco-Christianity and Papal declarations of the need for environmental responsibility. As heaven receded, the call of earth became more clamant.

The increasing secularization of society in Christian and Muslim countries and among modern Jews has been accompanied by growing sensitivity toward the threats posed to all forms of life by industrialisation, urbanization, consumerism and human population growth. A greater interest in the conservation and enhancement of life on earth while it reflects secular and scientific values, has deep roots as well in religious values and beliefs.

10.1 The Benedictine Monastery at Reichenau

Reichenau is a small island on the edge of Bodensee (Lake Constance). It is situated along one of the main German-Italian travellers' routes in an area popular today with international and German tourists, alike for its beauty and historical associations.[1]

Dominating the island is a complex of religious buildings preserving traces of a Benedictine abbey whose origins date from the year 724 AD. After several centuries of rich monastic activity including schooling and religious education, the production of illuminated manuscripts, rich wall paintings and the accumulation of relics and liturgical objects, Reichenau Abbey fell into decline. Eventually it was secularised, first in the mid eighteenth century then at the beginning of the nineteenth. A small community of monks was re- established in 2001 and the churches

M. Skilbeck, *Loving and Studying Nature*, International Explorations in Outdoor and Environmental Education 10, https://doi.org/10.1007/978-3-030-80751-1_10

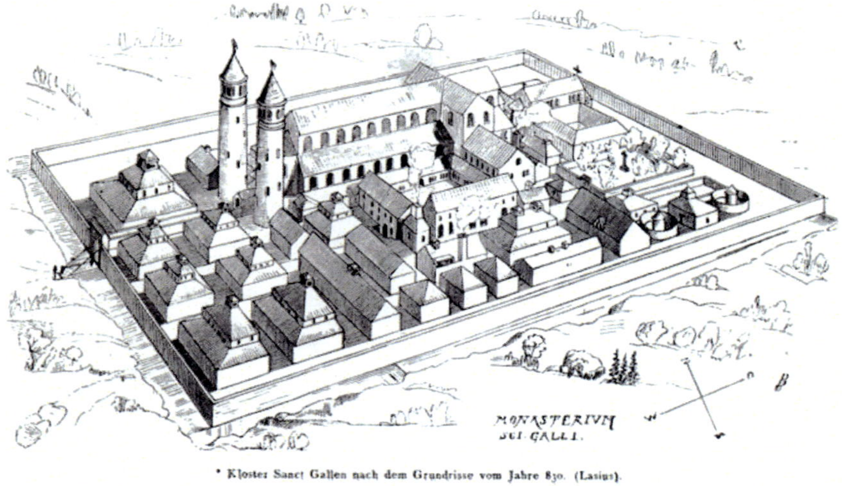

Fig. 10.1 Reconstruction according to the Saint Gall monastery plan, created in Reichenau in the early ninth century as the most appropriate layout for a Benedictine monastery. Drawn by Abbot Haito of Reichenau (763–836) for Abbot Gozbert of St Gall, this appears to have been a model, ideal layout rather than a design for a specific location. Walafrid Strabo was one of Abbot Haito's pupils at Reichenau. Image published 1876 by J.R.Rahn drawing on work by Karl Lasius. Johann Rudolf Rahn nach Lasius; upload by sidonius (talk) 12:57, 1 March 2009 (UTC), Public domain, via Wikimedia Commons

comprising the present complex are places of worship for modern-day pilgrims as well as locals.

Reichenau Abbey is surrounded by fields and glasshouses for vegetable and fruit growing. Nearby, a wetland nature reserve for migrating birds echoes the conditions that probably prevailed around the island when the monks first set up home there. Reichenau was declared a World Heritage site by UNESCO in 2000, as of global cultural significance.

The Abbey came to rival and for a time may have outshone the famous Benedictine Abbey of St Gall (in present day Switzerland) (Fig. 10.1) which was established in the middle of the ninth century. When the latter was under Magya threat early in the tenth century, the contents of its magnificent library were removed to Reichenau. Not all of them were returned, retained, one imagines, to strengthen its own holdings. St Gall, like Reichenau, is a UNESCO world heritage site. It, incidentally, houses a ninth century plan, regarded as the ideal layout for a Benedictine abbey.

10.2 St Benedict and the Early Benedictines

The typical mediaeval Benedictine abbey, monastery or convent, was often set in a remote place of natural, wild beauty that was thought to encourage meditation and engender an aura of mysticism. At first quite primitive, many monasteries grew into a complex of residential, largely self-contained and independent communal religious buildings established under the leadership of an abbot and with the authority of the Pope. The first foundation, early in the sixth century, was that of St Benedict at Subiaco in Italy, south east of Rome.[2] (Fig. 10.2)

Benedict moved in about 529 from Subiaco – regarded still as the most holy site of the order – to Monte Cassino, south of Rome and on the route to the Mediterranean. There, the famous and highly influential Benedictine rule was drawn up, as a guide to a virtuous religious life for those both within and without a religious house.

The Benedictine rule was not the first of the numerous affirmations of a hermetical or monkish Christian life. These lofty religious impulses and revelations were being experienced, if not articulated and recorded in any systematic way, as early as

Fig. 10.2 St Benedict of Nursia (480–547) depicted in an Eastern Orthodox icon. Unknown author, Public domain, via Wikimedia Commons

the third century in the Egyptian, Palestinian and other desert retreats. However, the Benedictine rule with its various offshoots became the model for the spread of religious houses throughout the Middle Ages and beyond. The requirements for its adherents are considered below.

At Subiaco, living as a devout hermit, Benedict appears to have attracted the notice of a group of monks who prevailed upon him to become their leader. The severity of his regime proving too much for the waning fervour of his followers (not an uncommon occurrence in the history of religious life), Benedict left Subiaco. After establishing several monastic settlements, he lighted on an ancient pagan site at Monte Cassino in the mountains south-east of Rome. Destroying the pagan temple there, he built what was to become the mother house of his order. Benedict appears to have wanted a life of piety and devotional reading above all, but over time the Benedictine order substantially enlarged this into a fuller, richer and at times more ambiguous way of life for the average monk.

Details of Benedict's life and character are hazy. They are largely drawn from an inspirational life composed by Pope St Gregory the Great which is deficient in biographical information. Instead of solid data concerning Benedict's early religious regime, we have numerous legends, many of them miraculous with echoes from the Old Testament such as drawing water from a rock. One of these legends, which links Benedict to the later Franciscans (Chap. 11), tells of a crow fed by him from his daily ration of bread. The crow snatched a poisoned loaf that Benedict was about to eat (served perhaps by one of his disgruntled followers). This was a popular story at a time when kindness to animals (horses and some domestic pets apart), was widely regarded as ludicrous while belief in miracles was actively fostered by the church.

10.3 The Benedictine Rule and Its Practice

The famous Benedictine rule, written at Monte Casino, was not at first for monks but for lay people wishing to follow a path to God. Clerical status was later pressed on these people. While no specific tasks were laid down, as in other orders, such as preaching, combating heretics, freeing slaves, missionising, teaching etc., the rule aimed at helping one and all to live a good and righteous life, in the imitation of Christ, as in *Matthew 19:21*, "If thou wilt be perfect, go and sell that thou hast, and give to the poor, and thou shalt have treasure in heaven: and come and follow me." Reduced to the three simple directives of obedience, work and prayer, the rule became elaborated and variously interpreted over time and in different settings.

The monastic life under the rule went through waves of devotion, backsliding and reformist zeal such as tenth century Cluniac reforms under the Abbot Berno. Chastity was not originally required and poverty was for individuals not the house as such. Indeed, many of the Benedictine and other religious houses grew into great landed estates and with the profits therefrom, together with pilgrims' gifts and the donations of reprobates seeking salvation, magnificent buildings were erected.

Unlike some others, the Benedictine order did not entail extreme asceticism; there are many instances of indulgence at the expense of devotion. Some mediaeval scholars have made a point of producing detailed accounts of the moral and intellectual shortcoming of the monks up to the time of the Reformation.[3] As we see below, monks and clerics generally were often the subject of scandalous popular tales.

Productive work by monks, especially in field and on the farm, is often seen as one of the revolutionary and progressive aspects of monastic religious life during a period in history when such work was almost universally carried out by slaves and peasants. For the Benedictine rule, open air work was conceived, at least in principle, as the natural condition of man and as necessary for his well-being. An active, practical interest in agriculture, horticulture, the processing of food and other natural items was encouraged. Some scholars have claimed that whatever may have been valued in principle by the monks in the early period, as time passed they were prepared to see much of the practical work carried out on their grand estates by the peasants, themselves taking to more sedentary pursuits.

The first century BCE Latin poet, Virgil, in his *Georgics* had extolled a productive rural life, not for peasants alone. In the ninth century, virtuous Roman, Benedictine monk and Virgilian poet, Walafrid Strabo, whose work in the abbey garden we discuss below, carried the ideal into Christian practice as well as expressing it in beautiful poetry. Walafrid, as we shall see, had no hesitation about plunging his hands into the soil which the aristocrat Virgil would most likely have assigned to his slaves.

We do not know how many monks followed Walafrid's practical example. It seems that despite the language and sentiment of the rule and the healthy early practices, it was rare, at any rate during and after the high Middle Ages, for monks to work in the fields, as distinct from their own little plots attached to their individual cells. Some Benedictines, in their labours, saw themselves as fellow workers with Jesus Christ (who was not, however, reported as having cultivated the land).

10.4 Culture and Learning in the Monasteries

The mediaeval monasteries are often celebrated as centres of art and learning. Illuminated manuscripts were copied in the scriptorium and ecclesiastical texts were pored over and commented on in the library. Relics and other precious didactic and religious objects were housed. Schooling was provided by the monks, not only for their own oblates but also, at least in some of the monasteries, for the sons of the well-to-do. As wealth accumulated, splendid Romanesque and Gothic architecture and sculpture featured in the rural monasteries as well as in the urban bishoprics. In both religious art and society generally, the value of earthly, human life was gradually being affirmed and elaborated in various secular ways.[4]

But just how widespread and systematic were these cultural endeavours? Not all monasteries enjoyed the immense fortune in land, property and domestic culture of

Cluny, the most famous of the Abbeys. There appears to have been a considerable variety of cultural opportunities and practices. Latin, for example, which had become the principal language of mediaeval Christianity, was studied and taught. However this was usually for rather narrow pious and liturgical purposes. The classical ideal of grammar, rhetoric and the other liberal arts was eventually revived in the great cathedral schools and universities during the high Middle Ages but in the earlier Christian centuries, they were viewed with suspicion – as their spread had been by St Augustine and other theologians (Chap. 6). However, some religious leaders, themselves highly educated in the Latin classics, advocated the teaching of the liberal arts and other secular knowledge, law and science, for example. One such is the Roman statesman, Cassiodorus (c.485–585), who in retirement founded the Christian monastery of Vivarium on his family estates.

The conflict between those for and against the pagan classical learning, literature and poetry is dramatically illustrated in the writings and life of St Jerome who translated the *Bible* into Latin. He wrote in his *Epistles*, "Wretched man that I was! I fasted and I read Cicero. After passing sleepless nights and shedding bitter tears at the thought of my sins, I took up Plautus (pre-Christian Roman comic dramatist). If at times I came back to myself and tried to read the prophets, the simple careless style in which they were written repelled me at once." In a dream, Jerome stood at the pearly gates, there to be asked "Who art thou?" "A Christian," he replied. "It is false," came the response. "Thou art not a Christian, thou art a Ciceronian. Where thy treasure is, the heart is also."

Pope St. Gregory the Great (c.540–604) whose list of seven deadly sins (pride, gluttony, lust etc.) became the standard, admonished his bishops on learning that Latin grammar was being studied. He wrote, "The same monk cannot sing the praise of Christ and the praise of Jupiter" (*Epistles XI*). He himself, however, as a young Roman scholar had excelled in grammar and rhetoric among other secular studies.

The risk of teaching the pagan classics was the substitution of "learned ignorance" for "spiritual life". What was important was to learn to read by speaking the words of a religious text, "to pronounce the sacred words in order to retain them, both the audible reading and the exercise of memory and reflection which it precedes are involved."[5] The processes of speaking, memorising and reflecting were of course designed to lead to pious Christian habits and actions. Yet pagan culture was not and could not be stamped out. For example, it has been suggested that the populace saw the multiplication of Christian saints as proxies for the gods of polytheism.

Monasticism became progressively and inexorably tied to Latin learning – however variable in quality – and pagan Latin texts were gradually recovered, read and discussed with the more scholarly monks evermore open to classical thought. Nevertheless, debates continued to rage between the purists such as the Virgilians and those advocating the Latin of the Church Fathers, coloured by their anti-pagan views.

The monastic life, apart from the Latin disputes, may have appeared at least on the surface as calm, under the clear, definite Benedictine rule, unambiguous and ideologically as well as socially placid. It developed into a triangulation of prayer with spiritual meditation, practical work, and study and reading of religious texts and copying of manuscripts all superimposed on poverty and abstinence. However, diverging tendencies reflected the lives of different communities and individuals within them. There were mystical desires – to contemplate and revel in the spiritual beauties of the celestial homeland.[6] The continuing quest for a precise and correct interpretation of the *Scriptures* and *Gospels* engaged the scholars. The call to the practical and to the continued developing of techniques of farming, brewing, wine-making, horticulture, and gardening appealed to others. And there were those who enjoyed managing and administering large estates with their communities of sometimes several thousands of members.

10.5 Contact with the World

Outside the monasteries, the lives of monks and nuns were at times pilloried or mocked by writers, including Boccaccio in Italy and Chaucer in England. Such practices as papal canonisations, the worship of relics, corruption, debauchery, avarice, gluttony and the ignorance and gullibility of monks and secular clergy were ridiculed in the popular stories that Boccaccio both drew upon and himself created in *The Decameron*. In *The Canterbury Tales* of Chaucer, the venery of friars was exposed by the Summoner (bailiff) in the tale he told in revenge for the Friar's exposure of the cupidity of mediaeval bailiffs. These stories excelled in vulgarity. For example, the Summoner retails a dream wherein a friar, finding himself in hell with an angel looks around for friars who by their misdeeds should be there in number. Satan is commanded by the angel to lift up his tail. From his "ens" the sinful friars pour out in their thousands, swarming like unwholesome bees through hell.[7]

Just as the Catholic Church, with its structure of parishes, bishops and an ever closer integration with secular life, adjusted its understanding of religion to the exigencies of economic, political and social affairs as well as to the refinements of the theologians, so the largely independent monasteries, generally answerable only to the Pope, developed ways of living ranging from the virtuous to the illicit and carnal. These included succumbing to the temptations and sins to which secular Christians were exposed. Such monastic lapses had more in common with everyday social life than with the self-inflicted torture of the desert anchorites from whom the monks had evolved.

Both secular Christianity and monasticism thus developed throughout the Middle Ages, not only in response to theological advances such as those of Boniface and Aquinas and the debates in the schools attached to cathedrals but by various assimilations with changing social and economic affairs in the wider world. Nevertheless, Christianity, over the mediaeval centuries, continued to mould attitudes and reinforce beliefs. Even when pilloried for misdoings, monks and priests were not only powerful symbols of divine authority, they wielded temporal power at times exceeding that of princes and emperors. Indeed, the religious image of man as a superior being, akin to the angels, was constantly strengthened by the ways in which religion drew toward secular life.

By the time of the founding of the Reichenau Abbey on the shores of Bodensee (Lake Constance), the Benedictine system had been functioning for some 200 years. Great houses were being established, and the order, intended at first as a purely local, domestic affair, had spread beyond its native Italy. The Emperor Charlemagne in his empire-building campaigns promoted the order as the norm for Europe (Fig. 10.3). It is not surprising that St Benedict has been declared by the Catholic Church to be the patron saint of Europe and, perhaps alluding to the European-wide Erasmus (study abroad) educational scheme, of Europe's students as well.

Fig. 10.3 Emperor Charlemagne (742–814) Frankish king and founder of the Holy Roman Empire at whose court Walafrid Strabo spent several years as tutor to a member of the royal family. Albrecht Dürer *Emperor Charlemagne*. c1512. German National Museum, Nuremberg. Albrecht Dürer, Public domain, via Wikimedia Commons

Accepting (at least in principle) the constraints of celibacy and personal poverty, mediaeval monasticism could provide a secure, enjoyable and – in the more liberal institutions – convivial life. For men and women who had pious aspirations, sought the security of institutional life and thrived in the company of like-minded people awed by the prospect of eternal salvation; living a reclusive monastic life had great appeal in times of political and social disorder.

The monasteries, until they became princely estates and unless they succumbed to fleshly temptation, reflected a Christian image of piety, love and devotion to a freely chosen holy life. By contrast the imperial powers, when converted to or brought up in Christian ways, saw the Christian community as part of their political domain. Conversion of pagans to Christianity was an affair of state and could be achieved through power not love. It was reported against Charlemagne, the Christian emperor, (at whose court our Benedictine monk, Walafrid, was to serve) that he drove his compulsory converts through rivers at sword point forcing their baptism. The European monks and the Anglo-Celtic Christian saints in Ireland, the North of England and Scotland – Patrick, Columb, Cedd, Wilfrid, Morton and Chad – adopted more peaceable means in their missionising.

10.6 Walafrid Strabo: Monk, Poet, Gardener

At Reichenau, as at other abbeys early in the ninth century, boys entered a religious life as an alternative to the life of the impoverished and slavish peasantry from whom many were drawn (Fig. 10.4). Destined to live their lives under monastic order, trained to a liturgical and seasonal calendar and to obedience to the abbot, the father of the monastery, these boys were guaranteed bed and daily keep over their lifetime. One such, was the eight or nine-year-old, Walafrid Strabo, born in either 808 or 809 in the part of Germany called Allemania which includes the modern province of Baden-Württemberg and parts of France and Switzerland.[8]

At Reichenau, the young Walafrid Strabo learnt Latin, not merely the limited ecclesiastical kind but that of the Roman poets, notably Virgil. He was fortunate to enter Reichenau in the time of the so-called Carolingian renaissance, when literary arts were being fostered in Charlemagne's court (although the Emperor himself could neither read nor write) and the monastery at Reichenau was one of a cluster of developing monastic cultural centres, including St Gall, Fulda and Lorsch. With a library containing some 480 titles, including Roman texts serving as literary models, and learned teachers, Reichenau had much to offer an eager student with poetic aptitude and a developing taste for the Latin language (Fig. 10.5).

By the age of 20 and having been under the tutelage of Latin scholars at Reichenau and the nearby Fulda monastery, Walafrid was being acknowledged as a poet and a humanist, as well as a dutiful and pious member of the Abbey. In 829, aged 21 or 22, he was summoned to the Carolingian court where, among other duties, he tutored Charlemagne's grandson, the future Charles the Bald, King of the Franks. His Virgilian love of nature was already apparent in a poem he wrote soon after

Fig. 10.4 Island of Reichenau on Bodensee. Heinrich Murer's 1627 copy of *Der Chronik des Klosters Reichenau* by Gallus Öhem dated 1505–1508. Heinrich Murer after Gallus Öhem, Public domain, via Wikimedia Commons

arriving at court (*Versus de imagine Tetrici*) which includes joyful lines relating to a golden equestrian statue, "the pigeons flew in and out of their nesting in the nostrils and wide-open mouth of his warhorse, and, all around, trees, flowers and sky were radiant in the sun of springtime."[9]

After 9 years at court, Walafrid was recommended by the Emperor Louis to become abbot at Reichenau. This was an indication not only of royal favour but also of the role the court played in such appointments despite the principle that monks had the right to elect their own abbot. Walafrid did not have a long tenure as abbot; he was drowned in 849 at the age of 41 or 42 when crossing the river Loire, while on a diplomatic mission to Charles the Bald.

Throughout his life, Walafrid wrote devotional poems, hymns, biblical glosses, histories and other religious works for which he was far better known than for his nature poetry. A monk, poet and religious scholar, his pious, sunny disposition, learning and overall capabilities earned him the favour of the court and the admiration of his peers. It is not for this array of talents, however that he is remembered today, but as the author of the poetic text *Hortulus – On the Cultivation of Gardens* which is held by mediaeval scholars and garden historians to be the first book on horticulture produced in Mediaeval Europe (there may be older Eastern texts).

There had been Greek and Roman treatises on botany and farming, and poetical works dealing with cultivation of the land, such as Virgil's *Georgics* and Hesiod's much earlier *Works and Days*; Homer's *Iliad* included a description of the ancient garden of Alcinus but not of its care and management. The study of plants had been the subject of scientific as well as practical treatises in the Classical Age but it was

Fig. 10.5 Abbot Walafrid Strabo (842–849). Catholic abbot, Virgilian poet and master gardener. Dutch tile, eighteenth century. Cathedral treasury, Reichenau-Mittelzell. Courtesy Roman Catholic parish, Reichenau. Image: Theo Keller

Walafrid who, as far as we know, first produced in the West a manual on how to prepare and cultivate a domestic-scale garden plot, "Despite many references to cultivated gardens in antiquity, this is the first work that articulates what exactly the garden is growing, explains the benefits and properties of the plants involved and give some idea of how they are to be looked after."[10]

Walafrid mentions some 29 plants in *Hortulus*. Almost all of them were grown for their medical value, although he writes also of their beauty and horticultural interest to the attentive gardener. In addition to verses, for example on fennel, mint and gourds, there is advice on general cultivation of these and other plants useful in the kitchen and the still-room. As well, there are rhapsodical, semi-mystical passages on the religious symbolism of the rose and the lily.

English translations of *Hortulus* range from the earthy to the more sedate style of the Hunt Botanical Library edition, whose translator was Raef Payne, classics teacher at Eton College and whose text commentator was botanical art authority Wilfred Blunt. I have chosen for the first extract the more earthy translation of James Mitchell as giving more of the feel of the practical gardener. Thereafter I draw on both translations.

10.7 *Hortulus,* the First European Garden Book

This is how Walafrid writes in the opening section of his poem:

1. On the Cultivation of Gardens

Among the many signs of a quiet life, not the least results when a person dedicates himself to the art of Paestum *(town in southern Italy famous for temples and rose gardens)* and learns to attend to the concerns of randy Priapus (god of fertility represented in gardens by obscene statues). However the plot of land is shaped, whether the soil is crap and lies paralyzed with sand and gravel, or if it's rich and moist and produces heavy fruits; whether it lies high up on a hillside or rests on an agreeable plain lined with ditches or slopes – no matter how it lies, your land will certainly produce the fruits of our native plants – assuming your attention doesn't grow numb and falter, and you don't foolishly squander the gardener's resources, and as long as you aren't afraid of going outdoors and getting calluses all over your grubby hands, and dumping baskets of cow-plop onto the parched soil. I didn't learn about this from people's chatter, or from looking at a lot of old books, or spending long days doing nothing – I gained my expertise through hard work and experience.[11]

Walafrid was well aware that the process of gardening as distinct from its products is no bed of roses. The soil has to be dug over and fertilised; in a small open courtyard seeds, sown in the winter-broken soil, need to be gently watered in order to encourage growth in warming spring weather. Constant labour is required to ensure success.

Having got his seeds up and growing, Walafrid then proceeds to enumerate his plants, their care and uses; there are melons whose, "open shimmering white pulp delights the gums with its flavour, and isn't hard on your molars either." (Fig. 10.6) Then there is sage (Fig. 10.7), rue, the scented leaves of southernwood, clambering gourds whose, "fruits come to the table often with other fine foods, soaking up the fat in a hot pan". As for iris, "We crush bits of your root and dissolve them in wine, a good remedy for piercing bladder pains". Chervil likewise, which is especially effective "if your belly is plagued by intense pain", when pennyroyal and poppy leaves are added. For snakebite and bruises, there is crushed lily, while descent from sorrow into oblivion is achieved by chewing opium poppy seeds. Strong pains during urination and the rumbling of indigestion may be relieved by ingesting ground celery seed. Horehound has a surprising but most useful virtue, "If ever a vicious stepmother mixes in your drink subtle poisons or makes a treacherous addition of lethal aconite for you, don't waste a moment. Take a dose of wholesome horehound." Of fennel:

Fig. 10.6 Sugar melon. Image from *Flore des serres et des Jardins de l'Europe* v14 217a, 1861. Louis van Houtte, public domain, via Wikimedia Commons

IV. SALVIA	IV. SAGE
Lelifagus prima praefulget fronte locorum, Dulcis odore, gravis virtute atque utilis haustu. Pluribus haec hominum morbis prodesse reperta, Perpetuo viridi meruit gaudere iuventa. Sed tolerat civile malum; nam saeva parentem Progenies florum, fuerit ni dempta, perurit, Et facit antiquos defungier invida ramos.	There in the very front glows sage, sweetly scented. It deserves to grow green for ever, enjoying perpetual youth; For it is rich in virtue and good to mix in a potion, Of proven use for many a human ailment. But within itself is the germ of civil war; For unless the new growth is cut away, it turns Savagely on its parent and chokes to death The older stems in bitter jealousy.

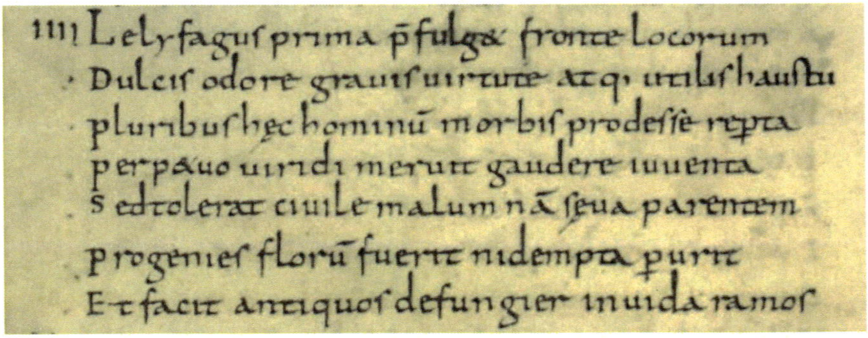
IIII Lelyfagus prima p̄fulget fronte locorum
Dulcis odore grauis uirtute atq; utilis haustu
pluribus hec hominū morbis prodesse rep̄ta
perpetuo uiridi meruit gaudere iuuenta
Sed tolerat ciuile malum nā seua parentem
progenies florū fuerit ni dempta p̄urit
Et facit antiquos defungier inuida ramos

Fig. 10.7 *Sage* – Section 4 of Strabo's *Hortulus*: Facsimile of Strabo's original manuscript in Latin; Latin transcription of Section 4; English translation of Section 4. Transcription and translation by Raef Payne. Courtesy of Hunt Institute for Botanical Documentation, Carnegie Mellon University, Pittsburgh, PA

They say it is good for eyes whose sight is clouded.
That its seed, taken with milk from a pregnant goat,
Eases a swollen stomach and quickly loosens
Sluggish bowels.[12]

A twig of pennyroyal behind the ear guards the gardener, when in the sun, from dizziness.
Finally, of the rose, "flower of all flowers", its strength and fragrance yield soothing oil. Nearby, the snow-white flower of the lily serves the poet as a symbol of "holy chastity" which shines so long as "the passions of illicit love do not destroy it." Walafrid was quite aware of the temptations monks and nuns faced. Rose and lily alike are to be praised and, presumably, cultivated for on the one hand representing the blood of martyrs and, on the other, providing "a radiant sign of shining faith." Walafrid's delightful poem ends on a high note of pious lyricism.

It is only in the final part that the poem becomes overtly religious. It might otherwise be read in the pagan terms of its Virgilian Latin, as a mediaeval contribution to the ensuing long line of herbals and as a practical guide to the seasonal tasks of the active gardener. While the poem is directed towards Walafrid's cultivation of a plot of land on the eastern or southern side of the Reichenau monastery and the uses that the monks could make of its produce, there are also references to broad acres, to horticulture more generally, of the kind that may be observed at Reichenau today. Except for a final section on the rose and lily, the poem contains no exhortations to virtue and piety other than the practical seasonal jobs of breaking the soil, controlling the weeds, fertilising and bringing plants to flower and fruition – virtues of temporal, rural life. Reference to the religious symbolism of the rose and lily come almost as an afterthought, albeit a passionate one.

The poem as a whole is a celebration of the cycle of nature and the living power of the plant world in its own right. God's gift of the earth is, as Walafrid tells us, bounteous and beautiful. It will remain so, provided it is properly cared for by good and capable people, that is by pious and sensitive gardeners.

Notes

[1] I am grateful to Ute Dickel for drawing my attention to Reichenau and for sending me material on the history of the monastery together with Strabo, W. (1966) *Hortulus* Trans. By Raef Payne, Commentary by Wilfrid Blunt. Pittsburgh, Penn: The Hunt Botanical Library.

[2] A readable, and rather adulatory account of the origins and development of the Benedictine movement from its origins to the end of the twentieth century is O'Donovan, P. (1984) *Saint Benedict of Nursia*. London: Collins. This is an English rendition of *Benedikt von Nursia*. Freiberg, Verlag Herder, by ecclesiastical historian and pastor, Walter Nigg.

[3] Notably the distinguished and controversial mediaevalist, Coulton, G.G. (1959, first in 1906) *Ten Medieval Studies*. Boston: Beacon Press; and (1961, first in 1938) *Medieval Panorama* 2 vols. London: Collins.

[4] Minne-Sève, V. (1991) *La France Romane*. Paris: Nathan: L'Empire des Moines, pp.44–59.

[5] Leclercq, J. (1957) L'*Amour des Lettres et le Désir de Dieu. Initiation aux auteurs monastiques du Moyen Age*. Paris: Ed. du Cerf.

[6] Coulton, G.G. (1961, first in 1938) *Medieval Panorama* op.cit. Vol. II "The Mystics"

[7] Chaucer, G. (1986) *The Canterbury Tales*. London: The Folio Society. Vol.II Prologue to the Summoner's Tale.

[8] For Walafrid I have drawn principally on two texts: Strabo, W. (1966) *Hortulus* Trans. By Raef Payne, Commentary by Wilfrid Blunt. Pittsburgh, Penn.: The Hunt Botanical Library; and Mitchell, J. (2009) *On the Cultivation of Gardens. A Ninth Century Gardening Book by Walafrid Strabo*. San Francisco: Ithuriel's Spear.

[9] Payne and Blunt *op.cit.* p.7.

[10] Payne and Blunt *op.cit.* p.12.

[11] Mitchell, *op.cit.* p.25.

[12] Payne and Blunt, *op.cit.* p.43.

Complete Listing for Wikimedia Commons Images

10.1 Rahn Kloster Sanct Gallen nach Lasius. https://commons.wikimedia.org/wiki/File:Rahn_Kloster_Sanct_Gallen_nach_Lasius.jpg Attribution: Johann Rudolf Rahn nach Lasius; upload by sidonius (talk) 12:57, 1 March 2009 (UTC), Public domain, via Wikimedia Commons.

10.2 Benedict of Nursia. https://commons.wikimedia.org/wiki/File:Benedict_of_Nursia.jpg Source: http://www.svetigora.com/node/976. Attribution: Unknown authorUnknown author, Public domain, via Wikimedia Commons.

10.3 Albrecht Dürer – Emperor Charlemagne. https://commons.wikimedia.org/wiki/File:Albrecht_D%C3%BCrer_-_Emperor_Charlemagne.jpg Attribution: Albrecht Dürer, Public domain, via Wikimedia Commons. The Yorck Project (2002) *10.000 Meisterwerke der Malerei* (DVD-ROM), distributed by DIRECTMEDIA Publishing GmbH. ISBN: 3936122202.

10.4 Plan der Insel Reichenau 1627. https://commons.wikimedia.org/wiki/File:Plan_der_Insel_Reichenau_1627.jpg Source: scan from *Die Bischöfe von Konstanz*, ISBN 3-922137-48-2. Attribution: Heinrich Murer after Gallus Oeheim, Public domain, via Wikimedia Commons.

Chapter 11
St Francis the Poverello – Little Poor Man of Assisi – And His Followers

Numerous legends and fanciful stories surround the life of the mediaeval friar, Francis of Assisi, who was canonised only two years after his death in 1226. Excepting Jesus and Mary, he is perhaps the most popular figure in the history of Christianity, "a faithful mirror of Christ", in the words of a modern Franciscan.[1] *He is famous for preaching to birds whom he treated as equals and for his response to the natural world – forests and meadows – as his place of worship.*

11.1 A Renewal of Asceticism

Born into a well-to-do Assisi merchant family in 1182, Francis was as a youth inspired by the ethical ideals of chivalry (Fig. 11.1). He sought glory in a military career, but in this was disappointed. After a year spent as a prisoner of war in a battle between the cities of Assisi and Perugia, he had a visionary experience of Jesus suffering on the Cross. This was a turning point; in his mid-twenties he took up a life of extreme poverty, prayer and itinerant preaching, closely modeled on that of Jesus as recorded in the Gospels.

Francis quickly attracted followers and disciples, at first four in number who in pairs went on missions, and then a further eight. In 1210, with twelve followers, Francis wrote a simple rule for which, travelling to Rome, he had the Pope's oral approval. A moving scene in Franco Zeffirelli's film *Brother Sun and Sister Moon* (2012) has the Pope falling in holy awe at the feet of the ragged monk from Assisi. The Rule, for which he was seeking Papal approval, was the founding document of the first of the three orders that Francis was to establish: the Men's Order of Minor Friars, the Women's Order of Poor Clares and the Third Order, for men not able to be itinerant preachers or women not able to lead a monastic life. The early Rule was later re-written to strengthen and intensify a way of life in close imitation of that of Christ and his disciples.

Beyond the established norms of mediaeval monasticism – chastity, poverty and obedience – Francis and his friars sought an itinerant life, divided between prayer, commonly in woodlands, and preaching, often in the open air – supported by work,

M. Skilbeck, *Loving and Studying Nature*, International Explorations in Outdoor and Environmental Education 10, https://doi.org/10.1007/978-3-030-80751-1_11

Fig. 11.1 Assisi in 1906. Photo: Messrs. Alinari and Brogi of Florence. In Heywood, W. *The Little Flowers of the glorious Messer St Francis and of his Friars*. 1906. Public domain

preferably manual, and begging. The goal was extreme asceticism and a fervid piety with emphasis on self-denial and poverty – a life guided by "the spirit of Divine wisdom". This life, starkly contrasted with that of other continental orders such as the Benedictines, bore some resemblance to the peripatetic monasticism and asceticism of the early Anglo-Celtic monks in Ireland, Scotland and England.

Not all of his friars were ready to accept the severity of Francis's harsh regime. Some sought a more intellectual or doctrinal calling and a more settled way of life. The Franciscan Order was soon to attract into its membership the philosopher-scientist, Roger Bacon, the Platonist critic of Aquinas, John Duns Scotus and the Aristotelian rival of Aquinas, the logician William of Occam, generating a debate not only between the everyday poverty of the mendicants and the esoteric knowledge of the scholars but also between the traditions of a simple, possession-free country life and an urban existence in universities and colleges. In time, this extended into fierce intellectual arguments over what was owing to God and what was reasonable in light of the call of nature and of the natural man.[2] The masses were captivated more by St Francis' religion of the heart than that of the philosophical mind (Fig. 11.2).

For a long time during Francis' life, and for a century or more after his death, the Franciscan movement suffered dissensions and disorder. The extreme austerity of

Fig. 11.2 Legend of St Francis, Sermon to the Birds. From a series of frescoes on the life of St Francis in the nave of the Upper Basilica of St Francis at Assisi. Painted around 1300 shortly after the canonization of St Francis, these 28 frescoes were traditionally attributed to Giotto, although that authorship is now disputed. Creator: St. Francis Master, Public domain, via Wikimedia Commons

his daily life, combined with total rejection of worldly pleasures that Francis imposed on himself, set a standard of quite extraordinary piety that few could match. The friars soon began to soften their message of salvation for popular consumption. Dante, alluding to the Book of Revelation, said that Francis was ready for the final seal (the train of events that must occur before God's response to the prayers of the saints). It was said that the five stigmata of Jesus's crucifixion were, late in his life, miraculously impressed on Francis's body. He himself told one of the friars that Christ in a vision said: "I have given you the emblems of My Passion so that you may be my standard-bearer."[3]

The symbolic gesture of the stigmata strongly impacted on his followers. The emotions of the credulous were stirred in a way that intellectual interpretations of the gospels by university philosophers and theologians never could. Francis himself remained hostile to the scholarly appurtenance of university, cathedral school, libraries and the material equipment on which they depended and in turn generated.

11.2 Fioretti: The Little Flowers of St Francis

The main source of what we know about St Francis is the *Fioretti – The Little Flowers of St Francis*, a work of piety written some hundred years after his death by a Franciscan friar, Brother Ugoline de Monte Santa Maria. Written in Latin (Actus), it was translated into Italian (*Fioretti*) and has appeared in numerous translations and editions ever since. Modern editions of *The Little Flowers* are a condensation of a large array of stories of varied aspects of the mendicant and ascetic Franciscan way of life.

The legends, folk tales and recollections transmitted orally recount events real or imagined. They include miracles, relating not only to incidents in Francis's life and that of other friars. Deeply felt religious convictions and aspirations for a kind of purity of Christian life feature throughout *The Little Flowers*. That life was attainable, if at all, only through the extremity of total submergence of the self in the life and actions of Jesus and his disciples.

It is this exemplary piety, presented in entertaining tales – some of them humorous – that largely explains the endurance of *The Little Flowers* as a Christian classic. For readers today, Christian or not, the appeal of the book is likely to be its combination of remarkable and fanciful stories and the immediate presence of the gentle, saintly character of Francis and his most devout friars. This at any rate is the image presented by film-maker, Franco Zeffirelli, in *Brother Sun, Sister Moon*. As the English mediaevalist G.G. Coulton put it, Francis and his early disciples evoked, "the sense of open air and sunshine in God's world rather than the devil's". This exultation of freedom and joy in the world was not to last, however, and Coulton records the Franciscan theologian, St Bonaventure (1221–1274), admonishing the faithful, "Keep thy heart with all diligence, and let it be given up to spiritual exercises only; suffer no images of earthly things to leave their impress thereon, that so it may remain stranger to all things created, and free to devote itself solely to the Creator."[4] Nature is evil.

St Bonaventure criticized actions by a friar such as twisting the end of his girdle around his finger, or carrying fruit or flowers in his hand as "a token of wantonness."[5] W.S. Gilbert seems to have had in mind to parody this solemnity when in the opera *Patience*, he had Bunthorne (a character based on Oscar Wilde) in a green velvet costume making his way along Piccadilly "with a poppy or a lily in his mediaeval hand".

What does *The Little Flowers of St Francis* tell us about Christian attitudes toward nature? First of all, we must understand what is not featured in the stories. There is among the friars around Francis himself no profession of arcane theological doctrines or meticulous arguments of the kind occurring at the time in the universities, cathedral schools and theological institutions. There is nothing in the early Franciscan way resembling the Scholastic philosophy whose leaders were constructing logical systems of proof of God's existence and a hierarchical order of nature. St Francis had faith that his piety and that of his followers would take them ultimately to heaven but this required no theological gymnastics.

As befits mediaeval stories of the supernatural, there are in *The Little Flowers* episodes of miraculous cures, ecstatic states, the raising of the dead, visions of Christ and conversations between him and the friars, levitation, a speaking statue, the measures required to keep demons at bay, the miraculous vintage from a priest's destroyed vineyard, and the conversion of sceptics and heretics. One of the more remarkable of these attempts at conversion was, when joining the crusaders in Egypt, Francis was said to have preached to the Sultan. If so, it was a magnanimous Muslim who did not have him beheaded.

Among the most famous of the stories in the popular imagination are peaceful Franciscan encounters with animals: Francis making nests for doves, preaching to the birds, and taming a wolf terrorising the people of Gubbio, and St Anthony converting heretics by preaching to a shoal of fishes. The prelude to those stories is an episode in 1209 when Francis and Brother Giles made their way through the mountains to the region of the Marches. This story, to which we return below, demonstrates the joy the friars experienced in beautiful natural settings and in knowing the virtues of birds, fish, animals, humankind, trees, stones, water (i.e. qualities and joys of the natural world).

However, perfect joy, as Francis was to explain to Brother Leo, comes only through, "conquering oneself and willingly enduring sufferings, insults, humiliations and hardships from the love of Christ."[6] Yet, there is, if not the highest perfection, a delight in the beauty of nature that cannot be gainsaid.

The world had been created good and beautiful by God. That is how Francis and his disciple, Giles, saw it when they came through a mountain pass into a superb "garden" – the Marches (borderlands) of Ancona and Fermo in the northeast of the Italian peninsula. The Marches was not only a region of great natural beauty, it proved to be fertile ground for conversion at an early stage of the Franciscan movement. St Francis happily recognised the world as good and beautiful, yet due to man's primordial sin, it was in need of redemption. So, finding a "paradise on earth" in the forests of poplars, oak, beech, the honeysuckle, clover, lilies, roses, the wheat and vines as they descended into the Marches, Francis and Giles anticipated fertile ground for conversion. The farmers and villagers, "embraced with extraordinary enthusiasm his [Francis's] message of peace through mystic union with God and good will toward all of God's creatures". In a very short time, the woods and hills of the Marches were dotted with "loci" (places), humble little friaries, abodes of prayer and self-denial from which Francis's followers ventured out on missions to the farmers and fishermen. Alluding to the New Testament, Francis prophesied to Giles, "Our Order will be like a fisherman who throws out his net and catches a great mass of fishes."[7] (Fig. 11.3)

Fig. 11.3 St Francis the preacher, as sculpted by Donatello. Photo: Messrs. Alinari and Brogi of Florence. In Heywood, W. *The Little Flowers of the glorious Messer St Francis and of his Friars*. 1906. Public domain

ST. FRANCIS
(DONATELLO)
From a statue in the Church of S. Antonio, Padua

11.3 Preaching to Animals in the Open Air

Reference to fishes and delight in the beauty of nature at this early stage of his mission introduces a theme that Francis and his followers developed in their preaching and way of life. The woods and open spaces, in which they often lived in small hermitages and worked, are frequently mentioned as sites for solitary prayer and preaching – holy places, as indeed they had once been for pagan worshippers. It is tales of preaching to animals that are among the best known and most popular episodes in *The Little Flowers*. There are others, and all, regardless of their specific content, have a purpose which relates to the distinctive attitudes toward nature of Francis and the friars.

One of the early stories tells of how Francis tamed the turtle-doves. He encountered a youth taking a number of turtle-doves to market: "Good youth, I pray thee give them to me, that birds so gentle, which in the Scriptures are likened unto chaste and humble and faithful souls, come not into the hands of cruel men who would slay them." Inspired by God, the youth handed the birds to Francis, who made nests in which the birds immediately laid eggs, bringing forth and multiplying in accordance with "the commandments of our Creator". Nor did they depart from the friars until Francis with his blessing gave them leave to do so. To the youth, Francis promised entry to the Order.[8]

By handing over his wealth – the marketable birds – the youth was expressing a readiness to live a religious life of poverty – as foretold by Francis. By nesting and multiplying instead of being slain, the turtle doves were not only beneficiaries of Francis' love and compassion, they were acting religiously by obeying God's commandment.

The famous story of how Francis tamed and converted a fierce wolf points to another religious lesson, one which was to be taken up in several stories.

While living in the city of Gubbio, Francis heard about a fierce wolf which terrorised the countryside, devouring animals and people alike. Putting his faith in God, he went out to meet the wolf who rushed at him with jaws wide open. Advancing, Francis made the sign of the cross over the wolf and commanded: "Come hither, friar wolf. I command thee in Christ's name that thou do no harm to me nor to any other." At this the wolf trotted as gently as a lamb and laid himself at the feet of Francis, who admonished him for "slaying the creatures of God without His licence". Instead of punishment, Francis proposes a peace treaty: "Friar wolf, inasmuch as it seemeth good unto thee to make and keep this peace, I promise thee that, so long as thou shalt live, I will cause thy food to be given to thee continually by the men of this city, so that thou shalt no more suffer hunger; for I know full well that whatever of evil thou has done, thou hast done it through hunger." It is not, then, natural sinfulness that causes the wolf to attack people but lack of nourishment for the natural body. In token of his acceptance of the agreement, and according fealty, the wolf held up his paw to Francis' hand, much as the lion did to St Jerome.

Preaching to the crowd that gathered to witness this remarkable event, Francis declared that such pestilences as a savage wolf are God's punishment for sin – the cause was not only the animal's natural hunger. Moreover, the jaws of a ravenous wolf, he said, are as nothing compared with the jaws of hell. At this, the people, "were fulfilled with so great joy and wonder, alike for devotion toward the saint, and for the strangeness of the miracle, and for the peace with the wolf, that they began to shout to heaven, praising and blessing God who had sent them St Francis."[9]

It was not only Francis who was able to bring animals into the circle of Christian devotion and symbolism. The celebrated preacher and Scriptural authority, St

Anthony, being at Rimini, preached to the fishes. Echoing the Old Testament rebuking of Balaam through the mouth of an ass, Anthony wished to bring the large number of heretics in Rimini to the Christian faith, "through the instrumentality of unreasoning animals, to wit the fishes". Hard of heart, the heretics to whom he preached, refused to listen so St Anthony decided to preach to the fishes, "as a preacher sent to them by God". "You fishes of the sea and river, listen to the word of God, since the faithless heretics refuse to hear it."[10]

At this, a great multitude of fish thronged toward the river bank, the smallest in front then the middle sized and, at the rear, the largest. Holding their heads out of the water, the fish gazed attentively at Anthony. He called on the multitude to give thanks to their Creator for the waters they swam in, places of shelter, sources of food, and many other blessings including serving as food for Jesus Christ. Reverently, the fish began to open their mouths and bow their heads. The people of Rimini, heretics included, flocked to witness this miraculous event. The heretics were converted and the faithful comforted as Antony expounded, "the true faith of Christ". He then dismissed the fishes who departed joyously; likewise the people who were converted and comforted.

Anthony's preaching to the fishes at Rimini paralleled Francis' famous preaching to the birds. Uncertain whether he should devote his life to prayer alone or to preaching as well, Francis sought counsel, whereby God's will was made known to him. So, after praying, he set forth with great fervor to preach. Disturbed by the twittering of swallows, he commanded them to desist, and they obeyed him. Going further, he observed great numbers of birds in trees along the road. He told his companions to await him on the roadside while he went to preach "to the birds my sisters".

He called upon the assembled birds to praise God, who had given them the means and the liberty to fly, the air to fly in, mountains and valleys for shelter and tall trees to nest in. Relieved of sowing and reaping, the birds could gather seed and, with feathers for raiment, they had need neither to spin nor to sew. Providing so many benefits, the Creator clearly loved the birds. In return, they were to give praise to God and avoid the sin of ingratitude. At this, the birds opened their beaks, stretched out their necks and opened their wings, singing praises to God. Francis, delighted in this display, "devoutly praised the Creator in them" for their attention and fearlessness. Finally, Francis making the sign of the cross over them, the birds were given leave to depart. Dividing themselves into four bands, they rose into the air, singing joyously. One band flew to the north, one to the south, a third to the east and the fourth to the west. Flying to the four corners of the world, "so the preaching of the Cross of Christ, renewed by St Francis, was about to be carried through all the world by him and his friars; the which friars, like unto the birds, possess nothing of their own in this world but commit their lives wholly to the providence of God."[11]

The Franciscan legends thus far discussed are relatively straightforward. If quite fanciful for modern readers and for the skeptical in mediaeval times, they are allegories of God's love of his creation of life which, regardless of any consideration of theological hierarchy, has its own virtue and value. As Brother John of Alverna observed, almost pantheistically, in a trance, that the Creator existed in all created things. At the same time, a theological doctrine is propounded, that life, whether animal or human, owes its creator gratitude and worship. As in the Psalms, all creatures have the ability and responsibility to praise God, just as he cares for them.

11.4 Were All Animals Truly Equal "Partners in Creation"?

There are, however, other animal stories in *The Little Flowers*, which give us pause to think that the early Franciscans at times had ambivalent attitudes toward their fellow creatures. Not all were "partners in Creation" – or at any rate not equal partners. The most holy Brother Simon, for example, did not care to be disturbed in his prayers by the cawing of rooks in the woods. He ordered them in the name of Jesus to depart. Obediently, they flew off never to return either to that spot or the entire district, that is, to their natural habitat.[12] Since Brother Simon had chosen to pray near a rookery, was it considerate of him to drive the rooks from their abode when he might have found another spot for his prayers?

A more complex issue arises in an incident in The life of Brother Juniper: How Brother Juniper Cut off a Pig's Foot Just to Give it to a Sick Man. Brother Juniper, one of the first and most elect of Francis's disciples, asked a sick friar if there was anything he wanted. "A pig's foot," was the reply. Dutifully, Brother Juniper rushed to the kitchen, took up a knife and pursued a herd of pigs. Cutting off the foot of one of them and leaving it maimed, he brought the foot back to the kitchen where he cleaned and cooked it for the ailing brother. The swineherd complained to his master who naturally protested to the friars. Hearing commotion, Francis came out and apologised, promising to make amends. The pig's owner rejected the apology and left, shouting threats and abuse. Francis chastised Brother Juniper and demanded that he pursue the man to make amends. Throwing himself down Brother Juniper offered an abject apology but did wonder why he should do so for what he believed to be a most charitable and loving act on his part. At first, in his fury, the pig's owner again rejected the apology but was soon brought to an understanding of the "charitable" act that he in turn apologised for his lack of understanding. He had the maimed pig slaughtered, dressed and presented to the friars. St Francis reconsidered the matter and while understanding the man's anger remarked, "My Brothers, if only I had a great forest of such junipers."[13]

As in the matter of the cawing rooks, the life of the unfortunate pig was made subservient to the friars' needs. Their understanding of (human) charity entailed the pig being brutally maimed and then slaughtered for the friars' kitchen. Would if not have been charitable to both ailing friar and pig to find an alternative gift and some other way of enabling the pig's owner to understand the nature of Christian charity?

11.5 In Praise of Nature: *The Canticle*

A fulsome expression of the spirit of life is to be found in a famous later "Additional Chapter" to the main text of *The Little Flowers*. This is *The Canticle of Brother Sun* in which Francis, when ill at San Damiano, praises God by praising his creation.

This Christian hymn, praising the forces declared by the ancient physicists to be the fundamental elements or forces of nature – sun, moon, stars, wind, weather, water, fire, earth – might seem as much a paean to the natural world as to God, its creator.

The Franciscan fondness for allegorising nature is nicely illustrated in the account of a visionary experience by Friar James of Massa to whom, "God opened the door of his secrets, and gave perfect knowledge and understanding of the Holy Scriptures and of things to come." In the vision, the friars of the world were held aloft in the branches of a great and beautiful tree whose roots were of gold and whose fruits were the friars. Christ, seated on a throne, summoned Francis, to whom he gave a chalice filled with the spirit of life. Francis was instructed to have the friars drink from the chalice. Some drank their fill, and became as resplendent as the sun. Others only sipped the spirit of life, or spilled it. They became dark and horrible. A great storm smashed the tree, throwing down those who had only sipped or spilt the liquid. A new tree, all of gold, sprang from the root of the original tree, bringing forth leaves and flowers and golden fruit, spreading, abroad "its beauty and fragrance and virtue".[14]

The allegory is clear; the devout could be certain that the Order would survive all vicissitudes, as indeed it has done. But what of the spirit of life? The earth is the theatre of life – of trees and friars alike – whatever may be its source. For the friars that source was and is God, to whom praise and gratitude are due. That is, we are called on to honour the source of our being, not to pollute, damage and repudiate it by our actions. There is a duty to establish and strengthen the best possible conditions for sustaining life in its richness and diversity. This is a percipient early ecological text (Fig. 11.4).

By contrast with the prudential morality of *Aesop's Fables*, the morality of the Franciscan stories is that of piety and reverence for the presence of God in nature. St Francis might be visualised as wearing over his rough garb a great flowing robe held at his tonsure by a circlet of profound faith. The robe captures the whole of nature in an image of creation, which enfolds the saint and which gives him the infinite joy so frequently affirmed in the stories.

In neither the *Old Testament* nor the *Gospels* is there a comparable depth of reverence for nature or delight in the presence of living creatures from worms to larks, that pervades both *The Little Flowers* and the *Fifty Animal Stories of St Francis as told by his Companions*. Raphael Brown, compiler and translator of the *Fifty Animal Stories* identifies four major recurring themes in the collection: (1) the birds and beasts display a remarkable, spontaneous confidence and affection for the saint; (2) they unhesitatingly and faithfully obey him; (3) he repeatedly urges them to praise God; and (4) he is deeply moved by compassion for them.

Fig. 11.4 Cypress (*Cupressus sempervirens*) said to have been planted by St Francis in 1200. According to legend, St Francis had come to Verruchio to found a new convent. When his helpers gathered wood for a fire, St Francis threw his own staff – a freshly cut branch of cypress – onto the fire. Next morning when the rest of the fire had burnt to ashes, the staff was still miraculously green. The saint observed that if the staff didn't want to burn, it should then grow, so he planted his staff in the centre of the new cloister. Image: Roberto Gelmetti. CC BY-SA 4.0, via Wikimedia Commons

The close communion Francis has with animals is, according to Brown, "a special grace granted by God to some souls who are so closely united to Him that they live, embody, and radiate a true Godlike peace". Intense love of God and, for God's sake, love of men and animals, that as Brown sees it, is the way to return to the mastery Adam and Eve exercised over animals before the Fall.[15]

11.6 Nature's Master?

Francis's relationship to the creatures depended on his special powers as a kind of benevolent master of them. That mastery is never questioned and reflects the grasp he and the Franciscan Brothers had of the biblical texts. Brown puts this quite unequivocally: "Animals were created by God to serve mankind. They had a so-called 'sensitive soul' or 'vital principle' that is not spiritual … Not gifted with reason and free will … Not immortal."[16] Animals, he says, do not have rights but men do have duties towards God regarding animals. Thus it is sinful to torture animals and degrading to the human soul to do so.

Friar Raphael Brown, however, had second thoughts on the matter of rights. Since animals are sentient, feel pain and psychological suffering, man's domain is limited, he is accountable to God for their treatment (but surely the principle of man's dominion always entailed accountability to God?).[17] Brown was writing not long before the time when philosophers, Peter Singer, Tom Regan and others, began developing an animal rights philosophy. The position of the Catholic Church during that era remained as it had been from biblical times. In 1950, Pope Pius XII while condemning cruelty to animals stated that, "the Creator has given the animal to serve man (*Genesis* 1:28), who because of his intelligence is essentially superior to the entire animal world."

Although the early Franciscans clearly held this view, the manner in which it is expressed is quite different. Francis and the Brothers with only a few exceptions – some already mentioned, others below – approached the animal world with love and respect and compassion, not through textual exegesis and exhortation but in their actions.

There are a few stories which demonstrate that Francis shared Jesus' capacity to temper compassion and forgiveness with righteous anger or to temper his affection with a moral reservation. For example, instead of taking up the biblical injunction to, "Go to the ant thou sluggard; consider her ways, and be wise" (*Proverbs* 6:6) Francis reproached the ants for their lack of trust in God's providence, "the excessive zeal they display in amassing supplies of food for the future".[18] Bees, however, were commended for building and filling their cells with honey reserves for the future.

On several occasions Francis chastised the birds for chirping too noisily. Sent off, the birds never returned. We are bound to ask of the *Fifty Animal Stories* as of *The Little Flowers* whether it was considerate to drive creatures from their chosen habitat even if, as when Francis was dictating a new version of the *Rule*, there seemed a good reason for silence. That episode was in the summer of 1223; in 1958, Friar Raphael Brown spent two hours at the spot where 700 years previously St Francis commanded the birds not to interrupt his important work. Friar Brown writes that wild birds were in the trees thereabouts, as he sat there. They could be heard only faintly; there was no nearby chirping.[19]

11.7 Unconditional Love?

The anger of Jesus against the scribes and Pharisees (*Matthew* 23) has its parallel in a couple of stories where Francis curses creatures and condemns their behaviour rather than forgiving it. A sow that killed a newborn lamb was cursed by Francis. It suffered for three days before dying. Then its carcass was flung into a ditch where not even a hungry predator would touch it. In another story, a clutch of young robin red breasts was given by the parent birds to the Brothers who had generously fed them. One, large and greedy, took its own food and prevented the others from getting theirs. Cursed by Francis, who predicted it would die a bad death, the bird climbed onto a water pitcher, fell in and was drowned. No animal would touch the corpse of the bird that Francis had cursed.

These tales of sinful creatures coming to a bad end, unredeemed by Francis's forgiveness, point to a familiar church morality concerning the necessity of repentance and penance for forgiveness. But they do not align well with the deep compassion and love of all creatures that generally characterised Francis's relations with the animal world.

11.8 A Veil of Piety Drawn Over Incipient Pantheism

Both *The Little Flowers* and the *Fifty Animal Stories* testify to a remarkable saintly life, now deeply embedded in folktales, miracles, legends and anecdotes of followers. The stories of St Francis express a religious nature philosophy in which ancient doctrinal elements, mainly drawn from the Christian Gospels, are blended with a fantastic moral fervour. The veil of religious piety through which the stories emerge screens an underlying pantheistic tendency while fully displaying the mystic unity of the universe. There is no reason to doubt that Francis was orthodox in terms of the prevailing official theology of the high Middle Ages, but the God of the stories invariably speaks and exists through nature, not only animal nature but in the majesty of forests and remote mountains and the beauty of flowers and landscapes. Reverence for an ever present God in nature and a compulsion to prayer and adulation both of and in the natural world provided a stage in which Francis and his followers played out their faith as others did in churches and monasteries. While the Catholic Church has claimed him as their saint of nature, his love of living creatures and the natural environment and his compassion for all creatures also serve as emblems of a modern universal nature ethic.

11.9 Patron Saint of Animals, the Environment and Ecology

It is in recognition of his teaching as symbolic of the spirit of life that Francis is today honored as the patron saint of animals, the environment and ecology. Francis, of all Christian saints, provides the exemplary mode of piety, the human image of Christ. He did a great deal to draw the virtues of love and kindness to the fore, while demonstrating self-sacrifice and discipline in the cause of profound faith and belief. These qualities were in stark contrast to the pride of monasticism and the princely lives of the prelates of the Church. A release was thereby provided from the logical entanglements of the Scholastic philosophers and the doctrinal disputes that had marked Christianity from its inception. He, and the early Franciscans, sought in their lives to show the meaning of piety and faith in action not words. As G.G. Coulton put it, "There, with but little actual quotation from the Gospel, is the true Gospel spirit."[20]

Nature, and particularly wild nature, forests, mountains, valleys and animals were – and are – to be enjoyed and loved. Stories of animals and life in the woods were directed at people's enjoyment of nature albeit often treated as allegorised religious texts. It is hardly to be expected that in Italy, in the period of the high Middle Ages, the Franciscans could see nature just in its own right, rather than a source of allegories designed to promote and cement the Christian faith. But they came closer to a love of nature in its own right than perhaps any other community at the time or for centuries afterwards.

On November 29, 1979 Pope John Paul II declared Francis the patron saint of ecology. American historian, Lynn White, had proposed this in his seminal article attributing the contemporary ecological crisis in part to Christian doctrines concerning man's place in nature. He also called for a new religious outlook grounded in the life of St Francis.[21] On World Environment Day 1982, Pope John Paul II enunciated the doctrine that man must protect and enjoy nature. He said that Francis' love and care for creation challenged Catholics, "not to behave like dissident predators where nature is concerned but to assume responsibility for it, taking all care so that everything stays healthy and integrated, so as to offer a welcoming and friendly environment even to those who succeed us." On World Peace Day, January 1, 1990, John Paul II said that St Francis, "offers Christians an example of genuine and deep respect for the integrity of creation". Upon his election as Pope in 2013, Cardinal Jorge Mario Bergoglio chose Francis as his papal name. Catholics and Anglican churches hold ceremonies blessing animals annually honouring Francis' Feast Day, October 2nd. These are important gestures but need to be followed by more specific directions and more concrete action.

Testament to the immense influence of the Franciscan legacy is the continuing flow of books and articles on St Francis and the early friars, paintings, such as the frescoes Giotto painted 70 years after Francis' death in the great basilica at Assisi and elsewhere in Assisi and Florence, and other frescoes by Italian artists, the naming of churches and educational institutions, painted windows and sculptures and cathedrals and churches and films such as Zeffirelli's *Brother Sun and Sister Moon.*

In the twentieth century, Hindemith and Poulenc, among other musicians, paid tribute to Francis, Francois Messiaen, celebrated for his bird-inspired and religious music, wrote the opera St Francis of Assisi and William Walton wrote the choral work *Canticle del Sol*. There are many others. The arts have shown a greater recognition of the Franciscan ideal than has western society at large. His unadorned, simple message has an enormous potential impact if taken into the heartland of the modern acquisitive capitalist world. Such inconsistencies and aberrations as occur here and there fade into insignificance against the force of the central idea: the integrity and holiness of nature.

In the centuries following the death of Francis and even during his lifetime, controversies raged in the Order, concerning the meaning and nature of the observance of poverty and the rightfulness of ownership of property. In the event, as with other religious orders, the Franciscans accumulated immense resources of great economic value in the form of magnificent buildings, vast tracts of land, sacred objects and works of art. Monasticism, in its numerous varied forms, has been a powerful economic agent, not only throughout the Middle Ages.

However, Francis' beliefs and values were maintained by strong forces within the Order known as Spirituals who observed Francis' dicta concerning poverty and ownership. Although the Franciscans did not rule the Church, among the Oder there have been notable political leaders such as Bonaventure, John Duns Scotus, Roger Bacon and four Franciscan and two Conventual-Franciscan popes. The present incumbent, an Argentine Jesuit, has identified himself with values of simplicity and humility while choosing to name himself Francis.

Notes

[1] Brown, R. (1958) *The Little Flowers of St Francis*. New York: Doubleday. p. 13.

[2] Le Goff, J. (1985) *Les intellectuels au Moyen Age*. Paris: Editions du Seuil. pp. 108–120

[3] Brown *op.cit.* pp. 210–211.

[4] Coulton, G.G. (1959) *Ten Medieval Studies*. Boston: Beacon Press. p. 44.

[5] Coulton *op.cit.* p. 49.

[6] Brown, *op.cit.* p. 60 .

[7] Brown *op.cit.* p. 18.

[8] Heywood, W. (1906) *The Little Flowers of the Glorious Messer St Francis and of his Friars*. London: Methuen & Co. p. 37.

[9] Heywood *op.cit.* pp. 35–36.

[10] Brown *op.cit.* p. 131.

[11] Heywood *op.cit.* p. 27.

[12] Brown *op.cit.* p. 136.

[13] Brown *op.cit.* pp. 219–222.

[14] Heywood *op.cit.* p. 78.

[15] Brown, R. (1958) *Fifty Animal Stories of Saint Francis as told by his Companions.* Chicago: Franciscan Herald Press. p. 10.

[16] Brown, R. (1958) *Fifty Animal Stories* p. 14.

[17] Brown, R. (1958) *Fifty Animal Stories* p. 17.

[18] "Sister Ant" in Brown, R. (1958) *Fifty Animal Stories* p. 28.

[19] "Don't disturb my dictating Sister Birds" in Brown, R. (1958) *Fifty Animal Stories* p. 94.

[20] Coulton, G.G. (1961) *Medieval Panorama 2. The Horizon of Thought.* London: Collins. p. 170.

[21] White, L. (1967) "The historical roots of our ecological crisis" *Science* March Vol 115 No 3767 pp. 1203–1207.

Complete Listing for Wikimedia Commons Images

11.2 Giotto – Legend of St Francis – 15 – Sermon to the Birds. https://commons.wikimedia.org/wiki/File:Giotto_-_Legend_of_St_Francis_-_-15-_-_Sermon_to_the_Birds.jpg Unknown source. Attribution: Creator:St. Francis Master, Public domain, via Wikimedia Commons.

11.4 Il cipresso del Convento Santa Croce. https://it.wikipedia.org/wiki/File:Il_cipresso_del_Convento_Santa_Croce.jpg Photo: Roberto Gelmetti. Questo file è licenziato in base ai termini della licenza Creative Commons Attribuzione-Condividi allo stesso modo 4.0 Internazionale (CC BY-SA 4.0).

Chapter 12
The Protestant Reformation, Consumer Capitalism and the Religious Society of Friends

With this chapter we reach something of a turning point in the enquiry into changing attitudes related to "the love and study of nature". We have moved from the personal experience of living in an inland city when one of Australia's most notable land restoration projects was underway, through reflection on ancient and modern man's treatment of wild animals, to a critique of the Judaeo-Christian doctrine of man's domination of the natural world.

By contrast with religious doctrines and creeds, we considered in the two preceding chapters the expression of religion in practical works and everyday life, observing a different, more earth-bound, understanding of man's place in the order of things.

In this chapter, we move to the revolutionary transformation of Western religious consciousness in the course of the sixteenth and seventeenth centuries Protestant Reformation and of implications some scholars have drawn for the modern ideology of capitalism. A key question is whether these religious, social and economic movements have combined to undermine or to reinforce the ancient Biblical licence to man to act as master – and exploiter – of the natural world.

12.1 The Collapse of the Mediaeval Religious Synthesis

Unity, both within the Christian Church and between its teachings and the socio-political order had been sought from early Biblical times but never fully attained. Against internally divisive heresies and schisms, the Church authorities had constantly to sustain and strengthen the bulwarks of orthodoxy. The splintering of Eastern from Western Catholicism, that eventually followed from Emperor Constantine's protective removal of the papacy from Rome to Constantinople in the fourth century, entailed rejection of Romish authority by the Byzantine (Eastern) confederation of churches.

For the Western Church, gradual permeation of forms of social and political authority did not proceed smoothly. It involved running battles with princes, kings

M. Skilbeck, *Loving and Studying Nature*, International Explorations in Outdoor and Environmental Education 10, https://doi.org/10.1007/978-3-030-80751-1_12

and emperors which sometimes favoured the religious and at other times the secular authorities. In the fourteenth century, rival claimants to the papacy were established, respectively in Rome and, under the patronage of the King of France, in Avignon. For some 70 years there were two popes. In these turbulent times the Church paid too little attention to chronic scandals and abuses within its ranks, not least the opulent, even princely lives of some of the popes, cardinals, bishops and monks.

Out of this deep spiritual malaise, from the fourteenth century onwards, the reformers Wycliffe, Hus, Erasmus and Thomas More, among others, pressed in vain for the kind of reform that might restore spiritual integrity and social harmony (Figs. 12.1 and 12.2). There was nothing new about such calls. In the twelfth century Benedictine abbess (and musical composer), Hildegard of Bingen, had urged Pope Eugenius III to work for reform of the Church (Fig. 12.3). The fifteenth century reformers, with varying degrees of support (and opposition) from secular authorities, and with the aid of the Gutenberg printing press, were able to spread their calls for change across much of Europe and especially the northern and central states. This kept alive the spirit of discontent and hope for reform.

What, in the high Middle Ages, had seemed to many an impregnable fortress of faith, buttressed by God-given reason, came to be undermined as much from within the Church as by secular forces from without. Well before the year 1517, when the German Martin Luther (1483 – 1546) of the Augustinian order of Friars, Doctor of

Fig. 12.1 John Wycliff (1330-1384) English religious reformer. John Wycliff reading his translation of the bible to John of Gaunt. Ford Madox Brown. Painting mid nineteenth century. Bradford Art Galleries and Museums. Public domain via Wikimedia Commons

Fig. 12.2 The burning at the stake of Czech heretic, Jan Hus (c1369-1415). Diebold Schilling the Elder, *Spiezer Chronik* (1485). Mediaeval miniature in fifteenth century illuminated manuscript. Public domain via Wikimedia Commons

Theology and Professor of Biblical Studies, pinned his *95 Theses* to the door of the Castle Church in Wittenberg, there were significant and widespread rumblings of discontent (Fig. 12.4). These were to lead in the course of the sixteenth century to breakaways and the establishment of the independent Protestant churches and sects of northern and central Europe. Protestant anti-clericalism was to find an ally in emerging nationalism and long-felt political resentment of clerical power.

The immediate source of Martin Luther's critical questions in the 95 Theses nailed to the Wittenberg church door was papal approval of the Church's sale of indulgences, even to redeem time in purgatory of the souls of the dead. However, discontent ran much deeper, leading many to think of the Church as rotten to the core. Christianity had to be purified and retrieved from the clutches of its ancient vehicle, the Catholic Church. Such long-criticised practices as simony in filling

Fig. 12.3 Hildegard of Bingen receiving inspiration from the Holy Spirit. Miniature illumination from the Rupertsberg Codex of her *Liber Scivias*. completed 1151 or 1152. Unknown artist, Public domain, via Wikimedia Commons

Church benefices, nepotism and family patronage, moral delinquency of priests and monks, spiritual neglect of the people and a clergy widely perceived to be corrupt were among the charges levelled against the Church. The popes themselves did not escape condemnation. Under the Protestants, doctrinal changes were to occur but initially they were not the primary concern.

Luther's protest was as much emotional as intellectual, if not more so. He referred to the pope as "anti-Christ", "the Devil's spawn". Resentment, indignation and revulsion underlay his *95 Theses*. Luther's famous hymn, A mighty Fortress is our God, A bulwark never failing ... His Kingdom is forever, appeals to the resolute heart. Not for nothing was it known as, "the battle hymn of the Reformation", nor is it surprising that, among many other composers, Johann Sebastian Bach used its stirring melody in several of his compositions including the Christmas Oratorio and Mendelssohn in the Reformation Symphony. Christians of all denominations, Catholics as well, have included it in their hymnals, to the present-day.

Fig. 12.4 Martin Luther (1483-1546), Catholic monk, scholar and theologian; leader of the Reformation. - Workshop of Lucas Cranach the Elder *Martin Luther*, 1532. Statens Museum for Kunst, Copenhagen. DSC08170. JPG. Photo: Daderot. Lucas Cranach the Elder, Public domain, via Wikimedia Commons

12.2 The Breakaways: Variations on Common Religious Themes

Historians have adduced many factors at work in the Reformation as it developed and spread. Starting as a debate over ways to reform the Catholic Church, the criticisms soon led to breakaways. Doctrinal matters inevitably came to play a part in establishing the new religious entities. For example, justification by faith alone and predestination, which underlined the supreme power of God in determining salvation, were proclaimed respectively by Luther and the French-Swiss lawyer and theologian, John Calvin (1509 – 1564), pastor, polemicist, author of *Institutes of the Christian Religion*, and their followers (Fig. 12.5). The doctrine of the priesthood of all believers and the importance of reading the Bible (in printed, vernacular texts) free of priestly mediation were widely proclaimed by the reformers, thereby challenging the authority of the Catholic clergy as intermediaries. This amounted to a revolution, not just a Reformation.

Each of these departures from Catholic orthodoxy was in turn modified. Thus the harsh logic of the doctrine of predestination, which under Calvin and his immediate followers meant that there was no place for good works as a means to salvation was softened, first by the identification of an elect – those who were to be saved – and then by acknowledgement that faith and good works could form an appeal to the grace of a relenting God. In the priesthood of all believers, some of the faithful had

Fig. 12.5 John Calvin (1509-1564), another leader of the Protestant Reformation, was invited to Geneva to establish a theocratic utopia – an authoritarian heaven on earth. Anonymous artist. Museum Catharijneconvent, Utrecht. RMCC s84. Cropped. Created c 1550. Museum Catharijneconvent, Public domain, via Wikimedia Commons

declared, for themselves, a better understanding than others of the Biblical messages. And while priestly mediation was attacked, the Protestant churches soon came to depend on ministers, pastors, synods, congregations and so on to mediate and to organise church matters.

John Calvin urged the establishment of a morally righteous, religiously guarded community under a Presbytery or Synod of ministers which, when combined with key elders, would ensure governance with what in effect was a moral and ecclesiastical dictatorship. He was invited to Geneva to oversee the implementation of such a theocratic utopia. The Puritans arriving in Massachusetts Bay from England and Holland early in the seventeenth century had similar aspirations.

The Reformation led to renewal of commitment to Biblical (as distinct from Church) authority, drawing on both Old and New Testaments. Thus Luther's hymn, A mighty Fortress, reflected the sentiment of Psalm 46. The sinfulness of the Fall was manifest to all the faithful not only in Church governance, but in the, "wretched, decaying state of the natural world".[1]

Among the reformers – Luther and his followers in the German states and Scandinavia; Zwingli and Calvin in Switzerland, France and the Netherlands; Knox in Scotland; Henry VIII; Elizabeth I and their ecclesiastical agents in England and Wales – there was considerable variation in practice, but also much common ground. On the surface this meant not only anticlericalism and hostility to the Pope, together with rejection of much Catholic ritual, worship of saints and "idols". Deep down was a desire to return, a reversion, to what was understood as the organisation and practices of the first Christian centuries.

In Luther's view, the gospel, as told by St Paul and transmitted by the most acceptable of the ancient doctors of the Church, St Augustine, rendered superfluous not only the apparatus of the Catholic priesthood but much of its complex, not to say, subtle theology. However, in their efforts to establish theocracies – not only in Geneva under Calvin but in Massachusetts under the Puritans and, for a short time in England, under Oliver Cromwell, "the Protector", – the Protestants moved in the direction of the mediaeval Catholic Church in seeking to claim secular as well as religious power. Christian theocracies were short lived and both Lutherans and Calvinists aligned themselves with secular authorities or were adopted by them. Islamic theocratic tendencies remain active in some modern Muslim states.

Eventually, strife between predominantly Catholic and predominantly Protestant states led to the European wars of religion, extending more than 100 years over the sixteenth and seventeenth centuries. Warfare further vitiated any prospect of a reunified Christianity. The Council of Trent (1545–1563) which had been called to overcome disunity in the Church, ended by rejecting Protestant tenets including salvation by faith alone. Through reform of various Catholic orders, adjustments to Church administration, and establishment by the Spaniard, St Ignatius Loyola of the Society of Jesus (Jesuits) immediately prior to the Council, the Church opened a powerful new movement, the Counter-Reformation.

The Protestant churches nurtured their own idiosyncratic doctrines and rituals while agreeing on general principles, such as the importance of direct communion with God, simplification of religious services, reduction in the number of sacraments, and the individualistic presupposition of the priesthood of all believers. Some, like the Church of England, with its 39 Articles of Religion dating from 1563 and Book of Common Prayer linking Church (Catholic) and Bible (Protestant), were, in the centuries ahead, to veer between traditional and liberal tendencies, together with an elusive quest for internal unity. Arguments over literal (fundamentalist) and liberal or metaphorical interpretations of the Bible came in time to divide Protestant churches both within and from one another, especially in the United States. The basic Reformation tenet of non-mediated Bible reading has been in practice a further factor in the failed quest of unity within Christianity as a whole. Had they been accepted as a needed impetus for reform within and by the Catholic Church, Luther's *95 Theses* might have provided a basis for internal reform of the Catholic Church and reaffirmation of religious unity. But that was not to be.

12.3 Did the Calvinistic Impulse Energise Consumer Capitalism?

To what extent did the Reformation advance or hinder unity between church (religion) and society at large (secular life)? The theocratic ventures mentioned above and the distinctive settlement of the Church of England are examples at least of the aspiration of this unity. However, over time, in these and other equations, religious authority declined in relation to secular power. There is, nevertheless, another aspect of this relationship between church, or religion, and society that, as some scholars have argued, calls for consideration. What relationship, if any, can be established between the nature and effects of the Protestant Reformation and the advancement of the now dominant, secular world ideology of consumer capitalism?

Early in the twentieth century, the distinguished German sociologist and legal scholar, Max Weber, suggested that modern capitalism has close affinities if not causal relations with ideas about personal and social life and economic affairs generated in the sixteenth and seventeenth centuries, during the Protestant Reformation and its aftermath, in both Europe and America (Fig. 12.6). These ideas, according to Weber, were mainly those of Reformation leader, John Calvin, and his followers. English economic historian and social critic, Richard Tawney, similarly found connections in Britain between the early stages of modern industrial capitalism and central tendencies in Protestant ideas about the conduct of life: "Capitalism was the social counterpart of Calvinist theology."[2] (Fig. 12.7). If, in the theology of Martin Luther and his followers, the avarice and selfish individualism of (traditional) capitalism were vehemently denounced, nevertheless idleness was sinful. By implication, work in accordance with Christian principles was commendable and disciplined work in the world was to become a religious duty, as well as a keystone of capitalism.

Tawney, in the above statement, was referring to modern capitalism which, like Weber, he distinguished from antecedent capitalist trends dating from antiquity, "Capitalism, in the sense of great individual undertakings, involving the control of large financial resources, and yielding riches to their masters as a result of speculation, money-lending, commercial enterprise, buccaneering and war, is as old as

Fig. 12.6 Max Weber (1864-1920) whose writings suggested links between Calvinist ideas and practices in the period of the Reformation and the development of modern capitalism. Photo portrait 1918. Unspecified, Public domain, via Wikimedia Commons

Fig. 12.7 Richard Tawney (1880-1962), English historian and political theorist. Memorial plaque on London house where Tawney lived during his lifetime. Photo: Megalit. CC BY-SA 4.0, via Wikimedia Commons

history. Capitalism as an economic system, resting on the organisation of legally free wage-earners, for the purpose of pecuniary profit, by the owner of capital or his agents, and setting its stamp on every aspect of society, is a modern phenomenon."[3]

Weber wrote of "affinities" between Calvinism and capitalism, the "psychological impulses" of the one giving direction to everyday life in the context of the other. He did not, as his intellectual biographer, Reinhard Bendix, wrote: "go substantially beyond the analysis of theological doctrines and pastoral writings. He sought to show that the inherent logic of those doctrines and the advice based upon them both directly and indirectly encouraged planning and self-denial in the pursuit of economic gain."[4]

On this point of economic logic, Richard Tawney was more reserved. He found much more of the indirect than the direct in the relationship between the Reformation and modern capitalism, "If it is true that the Reformation released forces which were to act as a solvent of the traditional attitudes of religious thought to social and economic issues, it did so, against the intention of most reformers."[5]

Calvin and his followers asserted the supreme role of Scripture, giving relatively little place to nature and reason as sources of revelation. From Scripture they took the basic idea of an all-powerful God who totally ruled over human affairs. Man was, as it were, utterly helpless unless redeemed by God, whose grace was inscrutable. It was man's duty to obey implicitly the will of God, thereby living according to Christian (Biblical) law. But in Adam's sin this capability was destroyed, to be redeemed through God in Christ, and then only as God willed.

As noted above, humanity was thus divided into two groups, those predestined to grace in the lap of God and those whose fate was eternal damnation, the latter greatly outnumbering the former. God in his infinite wisdom mysteriously made this pitiless distinction, whereby the small minority of the "elect" would be saved while the rest, regardless of whether they lived good or bad lives, would not. The elect included religious and social leaders who would naturally assume that, by virtue of their position, they enjoyed God's grace. According to the initial strict interpretation of Calvinism, the non-elect could do nothing to change God's edict. For the greater glory of God, they were nevertheless required to live virtuous lives of simplicity, frugality, piety and obedience to the social and religious order established by the elect, among whom were the magistrates. There are analogies in this line of thought to the class relations that for Karl Marx and Friedrich Engels were inherent in capitalism.

Weber argued that the state of anxiety, not say terror, engendered in the population at large by realisation of their possible (or likely) damnation would lead people to adopt the principles of the Calvinistic religion and social order. Work and material success, as well as religious piety might, they hoped, demonstrate worthiness, thereby signifying, if only to themselves, that they were in fact among the elect, the saved. According to Weber's thesis, the habits of mind and patterns of behaviour of the virtuous were at the very least consistent with, and perhaps also causative of, what he termed "the spirit of (modern) capitalism".[6]

Weber's argument is complex, ingenious, highly original and contentious. As with Tawney, he did not suggest that capitalism was created in the course of the Reformation; to the contrary, some of its elements such as private ownership of the means of production, profit-making, capital accumulation for purposes of investment, employed or slave labour, and financial institutions and procedures facilitating credit, could be traced to ancient times. But in the 16th and 17th centuries, new practices were adopted. These included the separation of business from household activities, sophisticated methods of book-keeping and accounting, establishment of joint stock companies and the stock exchange and, especially, more rational or systematic ways of organising work in a wage-earning economy. Legal changes recognised these economic developments and social and political principles increasingly recognised individual rights including private property. These developments assumed an interest by individuals in acquiring both worth and wealth. In Calvinist ethical terms, it was a duty for individuals to advance their own interests by achieving and demonstrating competence and proficiency in a "calling", that is a vocation. The idea of a "calling" was religious but it had significant implications for the individual's overall life. For example, while churchgoing was necessary, it was not sufficient. Material wealth acquired through regular, morally worthy work, was pleasing to God and therefore a duty.

Like much of Protestant reasoning, "wealth as duty" was not a new idea. The casuistry of Calvin's argument echoes early Christian debates over the advice Jesus gave to a rich young man, "If thou will be perfect, go and sell that thou hast and give to the poor" (Matthew 19.21). However, the theologian Clement of Alexandria (150-c.215 AD), favouring moderation, wrote that extreme poverty was not virtuous, rather that "excessive desire" had to be curbed. Literally following Jesus' advice, St Francis of Assisi had thereby risked ecclestiastical condemnation and lost the loyalty of a section of his followers.

Unlike many other students of modern capitalism, Weber looked for a driving force not so much in material conditions such as expanding trade, political changes or social evolution but in values and beliefs. The Calvinists aimed to create in their own lives an enduring image of God's righteousness, that is to say activities in everyday day life that would be pleasing to God. The organisation of a methodical life, within a vocational calling, would serve God's glory – for which the world solely existed. Thus disciplined work was sanctified, it tamed desire, led to moderation and alleviated the dread of predestination. Moreover, it enhanced self-confidence. Wealth, the fruit of working in a vocational calling, signified God's blessing. It was not to be squandered in enjoyment and frivolity but itself put to work through saving and investment. Thus work, conditional on meeting strict moral criteria, became a religious ethical ideal. Capitalism, on this Calvinistic model, was a form of pious self-denial – for capitalists and workers alike.

A distinctive mind-set either grew out of or was congruent with the teachings and practices of the religious reformers and especially the theology and the theocratic practices that Calvin and his followers attempted to establish in Geneva in the mid-sixteenth century. Although similar attempts to establish a Calvinistic theocratic regime were made in the early years of the Puritan settlement of Massachusetts Bay - part of the heartland of modern capitalism - neither Weber nor Tawney seems to have studied its relationship to proceedings in Europe, the source of the beliefs of the Puritan Pilgrim Fathers. Weber did, however, find significant parallels in Benjamin Franklin's advocacy – and practice – of a disciplined and prudent business ethic (Fig. 12.8).

The religious ideas of the reformers found favour, initially, among the "middling" sectors of society: craftsman, yeomen, small businessmen, traders and others who were of neither the aristocratic class nor the labouring poor. Benjamin Franklin, although not himself overtly religious, was for Weber a classic illustration of the Calvinistic work ethic and its capitalistic tendencies. His famous autobiography provides many examples of self-help and "getting on" in the world.

The virtues of this middling sector – the future middle-class – included hard, systematic work, thrift, prudence and a readiness and ability to organise economic activities and their working lives in orderly productive and profitable ways. Calvinism gave impetus and a religious licence to these worldly qualities. As population increased, business enterprises expanded, international trade flourished, and urban centres became larger and more numerous. Patterns of systematic employment and regular wages were established while financial institutions, if ever risk-taking, became more pervasive and powerful. Moreover, the capitalist system of industrial production and trade of huge quantities of material goods led to what was to become a virtually irresistible pressure to consume and to accumulate material possessions – modern consumer capitalism. And it is the forces within and

Fig. 12.8 Enlightenment thinker and an architect of the American Constitution, Benjamin Franklin (1706-1790) whose belief in self-help and "getting on" in the world linked him with the post-Reformation resurgence of capitalism. Benjamin Franklin by artist Joseph Duplessis 1788. Metropolitan Museum of Art. Joseph Duplessis, Public domain, via Wikimedia Commons

supportive of modern consumer capitalism that now constitute the greatest menace to nature, whatever benefits they bring in their train.[7]

The rise of Protestantism coincided with the great European voyages of discovery and establishment - first, of trading posts in Africa, South America and the Far East, and then of colonies. The enormous increase in trade brought great wealth to Europe, with opportunities for productive investment on an unprecedented scale. Economists, including Adam Smith, and philosopher and historian, David Hume, demonstrated that the old (mercantilist} idea of a fixed quantum of (tradable) goods had been superseded (Figs. 12.9 and 12.10). Investment and manufacture as well as improvements in land and resource use, had led to increased productivity – human consumption. Successful entrepreneurs demonstrated their social worth and virtue alike by generating profit. The energies of capitalism, therefore so far from constituting an alien secular force, could be aligned with the new religious energies that the Reformation was shaping and directing, "Both religion and secularism acquired a new energy from their mutual jostling."[8]

12.4 Continuing Human Dominance

While the above brief summary does scant justice to the meticulous research and finely reasoned conclusions of Max Weber and Richard Tawney, it does highlight their most interesting albeit controversial findings. In particular, it invites the

Fig. 12.9 The Scot, Adam Smith (1723-1790), pioneer of modern economic theory, best known for his advocacy of individual enterprise and free trade among nations. Known as "The Muir Portrait." c1800, artist unknown. National Galleries of Scotland. PG 1472. Scottish National Gallery, Public domain, via Wikimedia Commons

question as to whether the new directions in religious thought and practice and the emergence of modern capitalism made any difference to people's interpretation or view of the ancient religious and social doctrines concerning the centrality of humans in the natural order of things.

As we have seen, the general Reformist denunciations of the excesses of the Catholic Church, its centuries-long accretions of religious ceremony and liturgy, church ornamentation, monastic self-indulgence, and the opulent lifestyle of prelates, led to a demand for simplicity and modest living. Inspired initially by Martin Luther, the reformers wanted above all a more direct and personal relationship with God and they wanted people to live according to a responsible, individual understanding of the Divine Order. Christian principles were to be enunciated, adopted and acted on, free of the emotional and intellectual controls of an authoritarian Catholic Church. It is noteworthy that these ideas were contingent on individual and collective human agency, responsibility and consistent allegiance to the perceived grace of God. They proclaimed resurgent laymen-power in the world.

The Calvinist commitment to hard work, thrift and the simple life based on the satisfaction of necessities, avoidance of luxuries and the generation of profits from investment and employment of people pointed in the same direction as Luther's, but with greater force. Employed workers, free to enter into wage bargains, implied a level of individual understanding and confidence beyond that required of slaves and indentured labourers. The Calvinistic organisation and systematisation of work entailed new forms of social organisation and were to become a fundamental feature

Fig. 12.10 David Hume (1711-1776), Scottish empiricist philosopher, essayist and historian whose attacks on revealed and natural religion deprived him of university professorships in Edinburgh and Glasgow. Portrait by Allan Ramsay, 1776. National Galleries Scotland. PG 1057. Allan Ramsay, Public domain, via Wikimedia Commons

of factory production. The sober, responsible clean-living individual imagined in Lutheran and Calvinistic doctrines, was well-suited for diligent work in the emerging industrial organisations. For Lutherans and Calvinists alike, religion was a means of energizing and directing human capabilities in everyday life. The human species was to be empowered to rise to new heights of personal, social and economic achievement. But sin, as expressed in pride and avarice, had to be held in check in a precarious balance.

The Reformation was in no way a challenge to established religious beliefs and ideas about the status and agency of humans as articulated in the early books of the Old Testament and carried forward over the centuries of Christianity. These beliefs and ideas were indeed reinforced and enlarged and the links forged during the Reformation between religion and capitalism strengthened their grip on human life. That is to say, the rigours and the individualism of Protestantism and the energies released by modern finance capitalist endeavours through mass production and the organisation of labour further promoted the idea of man's God-given powers and mastery over nature. The Calvinist elect who were often known as "saints," might be thought of as a select group of human beings living on a higher or angelic plane, above and beyond the masses, and exercising more than ordinary human powers over nature and humanity. These special people were, like the Jews of the Old Testament, the "chosen" people. In the theocratic projects in Geneva and Massachusetts Bay, again like the ancient Jews, they saw themselves as the "chosen" (by God in his wisdom and majesty) leaders of a special "nation". Their nineteenth century analogues were "the captains of industry", less politely known as "robber barons".

As analysed by Max Weber, Richard Tawney and other scholars, Karl Marx among them, modern capitalism in its formative stages did nothing to challenge the prevalent view of man, although "fallen", as the master of nature. On the contrary, the image of the powerful entrepreneur or corporation, commanding vast financial and human resources, or of a unified proletariat battling to establish a communist paradise, suggested exaggerated human power and unlimited direction of the natural world and its resources, beyond anything hitherto experienced in human affairs. It seems not to have occurred to the apologists of rampant capitalism that its heroic figures might be among those children of Adam who had "fallen" the deepest.

"Avarice" was of particular concern to the religious reformers as it had been to religious leaders in the mediaeval church. Greed and avarice were widely recognised as ever-present risks which the relentless pursuit of profits could but exacerbate. The long drawn out assault by both Catholic and Protestant theologians on the greed of usury finally reached an accommodation in the time of the Reformation through acceptance of a "fair" rate of interest on loans. "Fairness" was a recognition of the labour of the lender, the principle of a share in profit earned from credit, the risks involved in trade and so on. Under capitalism, however, especially as it developed in the nineteenth century, such considerations provided no brake on extreme profiteering and few checks were applied to greed and avarice. Richard Tawney, in *The Acquisitive Society*, was acutely aware of just how the doctrine of "fair" return on investment could lead to the kind of exploitation and profiteering that the religious critics of usury had ever warned against.

Just as they struggled to balance the permanent tension in Christianity between the degradation and the exaltation of man, the early Protestant reformers, like Catholic theologians before them, sought to define and maintain a balance between a fair return for investment on the one hand and the human propensity to greed and avarice on the other. Over time the balance shifted markedly, as "the spirit of capitalism" came to prevail over "the ethics" of Protestantism. In this changing balance of power and influence, there has been a close coincidence and, even more, strong reinforcement of the idea of man's mastery of the natural world. It was as if the theology of Pelagius, against which Augustine had battled more than a thousand years previously had re-emerged: man could do anything, of his own volition.

The mainstream Lutheran and Calvinist churches and the various sects most often either fitted themselves to, or actively promoted capitalist economics and social organisation. Generally, they held in check the potentially radical social and political implications of doctrines such as "the priesthood of all believers" and they quelled politically radical tendencies within their own ranks. Some Protestant sects, however, as we see below, made more searching politically and socially challenging investigations into the meaning of the Christian faith. Despite religious revivals up to and including the present day, the ideal of a simple, humble response to guiding spiritual forces has been progressively eclipsed by the prospect of material gain and earthly comforts.

12.5 Radical Religion: The Religious Society of Friends

The Religious Society of Friends was one sect among several groups of English non-conformist dissenters who refused to accept the Anglican settlement of Henry VIII, Elizabeth I and their clerical agents. In a time of religious and political turmoil, the Friends, commonly known as Quakers (among other reasons, for shaking in "the presence within of the Lord") originated in small group meetings of Protestant sects (Seekers, Independents, Baptists and other sectarians who united as "Children of the Light" or "Friends in the Truth") in the north of England during the 1640s. They were hostile alike to the established Church of England and the royalist political culture which lay behind it. The movement then spread to the south and west of England, the Netherlands, the United States and, in the nineteenth century, Australia and New Zealand, parts of Africa and other countries. The Society has always been numerically small (today there are some 400,000 Friends worldwide) but socially and politically influential; for example, through engagement with campaigns opposing slavery and warfare.

The Society of Friends is of interest in this chapter mainly for two reasons. First their religion serves to illustrate the diversity of Protestantism by virtue of the repudiating of much of the traditional doctrinal basis and practice of Christianity in a quest for an ever more "pure and simple" form of religious experience. Second, in the modern development of Quakerism there is an even more radical subset of members, a liberal wing whose religion is either vaguely or overtly nontheistic. This wing is open to incorporating beliefs and practices from non or quasi-theistic religions, notably Buddhism, but also to taking inspiration from nonreligious sources including environmental and peace movements, ecology and even quantum physics.[9] Debarred access to the universities of Oxford and Cambridge, many Quakers nevertheless made contributions to science, including botany and to industrial development (as in the development of cocoa and chocolate). Some involved in the latter were implicated in slavery.

12.6 Quaker Beliefs and Values

The early leader of the Quakers was the Englishman, George Fox (1625 – 1691), who throughout his life was in and out of jail and frequently persecuted (Fig. 12.11). Quakers suffered imprisonment and even hanging until the eighteenth century when religious fires cooled from the zeal of the Reformation. Fox called not merely for a new sect but a radical, egalitarian, spirit-fuelled Christianity. Initially, Quakers were at times noted for crying in the streets, trances, nakedness, sackcloth garments and vehement itinerant preaching, but this phase passed. In the eighteenth century Voltaire in his *Letters on the English Nation* described the Quakers as possessing a "religious outlook almost Chinese in its rationality".

Like the early Continental leaders of the Reformation, Fox and others sought a model of a pre-institutionalised, very early church, free of priests, bishops and the apparatus of organised religion. As children of God, all people were said to have within themselves the power or light of Jesus Christ (the famous "Inward Light " or "Inner Light"). There is an echo in this of Thomas Aquinas's "sideways" glimpse of the divinity in his prayer, incorporated in the Catholic Benediction: "faith, our outward sense befriending, makes our inward vision clear" and "God, we beseech you, that we may so venerate the holy mysteries of your body and blood, that we may evermore perceive within ourselves the fruit of your redemption." Significantly this

Fig. 12.11 George Fox refusing to take the oath of allegiance at Houlker Hall, 1663. From painting by John Pettie, R.A. 1864. The oath was tendered to George Fox by the justices of the neighbourhood. Photo: Wilfred Whitten: Quaker Pictures, (The Friends' Quarto Series 1), Hicks, London, 1892, p 11. Public domain via Wikimedia Commons

"light" encompasses cooperation with and not mastery of nature. Faith in the universe follows, as a manifestation of spirit.

Fox had a religious experience in which he was filled with and felt inwardly the presence of Jesus. He believed that all people could have such an experience, thereby meeting God directly. It was for Fox - and remains incumbent on Quakers - "to always be open to the Light, from wheresoever it comes". It follows that individuals will find their own source of light and that these sources may be radically different from one to another. Outward forms of religion, doctrines, creeds, priests, churches and so forth Fox believed to be " anachronistic and apostate". Illumination would come to those who waited patiently, in silence and with minds directed by holy spiritual thoughts and emotions. This aspect of Quakerism had much in common with mediaeval Quietism, or contemplation. However, it was both a prelude and a companion to action in the world.

Quakerism has struggled from its inception to clarify and agree on relationships between the traditional Christian doctrine of man's fall from grace into a natural condition not oriented towards God but towards "the world," and the notion of a "light within". The virtue of this inwardness is its immediate appeal to and of the apprehended and appreciated God within the human heart. Exploration of these two poles has led Quakers over the centuries, but especially during and since the nineteenth century, in quite diverse directions.

Theology and philosophical enquiry have not been strong elements of Quakerism and the accents of individual religious experience have given rise to varied viewpoints concerning directions in which the "Inward Light " leads. Rejecting theology contingent upon divine prerogatives, Quakers have opted for descriptions, recognitions or personal experiences of universal human "spiritual conditions". At the same time, through regular meetings in "houses" (not churches) they have worked to achieve agreements and common understandings especially of desirable social actions. These actions go under the name of "testimonies", the most celebrated of which, as espoused by Fox, is "to peace". Given the depth of internal conflicts, as between mystical faith and quietist practice on the one hand, and zealous evangelicalism on the other, great effort has been required to achieve a more or less united front, especially in the United States of America.

Whereas many religious groups have creeds, and Christianity is replete with authoritative (and contested) doctrines, Quakers have testimonies. In a statement commemorating the 350th anniversary of the *Quaker Peace Testimony*, the Quakers in Britain stated that,

> Testimonies arise out of an inner conviction and challenge our normal ways of living,

- They exist in spiritually led action rather than rigid written forms
- They are governed by continuing spiritual experience and are not imposed in any way
- They require searching for ways in which the testimonies can become true for ourselves.[10]

Acknowledging that the way is not easy, but is eased, "with loving advice and a supportive community", the statement then listed and elaborated the testimonies in the following categories: Truth and integrity; Equality and community; Peace; Simplicity; The earth and environment.

The more conservative and evangelical wings of the Friends anchor personal recognitions and experiences in the *Bible*. The more radical liberal wing, as already noted, trusting to individual capacities which are spiritual, has reached out in other directions, most recently favouring scientific over recognisably religious explanations of human life, values and conduct. Regardless of what appear to be serious doctrinal differences among themselves, Quakers all share a belief in utmost simplicity – in speech, dress and manner and in a quest for detachment from material possessions. They also strive to achieve tolerance of diversity within their own ranks, in recognition of the mysteries of the workings of "Inward Light".

By the end of the nineteenth century Quakerism had as a whole come to embrace four main spiritual pathways:

1. Primitive, historical, the "Inward Light" provides direct engagement with God
2. Evangelistic, missionary, Wesleyan, Scriptural authority with minimal emphasis on inwardness and silence
3. Liberal, humanistic, anti-revivalist, interest in science together with criticism of traditional Christian doctrine
4. Evangelical, ecumenical, emphasis on silence and testimonies

In different parts of the world, different emphases on one or more of the above have emerged. If there is a mainstream, it is pathway number 4 above, but within national settings there may be a variety of strands. For example, 3 above has a strong body of adherents in parts of the United States and Great Britain. The import of this diversity is that modern Quakerism, while maintaining elements of the past (e.g. silence, inwardness, testimonies, active service) is flexible and open to a wide range of ideas about personal and social well–being.[11]

For Quakers, the presence of God, whether in the traditional form of a supernatural entity or supreme being, or as a metaphor for exalted spirituality, is to be experienced both privately and in social activity. Not only are hierarchies to be dismantled, men and women are conceived as equal and no-one is to stand over another as master or mistress. Hence Quakers were opposed to slavery and they campaigned long and hard for its abolition. Nevertheless, some Quaker merchants trading out of the port of Bristol for cocoa beans and sugar were among the slavers. Capital was required for this purpose to keep the manufacturies in continuous operation. So Quakers were among the early modern capitalists.

All warfare, according to the *Quaker Peace Testimony*, is an abomination. Among the reasons for persecuting Quakers has been their refusal to be enlisted by the state in any form of militarism. Testimonies to peace and nonviolence are perhaps what Quakers today are best known for. The testimony is not a formal doctrine or credal utterance, but an agreed direction of the action which at different times in different places will vary in content and working. This is further evidence of Quaker avoidance of ritual, and openness to changing codes of conduct.

12.7 Quakers and Nature

The deep reverence with which Quakers have always held the natural world was at first believed to come through divine revelation, directly to the heart. God, the creator, who invested the natural world with spirituality, and exists immanently in that world which was, therefore, to be revered. From late in the eighteenth century, there was a shift towards support for an active engagement in a more detailed scientific study of the natural world. Today, there is among Quakers a growing sense of responsibility for the well-being of the planet. This is not mystical but practical, a concern held primarily by those Quakers who identify with ecological environmentalism. The scientific flavour notwithstanding this is a religious as much as a social and scientific quest. Nature is still conceived, at least by those Quakers who remain fully theistic, as a temple that God created both for himself and for human and non-human creatures. The natural world became significant in part of the wider Quaker culture in the nineteenth century, since understanding nature came to be regarded as part of the divinely inspired search for order and truth.

As we shall see in what follows, Quakers in America have actively responded to the richness and beauty of the natural environment. In Australia, as discussed in Chap. 3, the Quakers Albert and Margaret Morris in Broken Hill and Edwin Ashby among others in South Australia, saw the natural world as a resource to be studied and treasured. The interest Quakers have long shown in natural history and science has been documented in recent years by Quaker organisations in Great Britain and the United States. Notable figures include Joseph Lister, pioneer of anti-septic surgery and President of the Royal Society; James Backhouse, botanist and missionary; physicist, John Dalton; astrophysicist, Jocelyn Bell Burnell; astronomer, Arthur Stanley Eddington and economist, Kenneth E Boulding . There are many others in the Quaker scientific Hall of Fame in numbers said to be disproportionate to the small overall size of Quaker membership.[12]

Several explanations have been offered for the scientific interests and achievements of Quakers. For Great Britain, it has been suggested that the Test Act of 1672, whereby Catholics and nonconformists were precluded from public office, both civil and military, and excluded from matriculating at the University of Oxford and graduating from Cambridge, led Quakers into avenues other than the learned professions, notably science and commerce. It has also been argued that it was their concern for practical matters that led Quakers to take an active interest in science and the social usefulness of its findings. Similarly, commercial activity was seen as playing an important role in meeting human needs. Free from religious dogma, Quakers became more enquiring in their relations with the world. It has also been suggested that the Quakers found in the study of God's creation the "book of nature", thereby coming nearer to God than could ever be achieved through the study of texts. Interest in modern scientific thought has led some Quakers towards the psychology of Freud, Jung and William James. In this respect, the Quaker mind "bloweth where it listeth". As well as psychology, Quakers have taken up social and cultural issues including feminism, poverty and environmental activism.

In a sense, the scientific, cultural and social reform interests common among modern Quakers are to be expected, inasmuch as Quakers have always shown great responsiveness to changing social conditions and cultural issues. John Woolman (1720 – 1772) argued in his famous *Journal* that over time there was a progressively greater apprehension of where "the light" was leading. For example, slaveholding Quakers would, he said, come to see the error of their ways (they did). For Woolman, this tendency may not have led him towards receptiveness to nontheistic religions and post-Darwinian "apprehension" of nature such as scientific ecology, but such developments are quite consistent with the idea of "progressive apprehension" over time.

Reservations about the *Bible*, including indifference to large parts of it, appear to have fostered a spirit of unfettered inquiry at least among the members of the Quaker Radical wing. At the same time there has been a resurgence of interest in the thoughts of early Quaker writers and, in both the United States and Great Britain, academic study of Quaker history and ideas where study centres and university programs have grown in popularity. Of particular interest is the frequency of reference to what is regarded as an authoritative text, *Apology for the True Christian Divinity* by Robert Barclay (1648-1690). Barclay's attraction may be that his thought transcended the doctrinal and organizational concerns of many Protestants in the seventeenth century. He called for a spiritual transformation of the world, not just spiritually-illuminated individual experience. At the same time he emphasized practical procedures, like experiencing direct revelation and a lay ministry.

Interest in nature, now becoming more widespread among Quakers, had a notable precedent in eighteenth century England and America in the form of a 30 year correspondence between a Quaker merchant in London, Peter Collinson, and a Quaker farmer in Philadelphia, John Bartram.[13] (Figs. 12.12 and 12.13) Collinson (1694 – 1768) was in many ways the epitome of Quaker interest in natural history which over the centuries became an important outlet for their intellectual, social and spiritual concerns.

Living in London, Collinson combined a commercial interest in the family haberdashery business with gardening and horticulture. Through both, he aimed to develop trade with the American colonies. With his Quaker wife, he set up house and a large garden in the Quaker stronghold of Mill Hill near to London. Collinson was deeply interested in botany but unlike some of his fellow Quakers who were denied access to the English universities, he did not enrol in Edinburgh where botanical studies were part of the medical degree and open to Quakers. But through his extensive horticultural activities including voluminous correspondence with botanists and plant collectors, together with his patronage, he was elected a Fellow of the Royal Society in 1728. Among those he influenced was fellow Quaker, John Fothergill, medical practitioner and naturalist, "whose efforts paved the way for all modern botanic gardens".[14] (Fig. 12.14).

Collinson's interest in and practical support for science is nowhere more evident than in his correspondence with Benjamin Franklin. Collinson received a letter from Franklin, dated 18 October 1752, recounting the famous kite experiment whereby, "the electric fire was drawn from clouds … when a thunder gust appears to be

Fig. 12.12 Quaker "brother gardener" Peter Collinson (1694-1768). Engraving by T. Trotter, Public domain, via Wikimedia Commons

coming on." Franklin thus made a significant contribution to scientific understanding by demonstrating the similarity of static electricity to lightning. Collinson, in relaying this letter to the Royal Society of London of which he was a Fellow (as Franklin was later), performed one of his numerous acts of patronage.[15]

Today, Collinson is best known for his collaboration with John Bartram which, starting with commissions by Collinson for the collection of seeds, roots, bulbs and plants, blossomed into warm friendship, and the exchange of horticultural and botanical knowledge. An avid gardener, like Collinson, Bartram also carried out hybridisation experiments comparable with the famous Fairchild's Mule, a cross in 1717 between the carnation and a sweet william made by nurseryman and author, Thomas Fairchild. For his botanical expertise, the fruit of practical experience and collaboration with Collinson among other collectors, Bartram was appointed the King's botanist in America, a paid position.

The relationship between Collinson and Bartram tells us a lot about the scientific, commercial and natural history networks in which Quakers played leading if often understated roles. There was a rich mixture of conviviality and curiosity about the natural world, combined with love and appreciation of its vegetable products: "Charleston doctors, and lady gardeners, explorers in wild hinterlands, aristocratic London collectors, compassionate Quaker philanthropists, Virginia slaveholders

Fig. 12.13 John Bartram (1699-1777), "brother gardener" to Peter Collinson. Nineteenth century sketch by Howard Pyle. Public domain, via Wikimedia Commons

agitating against British exploitation, Pennsylvania visionaries of a new world in the making – all were linked together by this wise and kindly haberdasher."[16] The name of Benjamin Franklin, although not a Quaker, might been added to this list as he was a friend of both Bartram and Collinson; Bartram joined Franklin's Philosophical Society of Philadelphia as a charter member.

Fig. 12.14 Quaker doctor and naturalist, Dr. John Fothergill (1712-1780), protégé of Peter Collinson and "father" of the modern botanic garden. Portrait from the painting by Gilbert Stuart. Wellcome Collection. M0002817 Public domain. CC BY 4.0 via Wikimedia Commons

The readiness with which the traditional Christian "religion of the book"was progressively abandoned by many Quakers, playing a diminishing role in their thought, meant that the meaning and significance of their core belief in "Inward Light" and its relation to the world had to be investigated. It has been suggested that such investigation is, above all others, the source of Quaker interest in environmental issues in the twentieth century.[17] But even this explanation has been challenged. Australian Quaker and writer, Susannah Brindle, in the course of her researches, questioned a number of Friends but found no causal relationship between their faith and their environmental activities. It just happened that they were Quakers who were also environmentalists.[18]

Just how far Quakerism in and of itself disposes its adherents to take an active interest in the natural world remains a matter of some debate. On one point, however, we can be reasonably confident. The move by the early Quakers away from the Bible as a way to God, and its replacement by personal experience through the "Inward Light" has meant that their religion of itself did nothing to support or strengthen the Biblical basis of man's domination of nature. When individual

Quakers did in fact demonstrate such domination, whether intellectuals, scientists or entrepreneurs such as the Cadbury and Fry families, the ironmaster Abraham Darby, or President Richard Nixon, they did so as individuals acting primarily it seems from motives other than directly religious or perhaps taking a Calvinistic approach to their Quakerism. At any rate they were being practical and no doubt believed that they were doing things useful not only to themselves but to others.

Was William Penn (1644 – 1718), Quaker, landowner and seventeenth century founder and early proprietor of the Quaker colony of Pennsylvania, revering and investigating nature or seeking dominance over it? An active Quaker from his student days at Oxford, Penn, in establishing the American colony, aimed to build an ideal or exemplary community on Quaker principles. Thus, with others, he drew up a constitution to enshrine these principles. They include social equality (although, like Thomas Jefferson later, Penn was a slaveholder), women's rights, fair treatment of Indians, an incipient democratic political assembly and respect for useful learning.

This colony was to be a haven for persecuted or reviled Quakers in England, seeking an environment in which they could freely practice their religion and prosper through productive work. According to a later commentator, "The Quakers flocked over to this Country in shoals ... not a few of good substance who went vigorously upon every kind of improvement ... by Diligence and Frugality, for which this Harmless sect is remarkable ... they have in a few years made Pennsylvania a very fine Country."[19]

The early history of the colony is the best example of Quaker ideology forming the basis of a successful civil society. Penn was a tree planter and in his design for the layout of the city of Philadelphia – which he wished to name Silvania – he provided for a green belt and urban parks, anticipating the late nineteenth century English Garden City movement. Just as George Fox had urged that schools established by the Society of Friends should teach, "the nature of herbs, roots, plants and trees", settlers under Quaker guidance exchanged plants as a matter of course. William Penn himself urged, "studying and following nature and how to become good naturalists". In a fulsome tribute, Leighton elevated William Penn to the head of a long line of distinguished conservationists in America: "that he himself was the first real conservationist to arrive in this country and to lay down rules for developing lands for both crops and forests, is evident in all his plan." Of every five acres of cleared land, one was to be left covered – especially, "to preserve oak and mulberries for silk and shipping".[20]

Many Quakers, as well as Collinson and Bartram, have opened the way to personal, peaceful and reflective ways of study relating to nature. By their actions and example rather than through doctrine and Biblical exegesis, they have invested Quakerism with ideals, standards and values directed toward development through practical actions considerate of the natural world.

In modern times, a mystical quest for Arcady has taken Quakers into movements to conserve and restore nature. This is in a very broad sense a religious quest which seeks to return to "Innocence", to "Arcady," to be achieved by positive action in the post–lapsarian world of the biblical fallen man. A statement issued by the Religious

Society of Friends, London, calls for "creative responsibility" for nature: "The world is a wonderfully rich resource for our spiritual and material needs. We should treasure it and preserve its capacity to sustain and inspire. That, in turn, calls for a creative responsibility towards the earth we have inherited and for proper sharing. It means seeing 'that of God' in the natural world around us, and being moved by considerations other than commercial gain. Habitats and species are sacrificed to products and services which are often far from essential. The future is constantly sacrificed to the present and the needs of others to the wants of self. It cannot be right to leave the world poorer than we found it in beauty or in the rich diversity of life forms, or to consume recklessly in the knowledge that our actions are bound to lead to future tragedy."[21] This passage refers to an "emerging testimony to the environment" that has now become an established one, with close links to the peace testimony and the testimony to simplicity. While it can be difficult to pinpoint specific attitudes towards nature and the natural world that have been derived ineluctably from their religious values and beliefs, there is nested in the Quaker testimonies a drive towards a more peaceful, perfect, harmonious, fair and caring existence for all. This embodies a deep respect for nature, its loving study which leads to a desire to conserve and advance the beauty and dignity of the natural world.

Not all elements of modern Quakerism have become actively and vigorously conservation-minded, but it is a strong emerging trend. Moreover, it is deeply rooted in the early forms of Quakerism, and can thus be presented to The Society of Friends as a whole as an evolving testimony anchored in the traditions of Quakerism. But with declining membership of the Society, especially among young people, it remains to be seen whether the Society of Friends will become a significant force in the multi-faith Eco-Christianity which is now drawing support from both Catholics and mainstream Protestants.

Notes

[1] Thomas, K (1984) *Man and the Natural World. Changing Attitudes in England 1500 – 1800.* Harmondsworth, Middlesex: Penguin Books. p. 20

[2] Tawney, RH (1920) Foreword to Max Weber *The Protestant Ethic And the Spirit of Capitalism* (trans. Talcott Parsons) London: Unwin University Books p. 2.

[3] Tawney *ibid* pp. 1 (B) – 1 (C)

[4] Bendix, Reinhard (1980) *Max Weber, an Intellectual Portrait.* Garden City, New York: Anchor Books. p. 63

[5] Tawney RH (1926) *Religion and the Rise of Capitalism.* London: John Murray, p. 84

[6] Weber, Max (1920/2002) *The Protestant Ethic and the Spirit of Capitalism.* Introduction and Translation by Stephen Kalberg. Los Angeles: Roxbury Publishing Company. "Introduction".

[7] Most recently, criticism of capitalist economics treats it as the major cause of human-induced climate change: Klein, N. (2014) *This Changes Everything. Capitalism vs the Climate.* London: Allen Lane/ Penguin.

[8] McNeill, WH (1965) *The Rise of the West.* Chicago: University of Chicago Press. p. 641

[9] For Quaker tenets and ideas I have found particularly useful Angell, S.W. and Dandelion, B.P. (editors) (2013) *The Oxford Handbook of Quaker Studies*. Oxford: Oxford University Press. I am grateful for information and advice from the late Ms. Susannah Kay Brindle, Quaker and author of *To Learn a New Song – Quaker contributions to real reconciliation with the earth and its people*. (Backhouse Lecture 2000) and to Australian and English Anglican and former Quaker, Emeritus Professor David Aspin.

[10] Quakers in Britain (2009) *The Quaker Testimonies*. "Our beliefs" Produced by Testimonies Committee of Quaker Peace and Social Witness, this edition published by Quaker Books 2009. Contact: Friends House, London. Available online: https://together.woodbrooke.org.uk/quaker.org.uk_mirror/Quaker%20Testimonies%20leaflet.pdf (Accessed Feb 17, 2020)

[11] Spencer, C. D. (2013) "Quakers in Theological Context" in Angel and Dandelion, *op.cit.*

[12] Benfey, T. (1995) "The History of Science includes many who were sustained by Quaker tradition" *The Scientist*. April. https://www.the-scientist.com/commentary/the-history-of-science-includes-many-who-were-sustained-by-quaker-tradition-58559 (accessed Feb 17, 2020). Also, Morris GGP (2010) *From revelation to resource: the natural world in the thought and experience of Quakers in Britain and Ireland 1647-1830*. PhD University of Birmingham.

[13] Wulf, A. (2009) *The Brother Gardeners. Botany, Empire and the Birth of an Obsession*. London: Windmill Books. The "obsession" is the British love of gardening.

[14] Mabberley, D.J. (2019) *Botanical Revelation*. The Peter Crossing Collection. Sydney: New South Publishing. p.iii.

[15] Hurd, D.L. and Kipling, J.J. (1964) *The Origins and Growth of Physical Science* Vol.2. Harmondsworth, Middlesex: Penguin Books. pp. 192-3.

[16] Leighton, A. (1986) *American Gardens in the Eighteenth Century - 'For Use or for Delight'*. Amherst, Massachusetts: The University of Massachusetts Press. Chap. 5 Gardeners and Collectors, p. 106

[17] Cresson, O. (2005) *Quakers and the environment: three options*. Posted June 18, 2005 in Blog Posts. (Accessed Feb 17, 2020) http://www.nontheistfriends.org/article/2005/06

[18] Personal correspondence, 10 November 2011.

[19] Byrd II, W. "History of the Dividing Line – on the settlement of Virginia," quoted in Leighton (1986) *op.cit.* p. 16

[20] Leighton (1986) *op.cit.* p. 44

[21] Quakers in Britain (2009) *The Quaker Testimonies*. "The earth and environment." Produced by Testimonies Committee of Quaker Peace and Social Witness, this edition published by Quaker Books 2009. Contact: Friends House, London. Available online: https://together.woodbrooke.org.uk/quaker.org.uk_mirror/Quaker%20Testimonies%20leaflet.pdf (Accessed Feb 17, 2020)

Complete Listing for Wikimedia Commons Images

12.1 Ford Madox Brown – John Wycliffe reading his translation of the bible to John of Gaunt. https://commons.wikimedia.org/wiki/File:Ford_Madox_Brown_-_John_Wycliffe_reading_his_translation_of_the_bible_to_John_of_Gaunt.jpeg Source: repro from artbook. Attribution: Ford Madox Brown, Public domain, via Wikimedia Commons.

12.2 Spiezer Chronik Jan Hus 1485. Author: Diebold Schilling the Older. https://commons.wikimedia.org/wiki/File:Spiezer_Chronik_Jan_Hus_1485.jpg. Public domain via Wikimedia Commons.

12.3 Hildegard von Bingen. https://commons.wikimedia.org/wiki/File:Hildegard_von_Bingen.jpg Source: Miniature from the Rupertsberg Codex of *Liber Scivias*. Attribution: Unknown authorUnknown author, Public domain, via Wikimedia Commons.

12.4 Martin Luther by Lucas Cranach the Elder – Statens Museum for Kunst – DSC08170. https://commons.wikimedia.org/wiki/File:Martin_Luther_by_Lucas_Cranach_the_

Elder_-_Statens_Museum_for_Kunst_-_DSC08170.JPG Photo: Daderot. Attribution: Lucas Cranach the Elder, Public domain, via Wikimedia Commons.

12.5 John Calvin Museum Catharijneconvent RMCC s84 cropped. https://commons.wikimedia.org/wiki/File:John_Calvin_Museum_Catharijneconvent_RMCC_s84_cropped.png Attribution: Museum Catharijneconvent, Public domain, via Wikimedia Commons.

12.6 Max Weber, 1918. https://commons.wikimedia.org/wiki/File:Max_Weber,_1918.jpg Source: https://cdn.britannica.com/49/39749-050-E773E614/Max-Weber-1918.jpg. Attribution: Unspecified, Public domain, via Wikimedia Commons.

12.7 R.H. Tawney plaque. https://commons.wikimedia.org/wiki/File:R.H._Tawney_plaque.jpg Author: Megalit. Attribution: Megalit, CC BY-SA 4.0 <https://creativecommons.org/licenses/by-sa/4.0>, via Wikimedia Commons.

12.8 Benjamin Franklin. 1788. Artist Joseph Duplessis. Metropolitan Museum of Art. metmuseum.org https://commons.wikimedia.org/wiki/File:Benjamin_Franklin.PNG Attribution: Joseph Duplessis, Public domain, via Wikimedia Commons.

12.9 Adam Smith The Muir portrait. https://commons.wikimedia.org/wiki/File:Adam_Smith_The_Muir_portrait.jpg Attribution: Scottish National Gallery, Public domain, via Wikimedia Commons.

12.10 David Hume Ramsay. Portrait, oil on canvas. Author Allan Ramsay 1766. Source: National Galleries Scotland. https://commons.wikimedia.org/wiki/File:David_Hume_Ramsay.jpg Attribution: Allan Ramsay, Public domain, via Wikimedia Commons.

12.11 Fox Refusing Oath. https://commons.wikimedia.org/wiki/File:FoxRefusingOath.jpg Attribution: From an 1864 painting by John Pettie R.A., Public domain, via Wikimedia Commons. Photo: Wilfred Whitten: Quaker Pictures, (The Friends' Quarto Series 1) Hicks, London 1892. P.ll.

12.12 Peter Collinson FRS. https://commons.wikimedia.org/wiki/File:Peter_Collinson_FRS.jpg Attribution: Engraving by T. Trotter, Public domain, via Wikimedia Commons. Source: https://www.europeana.eu/portal/record/9200105/B8A06F6C99095784D89D08F83E871DB07B1954E6.html

12.13 John bartram00. Artist: Howard Pyle. https://commons.wikimedia.org/wiki/File:John_bartram00.jpg Attribution: Howard Pyle, Public domain, via Wikimedia Commons.

12.14 Portrait of John Fothergill (1712-1780), English physician Wellcome M0002817. Artist Gilbert Stuart. https://commons.wikimedia.org/wiki/File:Portrait_of_John_Fothergill_(1712_%E2%80%93_1780),_English_physician_Wellcome_M0002817.jpg Attribution: See page for author, CC BY 4.0 <https://creativecommons.org/licenses/by/4.0>, via Wikimedia Commons. Source: Gallery: https://wellcomeimages.org/indexplus/image/M0002817.html

Chapter 13
The Reverend Gilbert White and the Religion of Nature

Being of an ambitious temper, and strongly attached to the charms of rural scenery, he early fixed his residence in his native village, where he spent the greater part of his life in literary occupations, and especially in the study of nature. This he followed with patient assiduity, and a mind ever open to the lessons of piety and benevolence which such a study is so well calculated to afford. Though several occasions offered of settling upon a college living, he could never persuade himself to quit the beloved spot, which was, indeed, a peculiarly happy situation for an observer. He was much esteemed by a select society of intelligent and worthy friends, to whom he paid occasional visit. Thus his days passed, tranquil and serene, with scarcely any other vicissitudes than those of the seasons, till they closed at a mature age – June 26, 1793.[1]

13.1 Rethinking Nature and Religion

This chapter continues the theme of post-Reformation attitudes concerning man's place in nature, with, as its main subject, changes occurring in one country, England, in the seventeenth and eighteenth centuries. The tumultuous religious and political struggles throughout the seventeenth century as the conflicting forces of Catholicism and Puritanism destabilized the Anglican church, and the rival forces of royalty and republicanism vied for political power, eventually settled into relative calm. But, by the eighteenth century, long established ideas grounded in ancient religion and reasserted in the Reformation were being fundamentally challenged.

The impact of the Italian Renaissance was strongly felt, in political and social life, literature and science, as new images of the good life on earth were projected. Many long-held religious beliefs about the cosmos were shattered by the succession of physical scientists stretching from Copernicus and Galileo to Kepler and Newton. Philosophy took a new turn with John Locke whose attack on the theory of innate ideas gave rise to disturbing questions not only about what and how we can know, but also about the nature of the soul and Adam's legacy of sin. Both Newton and Locke were Christians but the impact of their thinking was to have a profound effect in the eighteenth century not only in scientific and philosophical thought, but on

M. Skilbeck, *Loving and Studying Nature*, International Explorations in Outdoor and Environmental Education 10, https://doi.org/10.1007/978-3-030-80751-1_13

religious orthodoxy, on how educated people generally understood who they were and how they related to the natural world, and on the organisation of social and political life.

Not authoritative, bookish knowledge, as in Mediaevalism and its single Catholic custodian, but direct, felt, personal experience of God, in and not apart from, the physical and social worlds, gave rise to a wide variety of religious beliefs. In addition to Quakerism, discussed in the preceding chapter, these ranged from the Deist notion that God, having created the world, left it to its own internal processes, through various kinds of natural religion whereby God was to be discovered by enquiry into the natural world, to atheistic rejection of the Biblical narratives and the very idea of a presiding deity. There was also, in the variety of understandings of nature, a view that the social and political ethos and literary achievements of ancient Rome provided a standard of the good life that it was "natural", i.e. right or proper, for moderns to adopt and follow.

Indeed, "nature", in its wide and diverse range of meanings became a shibboleth, such that it served as a central organizing idea in ethics, religion, Augustan poetry, political life, antiquarian studies and in natural philosophy (physical sciences) and natural history (botany, zoology and related horticultural practices, forestry etc.).

In this chapter, we approach English clergyman, the Reverend Gilbert White (1720 – 1793), not simply as a diligent natural historian and modest country clergyman, studying his own environment and corresponding with fellow natural historians, significant as that is but in the wider setting of a century in which ideas about nature and religious viewpoints about human relationship to the natural world underwent radical change.

Gilbert White, late in the eighteenth century, published a ground-breaking book on natural history. Based on his personal observations of animal (mainly bird) and plant life in and around his native village in the south of England, and correspondence with fellow naturalists, White's *The Natural History and Antiquities Of Selborne* (1788) has become a classic in its field. It is neither a scientific text nor a religious tract, although White's methods of observation and his ways of analysing and recording are scientific in the broad sense of the term and his writing is animated with a religious spirit of reverence for what, as an Anglican priest, he took to be the creation of God and expression of divine beneficence.

White lived in an age when social, political and intellectual life generally were absorbing ideas of the law-like structure of the universe as presented by Boyle, Newton and other physical scientists. Miracles and seeming disruptions of nature and its laws might occur but John Locke and other philosophers and religious leaders saw them as part of God' s design of an orderly – not a disorderly – universe. God's presence in the universe could be discovered, with sufficient effort, whether in the distant heavens or the human heart.

White saw the living world as a whole, each of its vibrant elements to be understood and respected, even the birds that carried off his fruit. Above all, he felt and expressed affinity with his fellow beings, which was also an affinity with God who dwelt equally in him and all the creatures. This meant not the ancient and essentially arrogant idea of mastery, but a partnership of equal sharing. It is for this radical idea that White has, by some of his admirers, been cast as the forerunner of modern ecology, or perhaps that should rather be Eco-Christianity.

13.2 Religious, Secular and Early Scientific Views of Man's Place in Nature

Diverging and conflicting points of view about humanity's place in the natural world had been gathering pace from long before the violent eruption of the Protestant Reformation and the Catholic Church's rejoinder in the Counter-Reformation. As already mentioned, John Wycliffe (1330 – 1384) in England and Jan Hus (c.1369 – 1415) in Bohemia led those reformers who, prior to the Lutheran challenge, criticised corrupt and venal practices in the Catholic Church and challenged theological orthodoxy, including the notion that the Church consisted only of the clergy. Some paid heavily; Hus was burnt at the stake in 1415 and Wycliffe's body was wrenched from the grave in 1420 to be burned, his writings having been previously thrown to the flames. While the unorthodox ideas of Wycliffe and Hus were condemned, both of them, in working to break the priestly Latin monopoly of the Bible, widened its readership by translation into the vernacular.

Notwithstanding the efforts of critics to reform Church practice and strip its doctrines of the accretions of mediaeval theology, the concept of man as inherently sinful, possessed of a corrupt nature and redeemable – if at all – only by God's grace, remained largely intact. Orthodoxy proclaimed that human nature and the natural world could never fully recover from the disobedience of Adam and Eve, their fall from grace and expulsion from the delights and refreshments of the Garden of Eden. The Day of Judgement awaited sinners, with eternal punishment for the dammed. Readers of the *Inferno* of Dante Alighieri (1265 – 1321) could be in no doubt about what lay in store. Dante's nine infernal circles showed that there were varying degrees of sin and of punishment appropriate to them; but all sinners were condemned (Fig. 13.1).

Unbelievers and scoffers might lampoon the clergy and avoid church attendance, pushing church discipline to the back of their minds, but neither in daily work nor in moments of leisure and relaxation could the threat of eternal punishment for wrongdoers be entirely banished. The hope of eternal life could not be denied to the most exalted of the faithful, but for most people the hereafter was a forlorn prospect and the earth, meanwhile, was for the many a vale of tears.

From the early Renaissance onwards, a more optimistic view of man's nature and destiny did, however, emerge. Not at first displacing Christian orthodoxy, Renaissance ideas and values were eventually to play a large part in undermining pessimistic assumptions and unflinching doctrines. The two principal sources of the Renaissance elevation of man and optimism about mankind's earthly existence were the philosophy of humanism and nascent physical science.

Fig. 13.1 Dante Alighieri (1265-1321) whose great poem *The Inferno*, reinforced the gruesome mediaeval Christian doctrine of eternal punishment in the afterlife for earthly sinners. The circles of purgatory are shown behind, the heavens above, the Catholic church on the right and descent into hell at the left. Domenico di Michelino, *Dante and the Three Kingdoms* 1465. Museo dell'Opera del Duomo, Florence. Domenico di Michelino, Public domain, via Wikimedia Commons

Humanism, in the Renaissance referred, initially at least, to the recovery of the literary, historical, political, philosophical and scientific texts and the artistic achievements of ancient Greece and Rome. Their value was not only literary, scientific or aesthetic; they were commonly treated as a moral basis for virtuous, fulfilling, personal and civic life. Combined with Christianity (cleansed of its arcane doctrines and corrupt practices), the ancient classics were translated and pored over to yield models of the good life on earth. These ideas first gained currency in the city states of Northern Italy, then throughout Europe and the civilised Christian world.

Two figures may fairly be taken to represent, or at any rate to symbolise, Renaissance humanism: Petrarch in Italy and the South, and Erasmus in Northern Europe (Figs. 13.2, 13.3). Francisco Petrarch (1304 – 1374), famous during his lifetime for the recovery of the letters of Cicero, was a cleric, scholar, poet and traveller. Nowadays his popular appeal rests less on his classical scholarship and piety than on his love lyrics to "Laura", a beloved lady who, loyal to her husband, remained poignantly beyond his reach.

Petrarch has tantalised scholars by ambiguities in his character and writing which foreshadowed many of the cross currents of religious and secular life in the

Fig. 13.2 Francesco Petrarch (1304-1374), humanist, scholar and poet felt the grandeur and beauty of the natural world yet admired St. Augustine's theology. Fresco portrait, Altichiero da Zevio, c.1376. Oratorio di San Giorgio, Padova. Public domain, via Wikimedia Commons

Fig. 13.3 Moderate social critic, teacher, theologian and classical scholar, Desiderius Erasmus (1466-1536), condemned Church abuses but remained a Catholic. Hans Holbein, Portrait of Erasmus of Rotterdam. Kunstmuseum, Basel. Hans Holbein, Public domain, via Wikimedia Commons

following centuries. A Tuscan by birth, with his parents he moved to the south of present-day France in 1309, when Pope Clement V shifted the papacy to Avignon, dividing the Roman Catholic Church. Petrarch spent many years in France, where he trained as a lawyer, but also travelled widely, enjoying the earthly pleasures of a tourist. A famous episode was his ascent of Mount Ventoux, the windy "Beast of Provence", known in modern times for its ascent by pedal-pushing Tour de France cyclists on their progress through Provence. A great admirer of St Augustine, Petrarch climbed the mountain with his brother and two servants, a copy of Augustine's *The Confessions* in his pocket. Scholars have argued about the meaning of the climb, as to whether it signified a turning of humanist feeling toward the grandeur and beauty of the natural world or a spur to inward reflection on the nature of the soul. Petrarch gave rise to the latter interpretation by berating himself for enjoying the grandeur of the view and then thrusting into his pocket for the *Confessions*. Or was the experience something again? As recounted by Petrarch, it may be read as an allegory of the humanist aspiration toward a fuller, better life in this world.

In his *Renaissance in Italy*, the nineteenth century English poet, critic and historian John Addington Symonds, found in the "revival of learning" begun by Petrarch no mere renewal of interest in classical literature. It was, "the emancipation of the reason in a race of men, intolerant of control, ready to criticise accepted norms of conduct, enthusiastic in admiration of antique liberty, freshly awakened to the sense of beauty, and anxious above all things to secure for themselves free scope in spheres outside the region of authority. Men so vigorous and independent felt the joy of exploration."[2] Herein lay the seeds of a renewed belief in adventures of the individual human spirit in its earthly domain.

In the North of Europe – the Netherlands, northern France, England, Germany and Scandinavia – it was the spirit of the moderate Catholic cleric and theologian, classical scholar, social critic and teacher, Desiderius Erasmus Rotterdamus (1466-1536) that, a century later than Petrarch, carried Renaissance ideas and aspirations far afield. Probably best known today for *In Praise of Folly* (1511 and many successive editions), Erasmus joined Luther in condemning Church abuses but remained a lifelong Catholic.

Like Petrarch, an inveterate traveller and independent scholar, for some years Erasmus was Professor of Divinity at Cambridge. It was in the house of his friend, Sir Thomas More, that he wrote *In Praise of Folly*, actually the praise of reason and moderation. Both men shared a distaste for superstitious religion and pedantic scholarship, while holding antique literature and philosophy in high regard. More paid homage to reason in *Utopia* (1516), a fictional story of an island community of temperate, pious and rational people living in peace and contentment. A devout Roman Catholic and high official under Henry VIII, More was not to experience the dignified death and associated rituals that he wrote about in *Utopia* but instead was beheaded, on a charge of treason.

Erasmus translated the *New Testament* into Greek and fresh Latin. His texts were drawn on by the team of translators-editors who produced the *King James Bible* for Protestant England, a literary masterpiece. Calling for moderation in religious disputes, toleration of differences and harmonious international relations, Erasmus heralded a new age of civility. He rose above the hatred, bitter rivalries and furious battles of the time. His spirit as well as his name is commemorated in the magnificent study-abroad scheme for students in countries of the European Union. As with

Petrarch, the humanism of Erasmus found expression in the image of this-worldly dignity and creativity.

As "ideal types", Petrarch and Erasmus, together with Francis Bacon, are representative of the literary wing of Renaissance humanism.

Other aspects of the Renaissance than the humanist revival must also be noticed. For example, Michelangelo Buonarotti (1475 – 1564) for supreme artistic expression (Fig. 13.4); the engineer-inventor-artist, Leonardo da Vinci (1452 – 1519), the Renaissance equivalent to the all-round liberally educated Greeks of Antiquity (Fig. 13.5); and Galileo Galilei (1564 – 1642), a voice for scientific observation and reason against religious dogma (Fig.13.6). Marxist literary critic, George Lukacs, in a study of "world historical" figures, gave the accolade to Leonardo in a swipe at elite culture, (Stephan) "Zweig's choice of Erasmus as the model type of humanist," he wrote, "is a distortion of the historical picture, for he disregards the militant type of humanist who was deeply involved in the problems of popular life. Engels in his enthusiastic analysis of the great men of this period puts forward Leonardo da Vinci and Dürer as the great types."[3]

We might add to Engels' and Lukacs' "militant" Leonardo, the image of the chivalrous knight, ready to defend his prince, to honour his lady, and to gracefully adorn the court in its periods of peace. However rather than conjecture about "representative types", we must pursue the enduring Renaissance images of humanity a little further. By numerous and varied efforts, including teaching in schools and colleges, humanists spread their liberating ideas across fifteenth and sixteenth century Europe and succeeded in creating a practical ideal of individual and collective human perfectibility on earth, combining moderate Christian faith, reason, good sense, toleration, social and civic virtue, practical capability and intellectual, aesthetic and moral cultivation. The mediaeval monkish ideal of withdrawal and contemplation was not entirely forgotten but absorbed into that of the active civic life.[4]

Fig. 13.4 Michelangelo Buonarroti (1475-1564). Sculptor and painter of the ceiling of the Sistine Chapel in the Vatican. From unfinished portrait attributed to Daniele da Volterra, probably ca 1544. Metropolitan Museum of Art. Daniele da Volterra, Public domain, via Wikimedia Commons

Fig. 13.5 Leonardo da Vinci (1452-1519), prince among "world historical" figures of the Renaissance, his vision rooted in the natural world. *Self Portrait* 1512 in red chalk. Biblioteca Reale, Turin. Leonardo da Vinci, Public domain, via Wikimedia Commons

Fig. 13.6 Galileo Galilei (1564-1642), Italian physicist and astronomer, was condemned by the Inquisition for championing Copernican heliocentric theory that the earth moves around the sun. Portrait by Justus Sustermans, 1636. National Maritime Museum, Greenwich. Justus Sustermans, Public domain, via Wikimedia Commons

Just as we turned to Petrarch and Erasmus to illustrate key aspects of Renaissance humanism, we may draw in two other figures, Vittorino da Feltre (1378-1446) (Fig. 13.7) and Giambattista Vico (1668-1744) (Fig. 13.8) to exemplify further developments, practical and philosophical respectively.

Fig. 13.7 Vittorino da Feltre (1378-1446), Renaissance humanist who thought the schoolroom should be a place of joy in the world. Portrait by Justus van Gent and Pedro Berruguete, c 1474. Musée du Louvre. Justus van Gent, Public domain, via Wikimedia Commons

From early in the Renaissance, the humanist ideals began to influence educational thought and practice, not only in universities: "The Court schools were the chief centres of higher humanist teaching in the early Renaissance."[5]

Best known and most influential of these Court schools was that established for the Marquis of Mantua by Vittorino da Feltre, who turned from university teaching to establish a new kind of schooling, as a number of scholars, including Bertrand Russell, have continued to do to the present day. Vittorino's pupils, girls as well as boys, were the children of the marquis and other noble families destined for future political and social leadership. They also included talented poor children, anticipating future meritocratic secondary education. Vittorino's curriculum was broad; religion was the keystone but studies ranged through mathematics, music, art, history, poetry to physical exercises and games. Due to a happy learning atmosphere, the school was known as The House of Joy. Vittorino's fame spread beyond Italy to Germany, France and England, where it had a considerable influence on both educational essayists and schoolmasters.

In the eighteenth century, the by then well-established humanist ideal of learning, directed towards a cultivated and virtuous worldly life, with an aura of rational piety, received intellectually robust and novel foundations from the Italian

Fig. 13.8 Giambattista Vico (1668-1744), historian, philosopher and pioneer of modern social science methods. Portrait by Francesco Solimena, eighteenth century. Francesco Solimena, Public domain, via Wikimedia Commons

philosopher and professor of rhetoric, Giambattista Vico.[6] Author of what was to provide an influential rationale for the substance and methodologies of modern social and cultural studies – *The New Science* (1725) - Vico had set himself an ambitious task worthy of his Renaissance forebears. This was, first, to question the rationalistic, mathematical circumscription that Descartes imposed on philosophy and, second, to provide a more ample, historically grounded framework. The new science of Vico was a blend of: Cartesianism (limited to its "proper" scientific domain – the mathematics and physical sciences of his time); the wisdom of ancient literature and philosophy; and the "divine wisdom of Providence". From these, and a profound awareness of the social conditioning of human action, Vico developed his ideas of historical development and methods of inquiry. These he expounded in a mass of technical detail which need not detain us. But something must be said about his deep interest in the study of concrete, human society.

Human knowledge, or the kind that interested Vico and his readers, the present author among them, was conceived as a function of social development over time. The knowledge of anything requires us to inquire into how it came into being and what it is as a product of human action (a constructivist theory of knowledge). History is "ideal", eternal, the manifestation of Providence, and "traversed in time by the particular histories of the nation, each with its rise, development, acme, decline and fall." This became a favourite theme of eighteenth and nineteenth century historians and philosophers of history, in Germany, France and England, lingering into the twentieth century; for example, in the works of Collingwood, Toynbee and Isaiah Berlin. But history is, equally, an adumbration of "the empirical phenomena of the world which are from human choice: the language, customs, and actions of people which make up human society." Hence Vico's concept of a "new critical art."

This was an art, or methodology, whereby the quest for knowledge and understanding of the human and social world was to be informed by "poetic feeling". Materialistic science and philosophy had their place, but their methods were inadequate for the human sciences. It was as if Vico were seeking to extend something of the "poetic metaphysics" that he found in the first stages of development of human society into his own times. Institutions, he observed, arose first from the immediacy of sense expression, "pure feeling, curiosity, wonder, fear, superstition, and the child-like capacity of human beings to imitate and anthropomorphize the world around them". This concept of "poetic feeling", as we shall see below, found several parallels in the Reverend Gilbert White's intuitive philosophy of nature. But it is most unlikely that White ever read, or heard of, Vico.

From the period of the Renaissance until the eighteenth century Enlightenment and including the scientific revolution of the seventeenth century, scholars and scientists generally were, like Vico, religious believers although of quite diverse types. Deism came to be favoured among those Protestants who, picking holes in Biblical Revelation, looked to study of the natural world for disclosure of patterns and designs, laws and regularities that might reveal a supreme being at work.

In the seventeenth and early eighteenth century in England, Deism attracted many adherents among the educated. In France, its ranks included Montaigne, Montesquieu, Voltaire, Rousseau and Robespierre, while in America were Franklin and Jefferson among others. Deism was not a creed or sect but a cluster of ideas and beliefs which ranged from hard edged materialism hostile to Christianity and especially clericalism, as in France, to a form of natural religion in which many theological claims were accepted but on the basis of reason and scientific enquiry, not biblical authority. Many Deists, scientists included, were content with Christian moral codes and social norms.

There were echoes in Enlightenment Deism of the ancient Stoic, Lucretius (*De Rerum Natura*) and the Epicureans, although it centred on one God instead of the pagan galaxy. But, whether Deist or not, the scientists did not look to the ancients; they turned their new astronomical instruments towards the material heavens and applied mathematical tools to the analysis of planetary motion, gravitation and the phenomenon of light. God, in creating the world, had endowed it with self-acting powers and then quietly withdrawn, leaving the physicists, mathematicians and chemists to unravel its complexities – all of them ultimately explicable – and to enlighten the educated public. The human intellect, not the Bible, was to engineer this general illumination. Alexander Pope's famous epitaph to Isaac Newton (Fig. 13.9) perfectly encapsulated the new faith:

> Nature and nature's laws lay hid in night,
> God said "Let Newton be" and all was light.

The scientists, scholars, artists and writers of the Renaissance, in glorifying human striving and achievement, would have shocked the early church doctors but

Fig. 13.9 Isaac Newton (1646-1723) mathematical genius and *Old Testament* scholar who elevated science onto a pedestal from which it has never fallen. Mezzotint by T. O. Barlow, 1868, after Sir G. Kneller, 1689. Wellcome Collection. Attribution 4.0 International CC BY 4.0 via Wikimedia Commons

they were at one with them in claiming a superior place to man in the hierarchical order of nature. Neither humanistic scholarship and artistry nor scientific enquiry cast doubt on this article of faith. But it was not only humanistic scholarship and scientific advance that celebrated human power.

From the fifteenth century onwards, a new phase of European endeavour gave further reason to this ancient faith; humans were conquering distance, reaching to the ends of the earth. The Portuguese, Spanish, Dutch, French and English voyages of discovery not only opened new lands and peoples for European exploitation, they generated new knowledge and wider understanding of human culture and the natural world (Chap. 15). It has been suggested that awareness of contemporary religions other than Christianity, in distant colonised lands, was a factor in the rise of Deism. This may have strengthened Deism but was not the cause of it. There is no doubt that discovery of plants and animals, landforms, geological features and oceans resulted in a vast increase of interest in natural religion, the life sciences and the study of nature, popular as well as scientific. Missionaries who accompanied or followed the voyagers and traders were important contributors to this bank of natural history knowledge, fitting it into the religious glory-creation.

13.3 Breaking with Rome: The Fortunes of the Anglican Church

The entanglement of religion, politics and international relations in early modern colonialism that was a marked feature of the century of religious wars which followed hard on the heels of the early Reformation, was not a novelty, but new interests were involved. A striking example is the struggle between England and Spain in the Elizabethan period. Family relations between the Continental dynasties (and most conspicuously those of France and Spain) and the wavering fortunes of the English throne did nothing to diminish international rivalries. With impunity, the privateer-pirate, Francis Drake (with covert investment in his voyages by Queen and Court) attacked Spanish ships off the colonial Central American coasts and brought home rich booty. He was one of the leading admirals in the rout of the Spanish Armada sent by Philip II (related, by marriage to her half-sister Mary, to Queen Elizabeth I). Had the Spanish campaign to invade England succeeded, Catholicism may have been re-imposed.

However, there were not only religious issues to settle but also power relations among the European nations and the dubious involvement of the papacy, in fomenting conflict or seeking to broker peace and territorial disputes. The politico-religious ascendancy in the Middle Ages of the Roman Catholic Church was a story of struggle between princes, kings, prelates and popes for supremacy. Struggles of this sort continued throughout the period of the Reformation and for centuries thereafter. Eventually, matters were more or less to settle in mutual recognition of authority in the respective domains of church and state.

In England, a distinctive pattern had emerged in the establishment of the Church of England under Henry VIII and Queen Elizabeth I. For the essentially secular reasons of marriage, divorce and royal succession, Henry's break with Rome and excommunication in the 1530s established what was effectively a monopolistic state church whose head was the ruling monarch. Henry still regarded himself as a Catholic, and his church as dating from the early mediaeval Christianisation of England.

The Anglican break was in reality inconclusive while appearing decisive. "Romish" elements continued, in the liturgy of the service, the doctrine of episcopacy, and the rituals of the high churchmen. The rival forces of republican puritanism and Royalist-Tory authority destabilised the Church of England as did the Roman Catholic political resurgence of James II and his followers in the seventeenth century. That century was witness to the most turbulent period of English history since the power struggles of royal contenders from the Conquest onwards, whose copious bloodletting was so brilliantly recalled by Shakespeare in his history plays.

The turbulence extended far beyond its religious sources to embrace: the reluctant ordering by Queen Elizabeth of the execution of her Roman Catholic cousin Mary; the defeat and destruction of King Philip's Spanish Amada; production of the *King James Bible* and the *Book of Common Prayer*; the accession of the tolerant Catholic, James VI of Scotland and I of England; the execution of Charles I; the Anglo-Dutch naval wars; the civil wars and the establishment of the Cromwellian republic; restoration of the monarchy; and, in fear of a Roman Catholic succession, the invitation to the Dutch Protestant, William of Orange, and his English royal wife Mary, to take the English throne. The fierce battles and repression of Roman Catholics in Ireland, the constantly changing politico-religious alliances and the rise and fall of high and low Anglicanism, of puritanism and dissent, and the steady eroding of faith by science and secular life are all indicative of a highly energetic but conflicted and fundamentally uneasy society. The great Fire of London in 1666, and lingering plague were reminders of the mediaeval past while Sir Christopher Wren's construction of the new Anglican St Paul's Cathedral, the diaries of Samuel Pepys and John Evelyn, the philosophy of John Locke and the science of Boyle and Newton pointed to a new and different future. However, religious calm was not to be part of this future. As late as the middle of the nineteenth century, the Oxford Movement signified the resurgence of High Church Anglicanism after the great popular waves of Low Church Methodism. Celebrated conversions to Roman Catholicism included Henry Edward Manning and John Henry Newman, both to become Cardinals in the liberated English Roman Catholic Church, and Gerard Manley Hopkins, who became a Jesuit priest and a religious poet on a par with his great forebears, the metaphysical poets, John Donne and William Herbert. These were forceful indications of fresh waves of diverse religious feeling and belief, in the supposed settlement of Anglicanism as represented, for example, by the authority of the bishops of the church (Fig. 13.10).

This extraordinary religious panorama, stretching over several centuries, suggests anything but a "settlement" of the issues thrown up by the Reformation. Yet out of the ferocity and bitterness there emerged a new and calmer order of belief and ideas which, while they veered from High Anglican orthodoxy, to Methodism, Deism and atheism and were hotly debated, did not, after the seventeenth century, lead to bloody internal warfare.

A growing force in the religious debates was the idea that the study of the natural world was a more reliable base for religious belief than the *Bible*, or the *Bible* alone. This crystallised in a clear-cut distinction that was drawn between natural and revealed (biblical) religion. Sometimes seen as an opposition, these were more often subscribed to as complementary; natural religion serving as a sure foundation for the faith required by revelation (much as, according to Aquinas, mediaeval reason was an indispensable buttress to ancient faith).

Fig. 13.10 Dr. Michael Ramsey, 100th Archbishop of Canterbury, senior prelate of the Church of England, 1961-1974. Unknown author, CC BY-SA 3.0 NL, via Wikimedia Commons

13.4 Natural Religion

Natural religion had affinities with natural science although by no means all natural scientists and their philosophical allies accepted that scientific observation and reason had necessarily cleared away all the Biblical and ritualistic fabric of Christian belief. For example, John Locke continued to believe in miracles, as evidence of the truth of Christianity, and the mathematical and scientific genius, Isaac Newton, a lifelong Christian (a Unitarian who challenged the orthodox doctrine of the Trinity), dabbled in alchemy, asserted the need for God to periodically re-energise the universe and held that God created the particles of his particle theory of light. Newton gave as much time and effort to studying ancient religious texts as to light, planetary motion and gravity. He spent many years calculating a chronology of the *Old Testament*. John Maynard Keynes described Newton as not the first of the age of reason but, "the last of the magicians, the last of the Babylonians and Sumerians, the last great mind that looked out on the visible and intellectual world with the same eyes as those who began to build our intellectual inheritance rather less than 10,000 years ago".[7] Nevertheless, Newton firmly believed that it was through the power of the observing, reasoning human mind that nature's secrets would in time be revealed. It was this perception that was to make such a deep and enduring impression on eighteenth century intellectual life and thereafter.

The emergence of natural religion in the seventeenth and eighteenth centuries was not, of course, uncontested. New questions about humanity's place in the order of nature were raised, since the study of the whole of creation as a source of religious understanding necessarily included "man" as one - but only one - of God's creatures. What is more, if all the creatures are of God, may they not in some sense be of equal worth in the eyes of God? The biblical "mastery" doctrine was not directly challenged but was, by implication, gradually being whittled away.

Natural religion had its heyday in England from the latter part of the seventeenth until the end of the eighteenth century. Like so much else in Western thought, its origins or antecedents can be traced to the classical world of Greece and Rome. Philosophers and scientific enquirers in the pre-Christian era had quite distinctive if not always definite explanations for the beginnings, functions and organisation of the natural world. For example, it was the creativity and the rule of the gods that determined man's place in nature, man being inferior to the gods but in some respects able to consort with them. Consorting or not, the gods remained extremely active in the human world, behaving capriciously and arbitrarily, as in their interference in Homer's Trojan War or Zeus' attendance on a not unwilling Danae. An alternative view was that the gods, having created the world, withdrew from its affairs. This Stoic and Epicurean doctrine was, as we have seen, paralleled in the austere Deist notion that God was remote from the world.

Plato and the Neo-Platonist, Plotinus, and their followers regarded the quest for knowledge and understanding of the natural world as a religious mission into the ways and requirements of the gods. Its stakes were of the highest order for the well-being and immortal destiny of the human soul. Indeed, for the ancient world generally, theoretical knowledge of nature was a matter of religion as much as science or philosophy. Aristotle, however, did separate the two in his natural history studies (Sect. 13.6). Since in the Judaeo-Christian tradition there was an authoritative and literally true compilation, the *Bible*, it might seem that there was no need to search out the workings of God in the natural world. For those who accepted the literal truth of the *Bible*, there appeared to be no need to go further.

However, neither human curiosity nor the practicalities of everyday life permitted matters to rest with literal Biblical interpretations. In everyday life – agriculture and animal husbandry, for example – questions evidently rose about weather patterns, climate, soil, suitability of crops, diseases and pests, water tables and flows, breeding, nutrition and so on. Neither the *Bible* nor God's omniscience and omnipresence appeared to provide all the answers to such practical questions. They were part of a different, albeit related, regime.

Important as such practical matters were for everyday life and livelihood, it was not primarily from them that natural religion developed as either a complement or an alternative to revelation. Its origins and purposes were to do with religious knowledge of the kind, "How is God to be known?" (once a heretical question) and "How is God's way to be understood?" (The *Bible* tells us but we would like the certainty that the new science promises).

Natural religion was thus worldly in the sense that it asserted God's presence in, not detachment from, his creation. God was not only present, he was accessible to enquiring believers. The cosmos and all elements of nature were manifestly God's creation. By the close and respectful study of nature, the ways of God, which were the ways of nature, might be understood. Love, worship and the closest possible personal relationship with nature were tantamount to the love and worship of God. If this might seem to veer dangerously close to pantheism, God was understood to be both in nature and independent of it (panentheism, not pantheism). As well, God revealed himself in the Bible. The role of natural religion was to fortify, not displace revelation. But it was not to be long before God slipped out of the study of nature altogether.

The rapid development of scientific enquiry and philosophical speculation in the Renaissance, the line extending from Copernicus to Galileo, Boyle, Newton, Descartes, Locke and Leibniz, was the most direct and powerful intellectual force behind what was "natural" in natural religion. But there were other forces at work and in order to understand the part they played, we must look into the various uses being made of the term "nature". As with "reason", it had many and varied uses.

Since the natural world included man, and man's beliefs and roles in the world, as serving God's purposes, knowing more about man was a central tenet of natural religion. The study of the natural world, in search of God, was necessarily also the study of man. Quite independently of the scientific and philosophical foundations of natural religion, the study of man as an individual and in society had been one of the principal objects of the Renaissance humanists, as we saw above. Huge changes in the pattern of social and economic life from the sixteenth century onwards led inexorably to the study of mankind, historically and culturally as well as anatomically, physiologically and psychologically, ethically and politically as well as geographically and linguistically.

In one of his famous couplets, the Augustan poet, Alexander Pope, tired of theological wrangling, captured the essence of the new mentality,

> Know then thyself, presume not God to scan
> The proper study of mankind is man.[8]

For the purposes of natural religion, however, God was to be scanned in the ways and workings of nature and it was man in his natural setting that Pope intended. But, as we shall see in the following section, "nature" for Alexander Pope and the Augustans generally was rather more than the physical world. Before coming to Pope's world view, however, we must mention other religious views and identify the two figures who seem to have introduced the naturalist Reverend Gilbert White to natural theology.

The Quakers, as we saw in Chap. 12, in the doctrine of the "inner light" had found a place for God in one domain of nature, namely the human heart. This idea of God's presence in man's inner nature, as it were in conscience, surfaced, or rather resurfaced, in the writings of two eighteenth century notables, Bishop Butler, author of the highly influential *An Analogy of Religion* and the moralist and social reformer, Lord Shaftesbury. They conceived religion and morality as "natural" inasmuch as they were congenial to man's true, or inner nature. In his heart or conscience man heard and could – or should - be directly and personally responsive to the word of

God, helped of course by a thorough knowledge and understanding of the *Bible*. As religious thought thus moved toward the works of man in his natural setting, that is, his normal, everyday life, it was inevitable that personal morality and individual conscience, social and sociable behaviour, altruism and charity would all be recognised as evidence of God's presence in human nature as a guide and judge, but not at the expense of free will. This was a very far cry from the doctrine of inherent sin, the hard determinism of Necessitarianism and the ineluctable sinfulness of human nature. These were aspects of Calvinism that the mainstream of Anglican religion was seldom if ever ready to welcome.

Before reflecting further on God's presence in human nature and the natural ordering of society, it will be useful to remind ourselves of the two wings of scientific observation and imagination that, with their origins in antiquity, from the Renaissance onwards were dominant features of Western culture. Most conspicuous and influential is the image, if not the precise detail, of Newtonian science: based on exact and prolonged observation of physical phenomena; instrumentation (Newton invented the first successful reflective telescope and worked out his colour theories using two lenses); detailed calculation and mathematical analysis (with Leibniz, Newton is credited with inventing Calculus). Discovering (or constructing) universal laws of nature was the great object of the scientific quest. The universe could be thought of as a perfect mechanism, an interpretation that William Blake derided in his magnificent image of a naked Newton on an algae-covered rock, dividers in hand and crouched over a scroll (Fig. 13.11). Blake also despised Deism, as in his poem, *Mock on, Mock on Voltaire, Rousseau*,

> Mock on, mock on Voltaire, Rousseau;
> Mock on, mock on, 'tis all in vain.
> You throw the sand against the wind,
> And the wind blows it back again.

13.5 Natural History

Newton and the physicists were "natural *philosophers*". The other version of science, akin to the feelings of Blake and the Romantic poets is natural *history*. Equally rooted in antiquity and with a belief in the value of close and detailed observation of natural phenomena equal to that of the natural philosophers, natural history was conceived as discursive and descriptive, not as a mathematical, science. It was avowedly subjective or intra-subjective, literary and aesthetic in tone, commonly a vehicle for the expression of positive and sympathetic feelings towards the natural world.

The subject-matter of natural history was life, and *feeling* for life, both animal and vegetable, together with the settings or habitats of life. Its study and practice have commonly had aesthetic and ethical overtones. In natural history, personal, direct sensory observations are the primary and most important technical apparatus. They may be facilitated, for example by binoculars, lenses, simple microscopes,

Fig. 13.11 Poet and artist William Blake (1757-1827) was among those not enchanted by eighteenth century Deism and Newton's materialistic physics. Detail from Blake's image *Newton* 1795. Tate Gallery, London. William Blake, Public domain, via Wikimedia Commons

tape measures, cameras, compasses and the like, but unlike the physical sciences which, according to Max Planck, by 1900 had become totally instrument-dependent for their data, natural history relied on the direct sensory experience of the observer. In this sense, Albert and Margaret Morris and the Barrier Field Naturalists were natural historians (Chap. 3). Feelings for the objects and creatures that are observed form part of the understanding; they are not incidental to it. Because relationships among individuals and groups and their environments were and are commonly studied, for example colonies of birds, it has been suggested that even in the eighteenth century, more than 100 years before the German Darwinist, Ernst Haeckel (1834-1919), coined the term, the naturalists were *ecologists* or proto-ecologists.

Among the English natural historians whose works were to have an impact on Gilbert White were silviculturist, John Evelyn (1620-1706); botanist, John Ray (1627-1705) and clergyman, William Denham (1627-1738). Evelyn was the author of a famous book on trees, *Sylva, or A Discourse of Forest-Trees and the Propagation of Timber*, an enthusiastic gardener, high-ranking Royal Commissioner and diarist. He encouraged fellow landowners to plant trees in order to keep the Royal Navy well provided with timber but loved nature as God's creation. John Ray and William Denham popularised the terms physico-theology or natural theology in their rationalist endeavours to demonstrate the existence of God and his works through observation of the natural world and reasoning about their findings.

For his *Wisdom of God Manifest in the Works of Creation* (1691) Ray took as a text *Psalm 104*, "How manifest are thy works, Oh Lord! In wisdom hast thou made them all." He asserted that the creatures of the earth, in their numbers and diversity, are "effects and proofs of God's omnipotence" and went on to make the teleological claim that they have been designed for a purpose, "their constituent parts adapted to function and use", all serving, "the public good." Ray's study of and attitude towards the natural world as an inspiration of Christian belief continue to enthuse modern Christians.[9]

It appears that, as in the *Old Testament*, a function of creation was to satisfy man's earthly needs and to minister to his uses. Accordingly, and in the spirit of Francis Bacon, for Ray the botanist, plants were to meet his needs as a scientist and, through his research and that of others, to prove of use and value to humanity. If not quite in the mental framework of Ray and other physico-theologians, this is also how plants were seen by governmental and private sponsors of the great botanical expeditions which in the eighteenth and nineteenth centuries brought the riches of the world to the museums, botanical gardens and herbaria of Europe. William Bligh's ill-fated voyage to transport breadfruit to the West Indies added a specific commercial purpose to John Ray's "public good". (Chap. 15).

Ray had an enthusiastic disciple in William Denham, Anglican clergyman and author. His teleological books *Physico–theology* (1713), *Astro-theology* (1714) and *Christo-theology* (1730) were to have an impact on William Paley (1743-1805) who argued for the view that the study of nature reveals evidence of design and a designer, God the great clockmaker, at work. Refuted by Darwinian evolution, the design argument has remained popular among Christian fundamentalists and some theologians.

Elected a Fellow of the Royal Society in 1703, William Denham extended his studies of the natural world from variations within species to atmospheric observations and estimates (reasonably accurate) of the speed of sound. In 1711–12 he delivered the Boyle lectures which had been endowed by the will of scientist and friend of Isaac Newton, Robert Boyle, to prove the truth of the Christian religion and refute atheism. Except for a short lapse, the Boyle Lectures have been delivered continuously from late in the seventeenth century to the present day. In this context, mention should also be made here of another distinguished series of public lectures on religious topics, established in 1888 by the will of the Scottish jurist, Lord Gifford (1840-87), in order to, "promote and diffuse the study of Natural Theology in the widest sense of the term – in other words, the knowledge of God." It will be recalled that William James's *Varieties of Religious Experience* and Alfred North Whitehead's metaphysical tour de force, *Process and Reality,* mentioned in Chap. 5, were first delivered as Gifford Lectures. As with the Boyle Lectures, the Gifford have kept alive something of the enquiring spirit of seventeenth and eighteenth century natural religion.

With its emphasis on scientific enquiry and, in the case of many of its adherents, relative neglect of revelation, natural religion attracted criticism not only because of suspected pantheistic tendencies whereby God was wholly absorbed in nature. At about the time, Gilbert White was preparing his correspondence and records for publication, in *The Natural History of Selborne*, the philosophical poet and closet revolutionary, William Blake (1757 – 1837) produced the first of his innovative illuminated relief etchings to accompany his aphoristic poetry: (1) *There is No Natural Religion;* and (2) *All Religions are One*.

In the first of these radical works, Blake, the apostle of imagination, mocked the prevailing Locke-inspired philosophy which he phrased thus,

> Man has no notion of moral fitness but from Education. Naturally he is only a Natural Organ subject to sense ... None could have other than Natural or Organic Thoughts if he had none but Organic Perceptions ... The Desires and Perceptions of Man, untaught by anything but Organs of Sense, must be limited to Objects of Sense.
>
> Conclusion: If it were not for the Poetic or Prophetic Character, the Philosophic and Experimental world would soon be at the Ratio of all things; and stand still, unable to do other than repeat the same dull round over again.

In order to break through the limiting, sensory "ratio" it was, for Blake, necessary to ignite the poetic vision or the inspiration of imagination. Instead of searching for God in scientific natural theology, Blake sought the infinite in all things, to be at one with God imaginatively and creatively, not by way of a sensory-based description and analytic classification of plants and animals. For long disregarded, Blake's views on the relationships between imaginative insight and the natural world began to attract quite a following in the twentieth century.

13.6 The Nature of Nature

As we saw above, from the Renaissance onwards "nature" was understood in diverse ways ranging from the cosmos, or physical universe and its physical laws to the hypothesised "natural" feelings of the human heart. "Nature" included violent volcanic eruptions and the sexual organs of plants as used by Linnaeus in his binomial system of animal and plant classification. "Nature" might also be the exotic world revealed in the reports of ocean voyagers, or the Rousseauist figure of the "Noble Savage"whose pure, pristine nature had been untouched by the vices of civilisation.

One unexpected use of the term "nature" was its reference to what was deemed to be right, fitting, or appropriate in human manners, taste, deportment, language and relationships. Augustan poets and moralists looked to the poets and writers of the era of the first Roman emperor, Augustus, for expressions of orderly, temperate behavior, to the refinements of the Latin language and to the poetic style, which they freely emulated, of Virgil, Horace and Ovid.

Alexander Pope's constant poetic theme was contemporary life, manners and taste - or lack thereof - and the absence of probity in public affairs (Fig. 13.12). His quest was for a resurrection of the imagined urbane Roman culture of the first century AD. Brought up a Roman Catholic, as a poet Pope showed no interest in the Christian metaphysics of his seventeenth century poet forebears, John Donne, William Herbert and Andrew Marvell but one thing he had in common with them

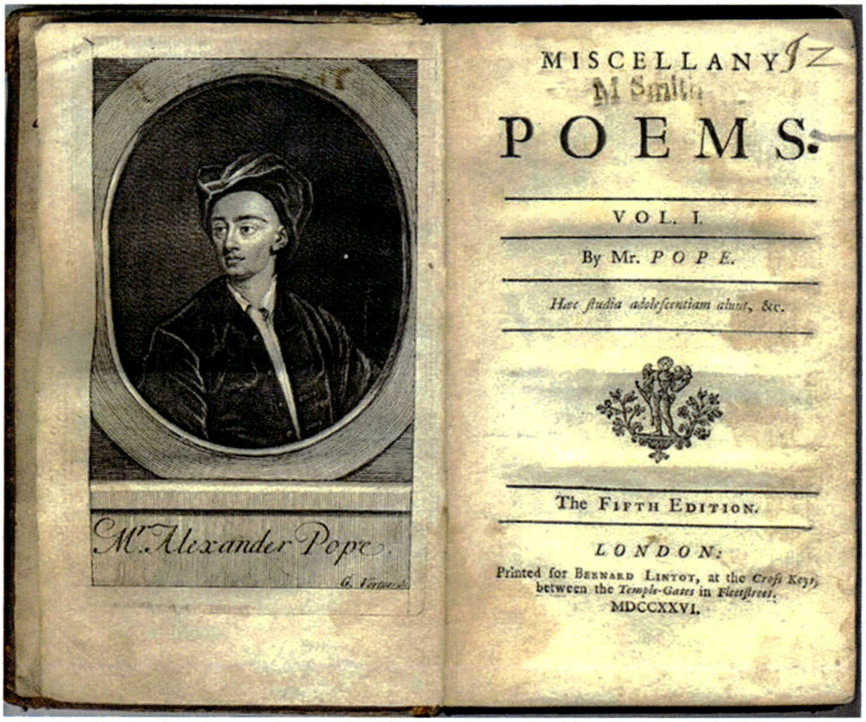

Fig. 13.12 Alexander Pope (1688-1744), philosophical poet and gardener. Pope's *Miscellany*, title page and G. Vertue frontispiece. Publisher Bernard Lintot London 1726, Scanner and Uploader Steven J Plunkett, Public domain, via Wikimedia Commons

and with the Elizabethan lyricists was a deep love of plants and gardens. An enthusiastic gardener, he took great pride in his own Thames-side Twickenham garden, its vistas and elaborate grotto. He was free of advice to his patron and friend, Lord Burlington, on the layout of a formal garden to complement Burlington House, a neoclassical mansion in the style of Palladio. It was to be "natural".

Known to all garden historians, Pope's *Epistle IV To the Right Honorable Richard, Earl of Burlington*, emphatically laid down rules for the correct "natural" design of an aristocrat's estate,

> Consult the genius of the place in all:
> That tells the waters or to rise or fall;
> Or helps th'ambitious hill the heav'ns to scale,
> Or scoops in circling theatres the vale;
> Calls in the country, catches op'ning glades,
> Joins willing woods, and varies shades from shades;
> Now breaks or now directs th'intending lines;
> Paints as you plant, and as you work, designs.
> Still follow sense, of every art the soul:
> Parts answ'ring parts shall slide into a whole;

Spontaneous beauties all around advance,
Start ev'n from difficulty, strike from chance;
Nature shall join you; time shall make it grow
A work to wonder at – perhaps a Stowe.[10]

Pope's lines on the art of laying out a garden and its planting express one element of his understanding of nature. The other is his normative view of classical Roman culture as a model for human conduct in society. In his *Essay on Criticism*, Pope is quite explicit in identifying nature with "Ancient Rules",

Learn hence from Ancient rules a just Esteem;
To copy Nature is to copy them …
First follow Nature, and your judgement frame
By her just standard, which is still the same:
Unerring Nature, still divinely bright,
One clear, unchanged, eternal light,
Life, force, and beauty must to all impart,
At once the source and end, and test of art
Art from that fund each just supply provides;
In some fair body thus the informing soul
With spirits feeds, with vigour fills the whole;
Each motion guides, and every nerve sustains;
But passion must be restrained
The winged courser, like a gen'rous horse
Shows most true mettle when you check his course …
rules of old, courses, not devised,
Are Nature still, but Nature methodised.

As for the application of these norms to Pope's interest in garden art, "nature" in the humanised, country landscaped park and urban villa was not at all to be taken as imitation of wild nature, any more than the miniature rocky peaks and pebble streams in classical Japanese garden art are to be taken as imitations of rugged mountains and flooding rivers. Instead, both for Pope and the Japanese masters, garden art is an evocation of idyllic landscapes of the mind. Nature is experienced by the senses as a stimulus, not to copying but to reflecting and to the evocation of poetic feelings. The making of the eighteenth century English garden, inspired by classical ideals of beauty, order and harmony, was the creation of a natural, that is to say proper, order, in the image of an ancient golden age.[11] As will be recalled from the previous chapter, for the creation of this ordered garden of the mind, North American forests, mountains and riversides were being scoured by collectors like Mark Catesby and John Bartram for plants and seeds.

Ideals and norms of nature reached ever further into the eighteenth century cultivated mind, inoculated with notions of politeness and gentility, orderliness and refined taste. These notions took visible form in the well-balanced Georgian architecture of London, Edinburgh, Bristol, Bath and Dublin, drawing on the style of the sixteenth century Italian, Palladio and the precepts of his Roman model, Vitruvius (first century BCE). The elegant turned and carved wooden furniture of Chippendale, Georgian silverware, marble fireplaces, ceilings and cornices in the Adams style, elaborate carved panelling and ornamentation by Grinling Gibbons, salon portraits by Reynolds, nostalgic rural scenes by Constable and the flattering portrayals of the landed gentry by Gainsborough, silk costumes, jewellery, bric-a-brac, and portrait miniatures were all expressions of polite taste, and the natural entitlement of the wealthy. These and other adornments displayed the cultivated taste and feelings of the aspiring middle-class as of the aristocracy. Polished in the course of that pilgrimage to art and good taste, the Grand Tour, this cultural ideal had its roots in an arcadian vision of idyllic "natural living", which it was the function of the poets and writers to deliver to their classically educated readers and connoisseurs. There is a double reflection in this ideal: that of rewards of success and achievement, relief from the pressure of money-earning business; and that of the gilded ancient past, source of standards that were fitting for those who were succeeding in the present.

A curious sidelight in this highly contrived image of nature is the art of representing flowers through collage. A very well-known and greatly admired society figure, was Mrs. Delany (née Mary Granville). Born in 1700 and living until 1788, Mrs. Delany, after a lifetime devoted to embroidery, shell work, painting, and gardening, turned to depicting flowers in paper collages made from hundreds of pieces of coloured paper. A prominent member of the circle of artists, scientists and collectors that gathered under the patronage of the Duchess of Portland, Mrs. Delany is a minor artistic figure but a very good exponent of the representation of nature in popular middle-class art.[12]

In London's National Gallery, one of the paintings most popular with visitors is by the young Thomas Gainsborough (1727 – 1788), *Mr and Mrs Andrews* (Fig. 13.13). Painted in about 1750, a recently married and elegantly clad young couple is portrayed in an idyllic, well-groomed natural setting. They are disposed at ease under a sturdy oak tree which has been interpreted by critics as the strong heart of England, whose gentry hold the country together like the branches of the tree.

Fig. 13.13 Thomas Gainsborough. *Mr and Mrs Andrews*. c1750. The rising bourgeoisie in an idyllic "natural setting." Thomas Gainsborough, Public domain, via Wikimedia Commons

Mrs. Andrews, demurely seated, is dressed in summery blue silk while Mr. Andrews nonchalantly stands alongside, leaning slightly towards his wife in the chair below. His white stockinged legs are lazily crossed and under his right arm, loosely held, is a long-barrelled hunter's gun. Sporting a hunting jacket and with his dog at heel, Mr. Andrews, if he can make the effort, is ready to leave Mrs. Andrews resting, and to head off for the shoot.

Behind the couple is an extensive, bountiful farm landscape. Neat, parallel rows of corn suggest use of Jethro Tull's novel seed drill and a bumper harvest, indicative of efficient modern agricultural practice. Against a distant background of low, well-wooded hills, sheep graze, doubtless on improved pasture grasses. To one side in the distance is the church perhaps suggestive more of their recent marriage than of piety.

The land is theirs, in part the dowry brought to the marriage by Mrs. Andrews. Its suggestion of prosperity and stability reflects the ideal of a carefully cultivated, well-managed country estate. The Andrews are not aristocrats nor old landed gentry; they are the beneficiaries of new wealth. Mr. Andrews's father was a prosperous businessman, into shipping, trade, finance and rented property. That is, he was a thriving capitalist. He not only bought an estate for his son, he found him a rich bride.

John Berger, an iconoclastic Marxist novelist, artist and critic has taken exception to Gainsborough's painting, on the grounds that it celebrates capitalist values and the wealth of a rising bourgeoisie, as it surely does. He also holds that the placing of Mrs. Andrews under the protective stance of her husband symbolises acceptance of the subservience of women in the eighteenth century patriarchy, as it may well do.[13]

There is much more that may be and has been read into this charming painting of the nouveau riche in their delightful and prosperous farm estate. Whatever varieties of symbolic meaning different critics have found, it can be agreed that this painting is an image of man's effective control and successful management of a large tract of fertile rural land. Nature has been harnessed to the advantage of a prosperous rising generation. Or can there be, perhaps, some doubt? Gainsborough has painted a looming sky which casts a dark shadow over the distant forests and hills. This may be a conventional painterly device, balancing light and shade, or perhaps there is a subconscious suggestion that the natural right to property proclaimed a century earlier by John Locke was contingent on economic and political circumstances, always subject to change. Historical materialist, John Berger, would have been happy with that interpretation.

The intellectual, artistic and broadly cultural ideals stemming from the Renaissance, and reaching a high level of sophisticated expression in the eighteenth century, combined with trends in religious thought to strengthen the image of the good life, here and now on earth. The various strands of the concept of "nature" that we have considered and others that we might have adduced, were brought together in a belief that human happiness and well-being are achievable when certain standards are met in a life that connects man positively and harmoniously with the natural world, as disclosed by science and religion and portrayed in art and poetry.

13.7 The Reverend Gilbert White of Selborne

Little-known to the wider public and treated as a minor figure in the world of letters, Gilbert White is an almost heroic figure in the spreading domain of natural history. Not merely admired, he is virtually sanctified in the circle of naturalists – field and literary – and conservationists that upholds the amateur tradition of observation and writing about nature in the countryside.[14] White is known to this community of naturalists above all for his masterwork, *The Natural History and Antiquities of Selborne* (1788). Further selections from his notebooks and journals have been published posthumously, and while they add to the details, they do not significantly deepen our understanding and appreciation of his methods of enquiry and the conclusions he draws from them. Essentially, the White we know today is the author of just one book, who lived in one house, The Wakes, and cultivated one piece of land, that surrounding his country residence in the Southamptonshire (now Hampshire) village of Selborne, a short distance east of Winchester.

As far as his biographers inform us - and the two I have mainly drawn on are Richard Mabey and P.G.M. Foster[15] - White gave relatively little weight to the momentous events of his time, including Britain's part in the Seven-Year War (1756-63/4) and, late in his life, the American Declaration of Independence and the ferment which was to lead to the French Revolution. Foster claims that by reading newspapers, journals, meetings with friends and family and so forth, White was quite familiar with the wider world and its happenings, but there is little evidence of his interest in them in his writing. White's attention when, after failing in his bid to become master of Oriel, his old Oxford College, he took up permanent residence in the family house, the former vicarage, at Selborne, seems to have been devoted almost exclusively to natural history, the doings of friends and family and letter writing with like-minded students of nature (Fig. 13.14).

Born in his grandfather's vicarage, in Selborne village, the house he was later to inherit, White in childhood and youth was privately educated; he then enrolled at Oriel College. Dating from the fourteenth century, Oriel is the oldest royal foundation at Oxford University. The reigning monarch is its official visitor; it was the last Oxford male college to admit female students, a reminder of its conservative history. During the Civil War times, Oriel was a high Tory, Royalist stronghold. Following the Royalist defeat, the Parliament removed five (almost one third of the total) fellows. Just how deeply White was infected by this High Church, Tory and Royalist culture is impossible to say but it seems improbable that it could have left no mark on him. In the nineteenth century, Oriel became the centre of Oxford's intellectual life, its resident fellows including leaders of the High Church Oxford Movement: John Henry Newman, later convert to Roman Catholicism and eminent Cardinal, and the high churchmen, Edward Pusey and John Keble, leaders of the High Church Anglican Oxford Movement. By contrast, it has been suggested that in the generality of Oxford Colleges in the eighteenth century, intellectual life may have been rather sluggish. From White himself, we know very little about his undergraduate studies or his academic interests but, as displayed in the BBC video of

Fig. 13.14 Gilbert White surveys his own estate and Selborne village beyond. Samuel H. Grimm, "North East view of SELBORNE, from the SHORT LYTHE" (Selborne, 1789) whose work (engraved by W. Angus and aquatinted) illustrated the first edition of Gilbert White's seminal work. Gilbert White (author); Samuel Hieronymus Grimm (artist), Public domain, via Wikimedia Commons

Michael Wood and Richard Mabey,[16] White as a student at Oriel was a devotee of rich food and good wine, a horseman and a hunter.

Elected to a fellowship which he retained for the whole of his life, White served for a year as a Proctor (university policeman) but seemed to give little if any attention to the academic life of the College. We may speculate, but his writings give no indication that he was interested in the political and theological culture of his Alma Mater over its long history. His early interest in the natural religion of Ray and Denham could have been taking him some way towards pantheism although there is no evidence that he was ever tempted by this heresy, unless it is a playful Virgilian couplet written in November 1763 when he would have been in his early 40s, *A Winter Piece, Dedicated to the Miss Batties,*

> Let festive glee th'enliven'd village raise,
> Pan's blameless reign, and patriarchal days;
> with pastoral dance the smitten swain surprise,
> And bring Arkady before our eyes.[17]

At Oriel, as a non-resident fellow, White was not expected to teach and appeared to have no duties to warrant the modest stipend he received from the College. It was but a useful supplement to his subsequent living as an Anglican curate and recipient of a small inheritance from his mother. All in all, his income was quite modest but adequate to support a country gentleman bachelor who never travelled abroad, had no family responsibilities and kept a simple household. College fellowships of the kind he held were expected to be relinquished once the recipient received an independent income but White did not observe this convention. Moreover, he followed the practice of pluralism by holding more than one curacy concurrently. This was a common abuse in the Anglican church, whose priests and bishops were not paid adequate salaries, unless they happened to be in well-endowed parishes or bishoprics.

In The Wakes, White's house and now a museum, stands a wax figure, neatly bewigged, and discreetly garbed with white stockinged legs and buckled shoes. It is a slight, unassuming and cheerful figure which, like White's quiet, uneventful life in isolated, rural Selborne, in no way prepares us for the remarkable scientific and literary achievements of his *Natural History*. Perhaps the dark, beady eyes and darting glances of the wax figure and the kindly expression provide a clue to the intensity of his observing eye and ear and the deep affection he showed for animals, especially birds, and for the whole of the natural world that surrounded him.

Periodically, White travelled, but no further than London and the south and south west of England, to visit friends and relatives. In his brother's bookshop in London, he made the acquaintance of two naturalists, his correspondence with whom was to form the substance of his *Natural History of Selborne*: Thomas Pennant (1726-1785) and Daines Barrington (1727/8-1800). A Welsh naturalist and antiquarian and Fellow of the Royal Society, like Evelyn, Ray, Denham and White himself, Pennant wrote several books on zoology, geology and botany, corresponding with many scientific and natural history notables of his day. His travels took him to France where he met the famous naturalist, Buffon, and the iconoclastic Voltaire. White corresponded with Pennant and letters to him form an early part of *The Natural History*. With Daines Barrington FRS, a lawyer, antiquarian and naturalist, White had a longer, more substantial correspondence, which forms the bulk of his *Natural History*. Even though Barrington's interests were largely antiquarian, it appears that White found him a more agreeable and useful correspondent on natural history than Pennant.

If this brief sketch of a modest, uneventful life suggests no more than the simple comforts and interests of a semi-retiring countryman, a minor cleric and an unambitious academic, then it is altogether misleading. White was all of these things but as Richard Mabey, doyen of contemporary nature writers in English, said in introducing the Penguin Classic edition of the *Natural History* and commenting on a letter by White to Daines Barrington about the flying manoeuvres of swifts on long summer days: "This is a new note in nature writing – not just because of the accuracy and percipience of the observations. White is plainly talking from his own experience … His swifts are real, not puppets – "bundles of responses" – all links in some taxonomic chain. They are living birds in a living and closely observed situation. With no more than a hint of anthropomorphism, White suggests that their lives have a richness and rhythm of their own. And by describing them on a summer evening in a particular English village he is able to awake in us the possibility of a human response to them."[18] It is just such a warm human response, and not simply scientific detachment, that Dutch epistemologist, Hugo Zwart, has called for, in our efforts to understand living creatures.[19]

Gilbert White so arranged his life that he was able to devote himself almost entirely to his great passion, the keen observation of, engagement with and reflection on animal and plant life, most of it in his own garden, nearby fields and woodland, together with the daily weather and seasonal patterns in his immediate environment. He was a paragon not only of ordered curiosity, but of delight in the immediacy and individuality of other creatures.

In the seventeenth century, following the humanist challenge to established authority and through the studies of the physical scientists, "curiosity" became a virtue. It had been, among the early Christian theologians, a vice, sinful even, and firm limits were set since heresy was always lurking. But events such as the establishment of the Royal Society, the writings of Francis Bacon, ocean voyages, travellers' tales and the aristocratic and gentry practice of filling cabinets with mineral and vegetable "curiosities", transformed a vice into a virtue.

Of itself, the idea of a book of correspondence, letters drawing on daily nature records, is not all that remarkable. Such observing and recording was, in White's time, becoming one of the favoured pursuits of the English country gentry and parson-naturalists. The study of local antiquities - the concluding and usually overlooked part of White's book - and the display of "curiosities"in cabinets had a great appeal. The best of them served as a foundation for local museums while the overflowing cabinets of London collector, Sir Hans Sloane, in 1753 provided the nucleus for the great collections of the British Museum.

Antiquarianism received a prestigious accolade when, in 1717, the Society of Antiquaries (founded in 1707) was granted a Royal Charter. Established interest in classical antiquities was demonstrated in the founding of the Society of Dilettanti in 1732. Through the Grand Tour, publications on excavations at Herculaneum (1738) and Pompeii (1748) by Lord Burlington and, later in the century, Edward Gibbon's *Decline and Fall of the Roman Empire* (1776 – 80), together with the purchase in 1816 of the *Parthenon Marbles* brought to England by Lord Elgin in 1808-12, all contributed to the idea that it was natural to look at the ancient past for standards of excellence and at the same time to feed curiosity about man's activities and achievements. Meanwhile, dogmatic Christianity was dissolving under the force of scientific enquiry and experience of the world of living nature.

It is not for his interest in antiquities, or even for any claim to uniqueness of his interest in the animal and subject matter of his observations, that Gilbert White is so highly regarded today. Others wrote about the flight and nesting of birds, hedgehogs and domesticated tortoises emerging from hibernation, the lives of insects and so on. But none has been thought to compare, during or before the eighteenth century, with White "whose beautiful and lucid journals ... spare, vivid, precise, yet energised by intense personal feelings about nature and his home spot",[20] have so enthused countless readers.

Such is the richness of Gilbert White's writing, it is exceedingly difficult to confine him by quoting just a couple of excerpts. One of the beauties of his *Natural History* is the continuous flow of observations, a rich potpourri of delight in diverse forms of life. In what follows I shall compare passages from White with an extract from the *De Anima* of Aristotle, the father of natural history, and another from *The*

Compleat Angler of Isaac Walton, whose study of fish and fishing deals with a popular form of nature enjoyment in the century before White's.

Gilbert White explains the importance of direct observation in the study of nature: "There is such a tendency in mankind towards deceiving and being deceived, that one cannot safely relate anything from common report, especially in print, without expressing some degree of doubt and suspicion."[21] His observations of nature are often coloured with allusions to contemporary country life or, as in the following instance, to notable landscape features: "These birds (daws) deposit their nests in the interstices between the upright and the impost stones of that amazing work of antiquity (Stonehenge): which circumstance alone speaks the prodigious height of the upright stones, that they should be tall enough to secure those nests from the annoyance of shepherd boys, who are always hiding about the place."[22]

White's environmental interests stretched from earth to sky, inanimate as well as animate nature, connecting everyday life to its earthern base. For example, he writes of fossil shells and soils and at some length on the characteristic local freestone: "This stone is of great request for hearth-stones and the beds of ovens: and in lining of lime-kilns it turns to good account; for the workmen use sandy loam instead of mortar; the sand of which fluxes and runs by the intense heat, and so cases over the whole face of the kiln with a strong vitrified coat like glass, that is well preserved from injuries of weather and endures thirty or forty years. When chiselled smooth, it makes elegant fronts for houses, equal in colour and grain to the Bath stone". (*Letter IV*) In other letters, White describes the habitats of hares, partridges and pheasants, quails, field mice and fallow-deer. All are presented as if fellow inhabitants of the earth, their lives vividly described, with high regard and feeling.

Gilbert White famously anticipated Darwin in his remarks about the environmental importance of earthworms: "The most insignificant insects and reptiles are of much more consequence, and have much more influence in the economy of nature, than the incurious are aware of; and are mighty in their effect, from their minuteness...... and from their number and fecundity. Earth-worms, though in appearance a small and despicable link in the chain of nature, yet, if lost, would make a lamentable chasm." (*Letter XXXV*) Continuing, White enumerates the many virtues of earthworms which include food for birds, and their work as "promoters of vegetation": "by boring, perforating, and loosening the soil, and rendering it pervious to rains and the fibres of plants, by drawing straws and stalks of leaves and twigs into it; and, most of all, by throwing up such infinite numbers of lumps of Earth called worm-casts, which, being their excrement, is a fine manure for grain and grass ..." These "hints", as White calls them, were given, "in order to set the inquisitive and discerning to work." "A good monography of worms would afford much entertainment and information at the same time, and would open a large and new field in natural history."

In another of his acute observations of the small creatures of which he was so fond, White wrote of the field-cricket, *Gryllus campestris*: "It is remarkable that, though these insects are furnished with long legs behind, and brawny thighs for leaping, like grasshoppers; yet when driven from their holes they show no activity, but crawl along in a shiftless manner." (*Letter XLVI*).

From White's numerous encounters with different species of birds, especially swifts and swallows, I will conclude these excerpts with a piece from *Letter XVI*, on a bird whose voice "country people" often mistake for that of a grasshopper: "The grasshopper-lark began his sibilous note in my fields last Saturday. Nothing can be more amusing than the whisper of this little bird, which seems to be close by though at an hundred yards distance; and, when close at your ear, is scarce any louder than when a great way off. Had I not been a little acquainted with insects, and known that the grasshopper kind is not yet hatched, I should have hardly believed but that it had been a locust whispering in the bushes."

The contrast between the manner in which Gilbert White studied and wrote about nature and that of the father of the genre, Aristotle, could hardly be greater. Not himself a field naturalist, as White inveterately was, Aristotle appears to have relied on teams of observers and collectors for his natural history materials. He writes, in *Parts of Animals*, that there is a purpose or telos in nature and that the parts (of, for example, an animal) have purposes or ends. Except that he eschewed metaphysical language, and that is a crucial difference, White might have written similarly, that the purpose or "end" of the flapping wings of the woodpeckers - who "fly *volatu undoso*, opening and closing their wings at every stroke, and so are always rising or falling in curves" (*Letter XLII*) - is to enable the birds to rise and fall in rhythmic motion.

In Aristotle's opinion, each individual and species is made for itself, its parts so organised as to ensure the good of the whole. Unlike White, the philosopher-scientist starts his enquiry with a lengthy discourse on *method,* a matter to which White gives practically no attention. He approaches his subject directly, almost as if he were alongside it in the meadow or on the bank of a stream. By contrast, it is difficult to feel personal engagement between Aristotle the observer and the objects of his scrutiny and analysis.

Having dealt in the introduction (Part I) to his *Parts of Animals* with the scope of the enquiry he is to undertake and a range of hypothetical matters, including the indivisibility of species and ways of classifying them, Aristotle eventually reaches the object of study, whether it might be "Bird" in the mass, as one might study generic "Man", or whether it should be species, "Ostrich, Crane and so on".[23]

To avoid, in his own words, a course that is, "somewhat unreasonable and long winded, because it makes us describe the same attributes time and again as they happen to be common attributes of many species", Aristotle proposes to describe "the common attributes of each group together", the groups being such as, "Birds and Fishes and others whose nature is to come into being and perish". He says that while our grasp of the category of "eternal things" is slight, since we live on earth among plants and animals, by taking sufficient trouble we can learn about them. True, while there is joy beyond measure in grasping the "excellence and worth" of eternal things, it is satisfying for us to have secure knowledge of living things on earth, "Moreover, because they are nearer to us and more akin to our nature, they are able to make up some of their leeway as against the philosophy which contemplates the things that are divine."[24]

Aristotle continues in this physico-philosophical strain with a determined interest in classification, purpose and causal relations. Thus he classifies the parts of animals, animal movements, common modes of generation and so on. Elements of his method and his assumptions of purpose and design continued to influence natural history over the centuries and remnants of this thinking lie beneath the records of Gilbert White. But there is a vast difference in the tone of voice and the way in which the eighteenth century naturalist engaged with living, sentient, individual beings. The categorised species of Aristotle are "the other" whereas for White the creatures are "thou". Aristotle is not drawn into personal engagement, his feelings are withdrawn, but Gilbert White, constantly, is almost part of what he is observing. For all that, Aristotle does make clear that his objects of investigation and use are akin to our nature, and writes of the immeasurable joy provided by Nature to "the eye of science" and "the philosophic spirit". We are together with all creatures and in embarking on their study we know, "that in not one of them is Nature of Beauty lacking". This message White, too, communicates to us.

It is in his methodical journal records and most of all those on weather and seasonal patterns that White's methods somewhat resemble Aristotle's. Familiar with contemporary proposals for the systematic recording of natural phenomena and events in the form of calendars, White was keen to make his records as systematic as possible. He studied Benjamin Stillingfleet's *Miscellaneous Tracts relating to Natural History, Husbandry, and Physick* (1762), a publication which included Stillingfleet's own calendar and one designed in Sweden. In 1767, White was presented by Daines Barrington with a similar journal which he enthusiastically used for his records. Each page of the journal was designed for a week's observations, classified into separate columns, on geological, and meteorological topics, trees first in leaf, first appearance and disappearance of birds and insects, and other themes. White's careful and precise use of the calendar and journal is the principal reason for characterisation of his methods as "scientific" in P.G.M. Foster's *Gilbert White and his Records. A Scientific Biography* (1988).

However, White's language is literary rather than scientific (although this distinction was not as strong in the eighteenth century as today) and his purposes were largely practical. Yet his records from the 1740s to the 1780s have proved to be valuable to historians reconstructing the conditions (climatic especially) under which horticulture developed in England.[25] He does not construct theories of animal behavior and the growth of plants, but does present phenomena with precision and feeling.

Despite his frequent appreciative references to John Ray, White did not care for some of the methods of those scientific botanists who were not, in that much used eighteenth century term, "improvers": "The standing objection to botany has always been, that it is a pursuit that uses the fancy and exercises the memory, without improving the mind or advancing any real knowledge." Speaking as a gardener, estate holder and farmer, White continues in Baconian mode: "Instead of examining the minute distinctions of every various species of each obscure genus, the botanist should endeavour to make himself acquainted with those that are useful. … The botanist that could improve the sward of the district where he lived would be a useful member of society; to raise a thick turf on a naked soil would be worth volumes of systematic knowledge; and he would be the best commonwealth's man that could occasion the growth of two blades of grass where one alone was seen before" (*Letter XL*).

On first and third January 1757, White wrote this characteristic page in his Garden Kalendar, his "two blades of grass":

> Jan:1st: Planted a row of Tulips, & Ranunculus's, given me by my Brothr Thomas, in the Border in the New-Garden next the street. Dug out the soil, & filled the trench with earth well-mixed with lime-rubbish.
>
> 3. Planted the Fir-Quincunx with five rows of Winsor-beans: dunged the ground, that was very poor, with ten wheel-barrows of very rotten dung. After the beans the ground to be trenched with Celeri.
>
> Levelled, & widen'd the Area of ye Melon-Ground; having made an under-ground Drain to prevent it's being flooded any more.[26]

Leaving Aristotle and the botanists, we turn to another naturalist, Isaac Walton. A successful businessman, devout Christian and biographer of the metaphysical poets, John Donne and George Herbert, Walton was a Royalist supporter in the Civil War and one of those who quietly retired to the country on the defeat of King Charles. His sympathies and associates were Tory and High Church.

In the form of a conversation between the author, as "Piscator" a fisherman, "Venator" a falconer, and "Aucups" a hunter, Walton's *The Compleat Angler* covers a week in a country retreat beside a river. Each of the three in turn extols the virtue of his craft but Venator and Aucups soon concede the palm to Piscator who proceeds to explain the craft of fishing and to expatiate in descriptive prose on the characteristics and qualities of a considerable variety of fish. The text is punctuated with poetry, anecdotes, songs and passages of religious moralising. In a tilt at capitalist enterprise, which he had left behind in retreating to the country, Piscator pities those Londoners, "of a sour complexion, money-getting men, men that spend all their time, first in getting, and next in anxious time to keep it ... poor rich men, we Anglers pity them." This was a theme that another critic of London money grubbers, Charles Dickens, was to take up in his widely read novels.

Although as a naturalist he speaks appreciatively of the fish he describes, Piscator's real interest in them is as objects for the fisherman and his cooking pan. From time to time, the trio enjoy the countryside as the gift of "the God of nature"; God is to be trusted and praised. In a closing passage, Venator declares that he, "will walk the meadows, by some gliding stream, and there contemplate the lilies that take no care, and those very many other various little living creatures that are not only created, but fed, man knows not how, by the goodness of the God of Nature."

Venator's pious observations touch on Natural Religion, but in the book as a whole interest in religion is a distant second to fish and fishing. All three protagonists are hunters and killers of animals and even though there is a school of thought that identifies hunting with respect, admiration and even love of the animal kingdom, Walton's message is that the fish in the river are there to serve man's hunting purpose and to minister to his enjoyment of the catch. The trust in God's Providence, in Venator's pious concluding remarks about contemplating the lilies, is trust in the provision that God has made for man's pleasure and comfort.[27]

As we have seen in this and previous chapters, in Western thought humans have almost always been at the centre of the scheme of things with the world projected as God's home for mankind. Even when, as weak and degraded creatures, humans were cast from the Garden of Eden, and in the state of nature depicted by Thomas Hobbes as, "solitary, poor, nasty, brutish and short", humans still ruled the roost. Gilbert White had no time for these images either of mastery or misery. For him, creation was a blessing and man, whatever his faults, an integral part of this blessing, as were "All creatures great and small." White's purpose was to convey an impression of the unity and the harmonious interrelationships of the living world. He expressed this optimistic outlook in the four distinct yet closely related activities to which he devoted his life and writings:

- Gardening and horticulture, healthy and meaningful practical pursuits suited alike to humble villagers and the gentry
- Observing, recording in daily journals and corresponding about the life-forms of his neighbourhood, and the overall environment, including uses of the land
- Reading and reflecting on literature, on natural religion and natural history, horticulture, Latin classics and poetry
- Writing, whether personally to friends and relatives (to whom there were periodic visits), or for ultimate publication in his *Natural History*

But the essence of Gilbert White is to be found not in generalisations about his "activities", it lies instead in an astonishing precision of expression of his daily observations of the life around him,

> Several swifts, Many swallows,
> Cut seven brace, & half of cucumbers.
> The great oak in the mead abounds with male bloom.
> Timothy [tortoise] is very voracious; he devours the lettuce.
> [May 7, 1788][28]

Gardening and horticulture, with chapped hands in the soil, was a lifelong occupation which White pursued in the manner of an applied scientist. At The Wakes, he had vegetable beds with frames for cucumbers and melons which he treated with loads of enriching farmyard manure, much as the monk, Walafrid Strabo, had done, and over which he closely watched, adjusting the glass lights according to the vagaries of the weather and the rate of fermentation (Fig. 13.15). He wrote of bounteous crops of cucumbers and sufficient melons for a village party held on the Hangar, the backdrop to his land, through whose wooded hillside he had cut a zigzag passage in the style of other landscape gardeners of the time. White had the common experience of marauding insect pests and diseases, part of the natural environment, for which not all (organic) treatments worked, but he kept trying. His horticultural interest was sustained not only by successful cropping but seed exchanges, the introduction of new varieties and recorded crop performance and persistence in the face of adversity.

Fig. 13.15 Gilbert White cultivates his vegetable garden. Note the glasshouse for cucumbers and melons, as well as White's pet tortoise. Artist Eric Ravilious (1902- 1942). Public domain

By the middle of the eighteenth century in England, country gardening and landscape design were coming into vogue for amateurs and professionals alike. As with the financial bubble created in the seventeenth century by the avaricious and reckless purchase of tulip bulbs, the excesses of huge investments in the "improvement" of their estates resulted in bankruptcy for some of the most ambitious landholders. Gilbert White showed an interest in this vogue but it was modest. He had a dressed stone wall built along the side of a ditch, the fashionable ha-ha designed to keep sheep and cattle from the garden while preserving the outward view. Unable or unwilling to afford a marble or lead statue, he had a local carpenter cut the figure of Hercules out of a flat piece timber and so erected in one of his meadows as to deceive the eye into creating of a three dimensional image.

In a life devoted to the love and study of nature, Gilbert White made a decisive break with one of the most persistent and earth-threatening ancient dogmas of his own Christian religion, replacing the arrogance of human's dominance by the communality of partnership with nature. As garden historian, Mark Laird, put it, he introduced the public to the "conversation of the animal kingdom".[29] In so doing, he opened new ways of knowing and understanding nature and succeeded in conveying to his readers loving insights into the lives of other creatures.

No theoretician, White was a pioneer in his study and love of nature, of the humanistic science of the singular and the personal as distinct from the general and the abstract. He studied and wrote about living creatures as individuals with an inner life whose meaning could be grasped in part at least through patient, empathetic observation and reflection. Gilbert White's lasting legacy, as summarised by Richard Mabey in a personal communication, is his "intuitive synthesis of vernacular science, poetic imagination and his broadening of the idea of community to embrace other species (which) remain pivotal in the history of both natural history and literature."[30]

The Catholic church has proclaimed St Francis, patron saint of ecology. Perhaps their Protestant co-religionists could come together to sanctify Gilbert White in a shared sainthood of ecology, joint spiritual fathers of twentieth century Eco-Christianity.

Notes

[1] The Works In Natural History by the Late Rev. Gilbert White, A. M. Fellow of Oriel College Oxford in 2 vols London. Printed for J. White, Fleet Street, By T. Bernsley Bolt Court 1802: Advertisement to volume 1 of a 2 volume set.

[2] Symonds, J. A. (1907) *Renaissance in Italy. The revival of learning*. London: Smith, Elder & Co, page 10

[3] Lukacs, G. (1962) *The Historical Novel*. London: Merlin Press, page 267

[4] Passmore, J. (1970) *The Perfectibility of Man*. London: Duckworth, ch.8 Perfecting by Social Action: the Presuppositions.

[5] Woodward, W.H. (1906) *Studies in Education During the Age of the Renaissance*. Cambridge: Cambridge University Press, page 14.

[6] In this section I draw on the online Stanford Encyclopedia of Philosophy entry on Giambattista Vico, by Timothy Costelloe, 2018 https://plato.stanford.edu/entries/vico/ (Accessed Feb 17, 2020) and recollection of my now distant studies of methods of inquiry in social science and theory of knowledge.

[7] Keynes, J. M "Newton the Man", an addendum to "Essays in Biography" (1933) (edited by Keynes, G. (1963). New York: W.W. Norton and Co. page 311.

[8] Pope, Alexander. "Essay on Man. Epistle II"

[9] In his sermon, "John Ray and our attitude to the natural world," which the Right Revd Cyril Tucker CBE gave at St Catharine's College, Cambridge in November 1988, he indicated that Ray stressed the importance of attitudes of caring and responsibility towards Creation. *St Catharine's College Society Magazine* 1989. pp. 30-31. https://www.society.caths.cam.ac.uk/Public_Magazines/1989r.pdf (Accessed Feb 17, 2020)

[10] Pope, A. "Epistle IV To the Right Honorable Richard, Earl of Burlington" in Fussell, P. (ed) (1972) *English Augustan Poetry*. Garden City, New York: Anchor Books. Doubleday& Co., Inc. p. 353

[11] Hunt, J. D. and Willis, P. (eds) *The Genius of the Place. The English Landscape Garden 1620-1820*. New York: Harper and Row.

[12] Laird, M. (2015) *A Natural History of English Gardening*. New Haven and London: Yale University Press. ch.6 Virtue. The Natural Pleasures of Mary Delany at the Duchess of Portland's Bulstrode; Hayden, R. (1980) *Mrs Delany. Her life and her flowers*. London: British Museum Press.

[13] Berger, J. (1972) *Ways of Seeing*. London: BBC; Harmondsworth: Penguin.

[14] Mabey, R. (ed) (1995) *The Oxford Book of Nature Writing*. Oxford: Oxford University Press. Passim.

[15] Mabey, R. (1986) *Gilbert White. A Biography of the Author of The Natural History of Selborne*. London: Century Hutchinson.; Foster, P.G.M. (1988) *Gilbert White and his Records. A Scientific Biography*. London: Christopher Helm.

[16] BBC Two (2007) *Gilbert White: The Nature Man*. Presenter Michael Wood with Richard Mabey. Video.

[17] Massingham, H. J. (Ed.) *The Essential Gilbert White of Selborne* Selected and Introduced by Mark Daniel with Wood Engravings by Eric Ravilious (1983) London Breslich and Foss, page 360.

[18] Mabey, R (1977) Introduction to Gilbert White: *The Natural History of Selborne*. London, Penguin Books pages ix –x.

[19] Zwart, H. (2008) *Understanding Nature. Case Studies in Comparative Epistemology*. Springer

[20] Mabey, R (ed) (1995) *op.cit.* Page ix

[21] Quoted by Mabey, R. (1986) *op.cit.* p. 53

[22] Quoted by Mabey, R (1986) *op.cit.* p. 58

[23] Peck, A.L. (1968) (trans.) *Aristotle: Parts of Animals*. London: William Heinemann Ltd.; Cambridge, Mass.: Harvard University Press. p. 95

[24] ibid. p. 99

[25] Laird, M *op.cit.* Pages 216 – 17, 258 – 59, 300, 314, 319

[26] White, G. (1975) *Garden Kalendar, 1751-1771*. Reproduced in facsimile from the manuscript in the British Library. Introdn and notes by John Clegg. London: The Scolar Press.

[27] Quotations are from *The Compleat Angler* by Izaak Walton. Available online with Project Gutenberg: http://www.gutenberg.org/ebooks/683 (Accessed Feb 17, 2020).

[28] Mabey, R. (General Ed.) (1989) *The Journals of Gilbert White* Vol.3 (1784-1793). Ed. Greenoak, F.. London: Century.

[29] Laird, M. *op.cit.* p. 25

[30] From email letter, Richard Mabey to Malcolm Skilbeck, 5 May 2016.

Complete Listing for Wikimedia Commons Images

13.1 Domenico di Michelino – Dante and the Three Kingdoms – WGA06421. https://commons.wikimedia.org/wiki/File:Domenico_di_Michelino_-_Dante_and_the_Three_Kingdoms_-_WGA06421.jpg Web Gallery of Art.Attribution: Domenico di Michelino, Public domain, via Wikimedia Commons.

13.2 Ritratto di Francesco petrarca, altichiero, 1376 circa, padova. https://commons.wikimedia.org/wiki/File:Ritratto_di_francesco_petrarca,_altichiero,_1376_circa,_padova.jpg Attribution: Altichiero, Public domain, via Wikimedia Commons. Source: unknown.

13.3 Hans Holbein d.J.046. Portrait of Erasmus of Rotterdam. https://commons.wikimedia.org/wiki/File:Hans_Holbein_d._J._046.jpg Attribution: Hans Holbein, Public domain, via Wikimedia Commons. The Yorck Project (2002) *10.000 Meisterwerke der Malerei* (DVD-ROM), distributed by DIRECTMEDIA Publishing GmbH. ISBN: 3936122202.

13.4 Michelangelo Daniele da Volterra (dettaglio). https://commons.wikimedia.org/wiki/File:Michelangelo_Daniele_da_Volterra_(dettaglio).jpg Source: Metropolitan Museum of Art, online collection (The Met object ID 436771) Attribution: Daniele da Volterra, Public domain, via Wikimedia Commons.

13.5 Leonardo da Vinci – presumed self-portrait – WGA12798. https://commons.wikimedia.org/wiki/File:Leonardo_da_Vinci_-_presumed_self-portrait_-_WGA12798.jpg Attribution: Leonardo da Vinci, Public domain, via Wikimedia Commons.

13.6 Justus Sustermans – Portrait of Galileo Galilei, 1636. https://commons.wikimedia.org/wiki/File:Justus_Sustermans_-_Portrait_of_Galileo_Galilei,_1636.jpg Attribution: Justus Sustermans, Public domain, via Wikimedia Commons.

13.7 Giusto di gand e pedro berruguete, vittorino da feltre. https://commons.wikimedia.org/wiki/File:Giusto_di_gand_e_pedro_berruguete,_vittorino_da_feltre.jpg Attribution: Justus van Gent, Public domain, via Wikimedia Commons.

13.8 Giambattista Vico. https://commons.wikimedia.org/wiki/File:GiambattistaVico.jpg Attribution: Francesco Solimena, Public domain, via Wikimedia Commons.

13.9 Sir Isaac Newton. Mezzotint by T.O. Barlow, 1868, after Sir Wellcome V0004247. https://commons.wikimedia.org/wiki/File:Sir_Isaac_Newton._Mezzotint_by_T._O._Barlow,_1868,_after_Sir_Wellcome_V0004247.jpg Source: Wellcome Collection gallery. Attribution: See page for author, CC BY 4.0 <https://creativecommons.org/licenses/by/4.0>, via Wikimedia Commons.

13.10 Michael Ramsey 1974. https://commons.wikimedia.org/wiki/File:Michael_Ramsey_1974.jpg Source: Dutch National Archives (ANEFO). Attribution: Unknown authorUnknown author, CC BY-SA 3.0 NL <https://creativecommons.org/licenses/by-sa/3.0/nl/deed.en>, via Wikimedia Commons.

13.11 Newton-WilliamBlake. https://commons.wikimedia.org/wiki/File:Newton-WilliamBlake.jpg Source: The William Blake Archive. Attribution: William Blake, Public domain, via Wikimedia Commons.

13.12 Pope 1726. https://commons.wikimedia.org/wiki/File:Pope_1726.jpg Attribution: Publisher Bernard Lintot London 1726, Scanner and Uploader Steven J Plunkett, Public domain, via Wikimedia Commons.

13.13 Thomas Gainsborough – Mr. and Mrs. Andrews. https://commons.wikimedia.org/wiki/File:Thomas_Gainsborough_-_Mr_and_Mrs_Andrews.jpg Source: The National Gallery. Attribution: Thomas Gainsborough, Public domain, via Wikimedia Commons.

13.14 NorthEast view of Selborne from the Short Lythe. Foldout frontispiece. https://commons.wikimedia.org/wiki/File:North_East_view_of_Selborne_from_the_Short_Lythe._Foldout_frontispiece.jpg Source: https://www.biodiversitylibrary.org/item/80773#page/10/mode/1up. Attribution: Gilbert White (author); Samuel Hieronymus Grimm (artist), Public domain, via Wikimedia Commons.

Chapter 14
Pathways to an Earthly Paradise – The Mogul Emperors

The spreading wide my narrow Hands
To gather Paradise.
Emily Dickinson I Dwell in Possibility

Life on earth, in the natural, Fallen, world, has been viewed by theologians and zealots on the one hand as but a dubious prelude to immortality in heaven, and by their more relaxed brethren on the other hand, as a more or less delightful sojourn amidst the bounty of nature. Such earthly pleasures as gardening, painting landscapes and flowers, studying and reflecting on animal behaviour, comparing observations with scientific colleagues and expressing deep feeling for the natural world as a means to understanding it were, as we have observed, all part of the repertoire of eighteenth century naturalists. For some of them, it was through close engagement with nature that they felt ever closer to God. As we shall see in this chapter, sentiments such as these had been expressed in earlier times by followers of another great monotheistic religion, Islam.

Relating to the natural world in such personal, emotional ways has continued to the present, in parallel or interaction with more systematic and theory-governed approaches, as scientific enquiry from the Renaissance onwards became more specialised and professionalised. A strong amateur tradition of naturalism, while it overlaps with professional, academic science, has its distinct practices, most notably in its investment of enquiry with subjective values and expressions of personal feeling. To this extent, we may think of this amateur tradition of natural history as a branch of the humanities, coloured with scientific ideas and practices.

14.1 Connecting with Earth Through Gardening

Starting with the Mogul Emperors in North India and Kashmir, in this and the chapter that follows we consider how, in diverse religious settings, gardens and gardening and the natural world more generally became a passionate interest and an enjoyable way of connecting with the earth. Heaven was not always put firmly aside but from the late mediaeval period in the West, it progressively faded into the

M. Skilbeck, *Loving and Studying Nature*, International Explorations in Outdoor and Environmental Education 10, https://doi.org/10.1007/978-3-030-80751-1_14

background as ever greater interest was taken in the earth, its place in the cosmos and its richly varied plant and animal inhabitants. Humans gradually became less a unique, distinctive species, at least in the eyes of science, not separate from the rest of nature but part of it and the wilderness of the Fallen world of Adam and Eve gradually became converted into something of "a green and pleasant land".

In 1859, publication of Darwin's *The Origin of Species* and, in 1871, his *The Descent of Man and Selection in Relation to Sex*, decisively closed the gap between humans and the natural world. *Homo sapiens* was but another species, cousin to others and sharing a common ancestor with them. The plant and animal life of the earth became of increasing interest and importance in the quest to understand the earth and live a fulfilling life in association with nature. Paradoxically, as the human species became assimilated by science into the natural world, driven by the economics of consumer capitalism, mankind assumed greater control of the earth and in the process ever more effective at manipulating its resources even to the point of large-scale destruction and elimination of other species.

By contrast, gardens and gardening have long been an integral part of the quest to relate positively and harmoniously to the natural world and to the felt presence of the divine in nature. With roots reaching back to ancient Egypt, Mesopotamia and China, gardening, as we saw in the previous chapter, had become, by the 18th century, a favoured interest or activity of the emerging European middle classes, the aristocracy and, as with Gilbert White, the clergy. Of necessity, vegetable gardening had long been a domestic pursuit of the farmworker and cottager, a means to augment meagre earnings. Whatever incidental pleasure this may have given the smallholder, sustaining the family food supply was usually a necessity. In the ancient Greek hierarchy of orchards and pleasure gardens, growing vegetables was an occupation for the poor and slaves.

Gardens and gardening also inspired the imagination, providing subject matter for literary allusions and images; for example, in the poetry of Marvel, Milton and Pope, as it was in Islam in the work of the mediaeval Sufi poets and other masterpieces of Persian literature. Many of the tales told in *One Thousand and One Nights*, written in the tenth century but set in the first, are located in exotic gardens. The Garden of Eden, variously a symbol of heavenly bliss and of humankind's sinfulness and fall from grace, has featured not only in the Christianised West but also, through numerous references to Paradise in the *Q'ran*, in Middle Eastern Islamic culture. Paradise conceived as a garden has for long had strong earthly as well as heavenly connotations. [1]

14.2 Diverse Images of Paradise

Enunciating the "paradise principle" in secular terms, garden historian, Ronald King, declared that: "Nature and man are one; whatever divides them invites failure." Humans, he went on, crave for a natural setting for their activities, but much of modern urban life denies this. Gardens can provide it.[2] The term "paradise", however, has historically had a variety of religious associations, albeit commonly with some reference to nature. For example, it referred to the entrance court, often planted as a garden, of Christian churches, as if stepping into church was like entering heaven. Similarly, monastic cloisters enclosing grassy areas suggested paradise, as, in St Augustine's understanding, did the early Christian church.

"Paradise" has been associated with plants and nature, since the Garden of Eden, the earthly paradise. As a compliment to garden makers, it has become something of a commonplace. In an essay in 1790, James Walker wrote of, "the paradise which the Earl of Moira has opened in the wilds of County Down"[3]. Another garden writer, who had a strong religious bent, John Daniel Sedding, conceived the garden as, "the memorial of Paradise lost, the pledge of Paradise regained"[4].

Dante's *Divine Comedy* is an early fourteenth century allegory of the ascent of the soul to Paradise, through the Inferno and Purgatory. The empyrium, "highest heaven", is the ultimate stage of the ascent to Christian Paradise, where the soul, so to speak, rests eternally in the lap of a loving, forgiving God. This image speaks to piety and faith which in Christianity are inseparable from Eden. By contrast, some two centuries later, the Netherlandish painter, Hieronymus Bosch, in the triptych *The Hay Wain* (Fig. 14.1), depicted humanity careering on an overloaded hay cart from the Garden of Eden straight into the fires of Hell, while in *The Garden of Earthly Delight* a great medley of humans, animals, monsters and bizarre creatures writhe about in what seems to be a parody of the Christian Paradise, or Eden.

"Paradise" is a term and a concept which, nevertheless, has origins remote from religiously inspired images of faith, love and hope rewarded – or sin punished. Garden historians cite Xenophon, according to whom the term originated in old

Fig. 14.1 Hieronymus Bosch (c1450–1516) *The Hay Wain triptych*. Humanity careers on an overloaded hay wagon from the Garden of Eden straight to Hell. Museo del Prado, Madrid. Hieronymus Bosch or workshop, Public domain, via Wikimedia Commons

Persian, where it referred to a royal park, a country hunting ground, a place of pleasure for ruler, Cyrus the Younger (nephew of Cyrus the Great), when on tour, away from the demands of courtly duty and the toils of everyday life.[5] Within or attached to the royal hunting parks there may have been gardens, or the park was itself conceived as a garden, as it surely was a royal pleasure ground. The account given by Xenophon refers to a well-organized space, straight lines, or scatterings, of trees, and to the importance of husbandry in the king's domains (Section 14.3). There is a continuity to trace from these ancient Persian "paradises" and preceding garden proto-types, to the modern use of the term in the imagery of actual gardens, its religious uses and the metaphors of poets, modern as well as ancient.[6]

The 18th century English poet, Samuel Taylor Coleridge (Fig. 14.2), had this idea of a garden – paradise - park in mind in his famous, opium-inspired and unfinished poem *Kubla Khan, Or a Vision in a Dream: A Fragment,*

In Xanadu did Kubla Khan
A stately pleasure-dome decree:
Where Alph, the sacred river, ran
Through caverns measureless to man
Down to a sunless sea.
So twice five miles of fertile ground
With walls and towers were girdled round:
And there were gardens bright with sinuous rills
Where blossomed many an incense-bearing tree

Fig. 14.2 Colleague of William Wordsworth, Samuel Taylor Coleridge (1772–1834) wrote his celebrated poem Kubla Khan, or a Vision in a Dream: a Fragment, under the influence of an opium-inspired dream. He said it was left unfinished because he was interrupted by "a stranger from Porlock". Mezzotint portrait of Coleridge by Samuel Cousins after Washington Allston. Public domain, after Wikimedia Commons

And here were forests ancient as the hills,
Enfolding sunny spots of greenery.

The poem finishes – or broke off - on a high note, inspired by the poet's mesmerizing vision of an Abyssinian maid playing on a dulcimer and "singing of Mount Abora". He yearns to hear her again since that would inspire him with poetic genius, dangerous as that was to an unwary imaginary audience,

And close your eyes with holy dread
For he on honey-dew hath fed,
And drunk the milk of Paradise.

Coleridge's poem might serve as a text for this chapter since it beautifully captures the same romantic spirit which infused the splendid gardens and buildings constructed over the centuries of the Mongol-Islamic Mogul Empire while evoking religious sentiments. Coleridge as a young man aspired to an earthly paradise of a communitarian kind. Together with the poet, Southey, and influenced by Plato's *Republic* and William Godwin's *Political Justice,* he planned to establish an egalitarian utopian settlement in Panama. For various reasons, including lack of finance, the scheme did not proceed. Perhaps the great poem, *Kubla Khan*, itself never completed, served as a substitute.

The actual Mongol place, Xanadu, is now a UNESCO World Heritage site, of "outstanding universal value." (Fig. 14.3) It is of interest as something of a forerunner of cultural developments centuries later under the Mogul emperors, themselves

Fig. 14.3 Xanadu (Yuan Shangdu) today, a UNESCO World Heritage Site in the Mongolian-Manchurian steppes (grasslands). Flaumfeder, CC BY-SA 4.0, via Wikimedia Commons

descended from the Mongols who invaded China. Situated north of the Great Wall of China, Xanadu was the legendary capital of Mongol warlord, Kublai Khan, grandson of Ghengis Khan and founder of the Yuan dynasty. The UNESCO text describes it as a unique attempt to assimilate the nomadic Mongolian and Han (Chinese agrarian) cultures. Religious debates at Xanadu led to dissemination of Tibetan Buddhism. Shaped like a square, the site was planned according to Chinese feng shui, in relation to the nearby mountain and river landscape. The city remains include pieces of the city wall, temples, tombs, a canal and other water works, together with a nomadic encampment. As we shall see below, several elements of Xanadu were to be echoed in Mogul times: cultural assimilation, religious debate and garden layout.

Coleridge's poem draws on an account of Xanadu by the seventeenth century clergyman, Samuel Purchas, itself based on the meeting Marco Polo is believed to have had with Kublai Khan (Fig. 14.4). The pleasure park at Xanadu was alike a symbol of authority, a dwelling and a place of respite and pleasure in a garden setting. Pavilions, fountains and flowing water of the kind still to be found in the great Persian and Mogul Gardens, were all part of Coleridge's imagery.

Fig. 14.4 Marco Polo served at the court of Kublai Khan, 1275–1292. 18th century illustration of Marco Polo in Tartar costume. Artist: Grevembrock. Grevembrock, Public domain, via Wikimedia Commons

14.3 Eden and Antecedents

Leaving Mongolian and Han Chinese ideas aside, it is to the Judaeo-Christian Garden of Eden and the evocations of paradise in the *Q'ran* that we now turn for further elucidation of the religious concept of paradise. That concept is of a state or condition of fulfilment, reward for a life on earth of piety and virtue, or as an expression of God's unconstrained, freely given grace. Eden is alike a natural and supra-natural phenomenon.

The *Old Testament* Eden (meaning "pleasure", "delight") in *Genesis* provides not only an image and an aspiration, but also an admonition and a warning. The image is of a garden, a place of beauty and leisure as well as bountiful produce and sensual pleasure. There is peace and contentment for Adam and Eve, in association with fruitful plants and animals: the natural world in its fullness and splendour is a peaceful home for all forms of life. The Garden of Eden thus represents an unsullied state of nature wherein humans could have lived in harmony with plants and animals, experiencing eternal bliss free of labour and of the disappointments and frustrations of everyday existence. If only they had not disobeyed God, the creator! This was an image of paradise in which, living together, all species and life forms had their rightful place. A similar impression was given by the animals all together with Noah and his family on the Ark. However, warnings were given and the sinning Adam and Eve were punished, for in effect despoiling the garden. This might be taken as a metaphor for what humanity is now doing to its natural home, the earth.

According to the final book of the *Bible*, *Revelation*, awaiting the last coming is the holy city of God, the new Jerusalem. Flowing through it is, "a pure river of water of life, clear as crystal" bordered on either side by the tree of life with a perennial yield of fruit and leaves, "for the healing of the nations". The city is of pure gold, within a wall of jasper garnished with precious stones. There is no more death, no pain. The image of Eden endures to the end of the world, but only the righteous will be able to (re-)enter.

The imagined topography of the Garden of Eden, both physical and symbolic, has been a major inspiration for the so-called paradise gardens that have continued to be designed and constructed to the present day. A place of seclusion, calm and beauty, adorned with ornamental and fruiting trees, the symbolic trees of knowledge, of good and evil, and of life. Eden was a garden that Adam was, "to dress and to keep," not to destroy, as he now seems to be bent on.

According to the *Book of Genesis*, a river flowed east out of Eden dividing into four and then, somewhat obscurely, compassing several lands and the River Tigris, before running towards Syria, (River Euphrates). Leonard Woolley, the archaeologist who led the excavation of the Mesopotamian city, Ur of the Chaldees, argued that the delta formed by the two great rivers, the Tigris and Euphrates, accords with the biblical story. Abraham, effectively the founder of Judaism, was a native of Ur. This was a region of fertile soil not only for the beginnings of agriculture but also for a variety of religious cults, each with its temples, shrines, groves, and specialised cultic trees (as of the trees of life and of the knowledge of good and evil in the Biblical Garden of Eden).

Since the *Old Testament* was composed, or recomposed, after their enslavement in Babylon it is likely that the Jews were familiar with the construction and garden-making practices of that region. These included the planting of groves of trees around tombs, such as that of Cyrus the Great, the Persian founder of the Achaemenid empire, at Pasargadae, and the Imperial custom of planting trees and gardens during travels in the royal provinces. Biblical Eden did not lack physical antecedents and analogues in the ancient Middle East.

Pasargadae, located in a temperate zone to the south of the elevated plateau of present day Iran (former Persia), was where Cyrus the Great established his capital and laid out the first known historical park-garden, in the sixth century BCE. Palatial and other monumental structures, including Cyrus' tomb, which still stands, were set in a verdant park. Apparently drawing on even earlier Assyrian and Babylonian practice, Cyrus established a central, irrigated royal garden within the large park designed to reflect the glory of his new empire which stretched from the Mediterranean to Central Asia. Cyrus drew on and integrated traditions and skilled workforces from throughout his empire in creating a radically new style of royal settlement on the green fields site of Pasargadae. Key features included the use of cut stone to construct buildings and water features, a formal geometric overall design, sophisticated hydraulic engineering, an apparently unfortified garden and park setting with few and relatively modest buildings placed beside gardens and water features.[7]

A system of principal and secondary axes in the form of decorative open limestone channels 25 cm wide with, at intervals of 13 or 14 m, deep square basins, provided precious water in the demanding summer climate. It appears that the garden layout formed two rectangles and that, in the portico of a palace building, a stone throne for the king was erected to enable him to view the garden. The cultivated "garden" may have been up to 60 ha. in extent, within a "park" of up to 150ha., the two together comprising the royal "paradise". Now a UNESCO World Heritage site, Pasargadae may well disclose further secrets in the future but for the present it suffices to regard the general layout as a foundation of an enduring tradition of landscaping and garden design in upper class and aristocratic culture. In the regular geometry, use of cut stone, noble architecture and irrigation channels, it anticipated many of the achievements of the later Persian and Mogul urban landscapes including their paradise gardens.

Recent archaeological research by Iranian-French teams is providing more detail and depth to field work carried out in the second half of the nineteenth and twentieth century which popularized the idea of a well-structured and watered royal garden[8]. Current research, including pollen counts, is yielding deeper understanding of both historic planting schemes and the indigenous and introduced vegetation on site in Cyrus' era: herbaceous plants and grasses, decorative plants, vegetables, shrubs and trees including fruit trees.

The hypothetical reconstruction by Farzin Rezaiean of the central garden at Pasargadae shows a lush setting against a barren mountain range (Fig. 14.5). While

Fig. 14.5 Oblique view, recreating central garden, of Cyrus's capital Pasargadae. Farzin Rezaeian's virtual images draw on recent archaeological research at this important site. Source: *Recreating Pasargadae* p.59. Courtesy Farzin Rezaeian

the archaeology has established the layout of the gardens and buildings, the question of how many and which mix of trees at this stage remains somewhat speculative. This image shows the distinctive type of "garden-palace" with scattered buildings not a single consolidated mass, as in later palaces. A large pool in the centre of the garden was fed by a stream (to the right of the image), and was controlled by a series of sluices (not shown in the virtual reconstruction). Huge dams constructed some 30 kms upstream on the Pulvar River and tributaries allowed the water level in the pool to be maintained at 1.5 metres depth, even at the height of summer. Water reticulated through the gardens in a series of stone-lined channels (Figs. 14.6 and 14.7).

While dams and surface canals were the means to bring water to the Achaemenid site of Pasargadae, the arid lands of Persia later gave rise to the deep qanat, a notable underground canal, to bring water from mountain slopes to dry plains.

Current archaeological research points to the development from the first millennium BCE in several parts of the Middle East of a "shaft and gallery" approach to digging relatively shallow underground canals, many of them quite short, just a few metres. These shallow systems often dried up over time and were abandoned.

Around the middle of the first millennium AD – the Sasanian into the Early Islamic periods – Iranians developed deep qanat technology whereby more permanent sources of water deep underground were able to be identified, tapped and delivered via gravity to areas suitable for settlement and farming – up to distances of

Fig. 14.6 Excavated stone-lined water channel with sluice gate, Pasargadae archaeological site. © Iranian-French Archaeological Mission at Pasargadae. Courtesy Rémy Boucharlat

some 70 kilometres, though generally much less. This enabled much denser permanent settlement and population growth in arid areas from this period.[9]

The deep qanat also uses a "shaft and gallery" system but has much deeper shafts, and often much longer galleries (Fig. 14.8). As all labour up until the nineteenth century was by hand using traditional tools (shovels, picks, baskets), these qanats, whose underground galleries or canals were stone lined to avoid excessive water seepage, are quite remarkable feats of engineering. They remained in operation for centuries with regular maintenance, although few are still maintained and utilized today.

From the beginning of settlement, water has been a precious element in dry and arid-land gardens and orchards, both a necessity and a source of delight. Hence the idea of the garden as an earthly paradise. The remarkable *qanat* system of water delivery was a brilliant achievement of water engineers in the Middle Eastern drylands. In an area of sloping land, as in the foothills of mountains, a shaft was sunk to the water table. A second shaft was sunk, the base of which was a little lower than that of the first shaft on the slope. Shafts were then sunk in a line between the two – to enable a tunnel to be dug, and so on. Spoil was lifted in leather bags to the surface

Fig. 14.7 Excavated stone-lined water channel, Pasargadae archaeological site in springtime. © *Iranian-French Archaeological Mission at Pasargadae.* Courtesy Rémy Boucharlat

while a worker or workers below were fed fresh air. By this means, a gently sloped channel was formed to enable a steady gravity flow of water - not too fast, not too slow - from source to gardens, orchards and fields laid out in a grid pattern. Since the channel was underground, loss through evaporation was minimized, but danger to workers increased. Water use was monitored and paid for, with the highest bidder, who perhaps had financed the excavation – an early capitalist - having first claim. Thousands of *qanats* were constructed over many centuries, a small number still in use. Requiring maintenance, they sometimes collapsed, were blocked by desert sands blowing into the shafts or were destroyed in warfare and disputes. They are a testament to the ingenuity of Middle Eastern people as well as the life-giving importance of water.

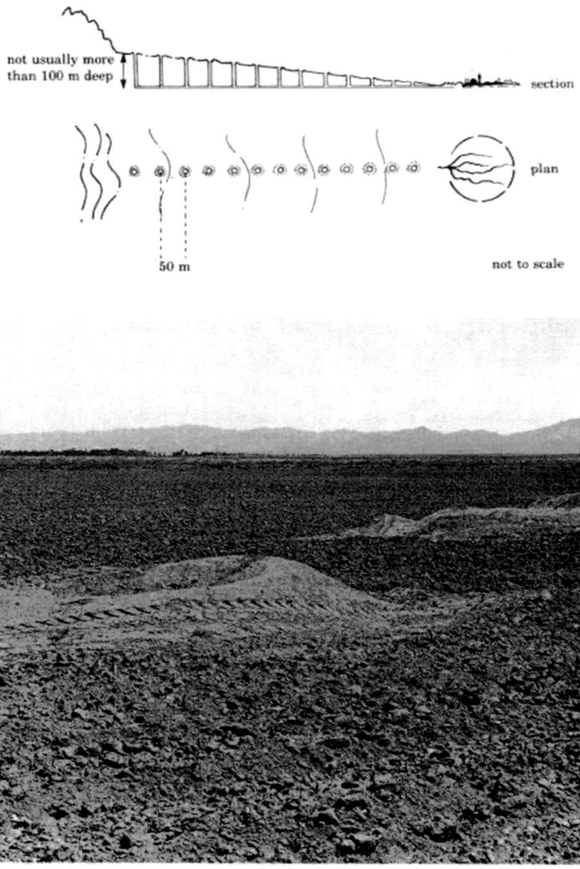

Fig. 14.8 Qanat. Transporting drinking and garden/ plantation water from the foot of mountains to settlements across the desert, a system in use for thousands of years. Plan and section of the construction of a typical qanat; qanat holes punctuate the central Iranian desert landscape. From: Brookes, J. (1987) *Gardens of Paradise. The History and Design of the Great Islamic Gardens.* London, Weidenfeld and Nicolson. p.29. Courtesy The John Brookes Denmans Foundation Trust

The imperial style of garden and palace construction included "paradises" of several hundred hectares in size as well as much smaller ones. As noted above, recent archaeological projects have revealed varied ways in which irrigation channels supplied life-giving water. Simplified and on a greatly reduced scale, they reflected (or perhaps grew out of) domestic building practice. By the second century BCE, in the hot dry lands of the Middle East, the form of house and garden in urban centres was a single, unified layout with water channels, troughs and tanks, trees and shrubs. People needed shelter from the heat and dust of summer and plants had to be irrigated. Whether very small and simple or large and opulent, the method of constructing a single, unified entity of enclosed house and garden has continued to the present with a garden, or even a single plant to provide relief. For the Swiss "wisdom" scholar, Titus Burckhardt, the modest courtyard house was not merely a device for privacy and a means to capture the downward flow of cool evening air, it had symbolic meaning, "The inner court of a house is an image of paradise. When it contains a fountain and water courses which gush forth to water trees and flowers, it does in effect recall the descriptions in the *Q'ran* of the abode of the blessed."[10] We return later in the chapter to this enduring principle of the unity of dwelling place and garden setting. Meanwhile there is more to say about relations between the earthly and heavenly paradise, both religious and secular.

14.4 Islamic Paradise

Paradise as the ultimate reward and resting place of the faithful plays a major part in Islam, the religion inspired by the message the Angel Gabriel is said to have brought to Muhammad, the son of a merchant family in Mecca, Arabia. Born in 570 AD, last of the prophets, he was believed by the faithful to be a descendant of Abraham. The message he received was uncomplicated but powerful: obedience to Allah, the one and only God (hence no Trinity as in Christianity) and observance of the holy edicts as revealed by Gabriel. These edicts were, and are: profession of faith; daily prayers; almsgiving; annual fasting (in the month of Ramadan); Hajj (pilgrimage at least once in a lifetime to Mecca). God can forgive sinners but they will be punished. There is Paradise for the faithful and there is hell, and deadly fruit therein, for the sinners. These edicts and promises, so similar to those of Judaeo-Christianity, were elaborated and detailed over time such that Islam became the basis of law, government, administration and community life, much as Christianity did over the Mediaeval centuries. Blood was (and still is) copiously shed in internal struggles over legitimacy (notably Sunni- Shi'a) and in the course of military conquests.

For Muslims, there are numerous images in the *Q'ran* of Paradise and much more enticing detail than is to be found in the Judaeo-Christian *Bible*. Moreover, there is a greater sense of sensual pleasure to come, including details of the exciting of different senses. Westerners have become familiar with the more lurid of these through the promises of almond-eyed Houris reportedly given to young suicide bombers in the jihad, or "holy war," being fought in recent times in the Middle East. The delights include reclining on luxurious couches in garden pavilions where the richly clad faithful are waited on by beautiful youths with vessels of silver, cups of crystal. The four rivers of Eden flow, with water, milk, honey and wine for all the blessed. Perfumed, those who repose in Paradise are dressed in silken garments, ornamented with gold, silver and jewels. The fruit of thornless trees is there for the plucking as the faithful enjoy balmy air in gorgeous surroundings. A verse in the *Q'ran* entices the slackers, "Ye prefer the life of the world; But the hereafter is better and more enduring."

Chapter 55, *Surah Ar-Rahman* ("The Merciful") of the *Q'ran* identifies four walled gardens in paradise, with springs, streams, the four rivers of Eden, ornamental and fruit trees, shelters, pavilions, castles, tents and houses. Not cities, not "Jerusalem", the structures of the Q'ranic paradise are spread over green, watered fields, houses are of gold and silver bricks in a landscape "as large as heaven and earth", an idyll for nomadic pastoral and agricultural peoples who had toiled lifelong in their dry fields and pastures. Likewise, these heavenly delights were for the faithful and virtuous leaders and rulers who in life on earth had borne the heavy burdens of political affairs, public administration, commerce and military command, not forgetting the strife of internecine family squabbles. They too were entitled to aspire to an afterlife in Paradise, of peaceful and calm delights.

The representations of the delights of Paradise have been likened to people's recollections of dreams unrealised, a drifting perhaps of the unconscious mind, or of delightful images and feelings in a reverie. Nature, in the form of animals, birds and flora, expresses and symbolises life. It is represented in paintings, tiles, stone carvings, jugs, vases, plates, bowls and carpets (Fig. 14.9). All are reminders to the faithful of God, the source and origin of life, and therefore of the beauty and delight of Paradise to come.[11]

The layout and furnishing of the Q'ranic Paradise and the deep, direct appeal to the senses, to bodily ease and refreshing water, speak to the conditions of everyday life in the hot, dry and dusty conditions in the arid areas of the Middle East, North Africa and Asia into which the Arab conquerors carried Islam following the death of Muhammad at Modena in 632 AD. Water, in these lands, signified not only survival but life itself. The pre-Socratic philosopher and cosmologist, Thales of Miletus, treated water as the source and principle of life, while Titus Burckhardt symbolically interpreted water in Islamic culture as "a kind of image of the soul in its fluidity and purity."[12]

The Q'ranic Paradise also reflects social and economic conditions in the countries conquered by the Arabs following Muhammad's death, the contrast between the basic if not altogether squalid life of most people, their labour and poverty and the opulent, luxurious lives of the favoured few. For the latter, as shown for example in Persian and Mogul miniature paintings on display today in leading Western museums and galleries, there were airy pavilions and palaces, ornate gardens, and garden parties, clean, flowing water, wine, milk and honey, endless leisure time, servants and enjoyment of the arts of story-telling, poetry, music, dance and painting. Even if some of this opulence was magnified by the artists and the subject of extravagant, secular tales spun by storytellers, as in *One Thousand And One Nights*, and even though conflict and warfare were a normal part of the lives of nobility (as of the foot-soldiers), it is not difficult to see that one of the great attractions of Islam, as conveyed in the verses of the *Q'ran*, was its promise to the faithful, of all classes, of eternal rewards in a verdant paradise of peace, love, friendship, leisure and ceaseless enjoyment.

A description by his sister, "Rose-Body", of festivities of Mogul Emperor Humayun's family, even including the time when they were exiled and limping toward safety in Persia, gives some idea of the sumptuousness of the lives to which

Fig. 14.9 An eighteenth century Kurdish garden carpet depicting a birds' eye view of the four rivers of Eden. A garden pavilion is indicated by the central cartouche. Bordering the water channels are stylized trees and flowers, possibly including cypresses and fruit trees. Garden squares fill the four corner beds. Metropolitan Museum of Art, New York. The James F. Ballard Collection, Gift of James F. Ballard, 1922. Accession Number: 22.100.128. Public domain

middle eastern royal families and nobility were accustomed: luxurious hangings and carpets in tents set up in a garden, feasting, poetry recitals, picture books of miniature paintings, games, visits to beauty spots and family gatherings, gifts and splendid and costly accoutrements.[13]

14.5 Other Paradises

We have touched on explicitly religious notions of paradise in the Judaeo-Christian and Islamic traditions. Dreams of paradise, of a happy and fulfilling afterlife, a coming golden age, are part of many cultures and belief systems. They include the Fields of Elysium in the Isles of the Blessed in the writings of Homer, Hesiod and Virgil and their physical revival in eighteenth century England; for example, the Elysian Fields of the grand gardens of Stourhead and Stowe. Similarly, today's Anglo-Italian paradise garden at Ninfa, near Rome, has an "Elysian quality" due to its serene beauty, charm and sense of restful enjoyment.[14] From Homeric times onwards, the image of Elysium on the banks of the river Oceanus has been represented – and realised in practice by garden makers - as one of beautiful meadows, bordering a watercourse, fanned by calming winds in an equable climate. Both Homer and Hesiod and Greek and Roman poets after them had the shades of their heroes enjoying life without pain, partaking in ease and comfort of the fruits of the earth.[15] Wealthy European landowners saw these images as worthy models for their country estates.

Ironically, the Champs-Elysées, the grand boulevard in Paris, is no longer an Elysian field - nowadays choked with noisy traffic, or angry *gilets jaunes* (protesters in yellow vests). It is heavy with pollution while pedestrian crowds bustle along broad pavements and the traffic roars past in endless streams. This is not what the ancient poets had in mind, but modern cities are scarcely poetic paradises. The Mayor of Paris, however, has declared her intention, following the summer 2024 Olympic games, to return the grand boulevard to an urban garden, as once it more or less was.

In the fanciful eternal domains of the blessed and the favoured, the after-life was to be enjoyed, whether by heroes slain in battle, the virtuous, or simply those who felt themselves to be among the deserving. When the seekers came from hot, dry lands, they looked to find in paradise shady, well-watered groves, cooling air and perpetual spring weather, with unconstrained opportunity to enjoy what they either happily experienced or were denied on earth. As we see below, those from cold, wet northern lands had different aspirations.

In the Far East, there were also Isles of the Blessed, the immortals, which, in order to reach, one hopeful Chinese emperor organised a voyage. It was unsuccessful. For the Taoists, in the Eastern Sea off the coast of China, lay the prospect of immortality. Through diet, exercise and ethical behaviour, the devout might achieve longevity, if not immortality. Crucial to the endeavour was working with nature, the "isness" of things, not struggling against natural forces and inevitability but accepting them. These Eastern Blessed Isles were seen as places of natural beauty: mountainous, lush in verdure, with beautiful flowers, and charming, peaceable animals. Water in plenty was life-giving, pleasure pavilions lined the shores. Legends had the Blessed Isles secured by great turtles (hence the representation of turtles in rock formations in the classic Japanese gardens). In later Taoism, the Isles became allegories of a spiritual quest, symbols of what could be attained through the Taoist discipline.

By contrast with such dulcet, peaceful, restful domains and conditions as we find in Islam and Taoism, in Norse mythology, Valhalla is a rough place. Presided over by the God Odin, in whose huge golden hall, enjoying a rowdy, raucous afterlife, are half of those slain in battle. The other half have joined the goddess Freya in her field. For these valorous warriors, brought to the afterworld by the Valkyries, there are boisterous field games, bloody fights with weapons, feasting on roasted boar, drinking wine and mead. Servants and the enjoyment of women as either experienced or sought for in life on earth are available to the warrior heroes in Valhalla. Like other versions of the afterlife in paradise, we find here a replica of a way of earthly life, intensified, of what was esteemed, envied or gave its inhabitants most pleasure and satisfaction in mortal existence. Valhalla, like the Taoist, Greek, Islamic and Judaeo-Christian paradise, was an image of idealised, earthly existence.

A popular retelling of both Scandinavian and Greek myths in late Victorian England was the long, narrative poem by William Morris, significantly named *The Earthly Paradise*. This poem, divided into 12 sections corresponding to the months of the year, recounts the travels of a group of mediaeval wanderers who sailed from the Northlands in search of everlasting life. Failing in this endeavour, after many vicissitudes they come at last as old men to, "some Western land, of which they had never before heard.' Shrivelled and grey, in "a nameless city in a distant sea" (a Greek colony) where people worship the ancient gods, over feasts they exchange their Norse legends with the Greek myths of their hosts.

Initially inspired by tales of "gardens ever-blossoming across a western sea where none grew old" the Northern travellers went ultimately in search of an earthly, not a heavenly paradise even though they were at the outset in quest of immortality. But they have no charts, they don't know where to go. Thinking to have "reached the gates of Paradise" they become bitterly disappointed over 30 years of wandering. Believing someone who had, at the outset, promised eternal life, they are brought to understand its impossibility - it is a delusion. But there is consolation in coming to know truth and in the enjoyment of exchanging myths and stories which, in the poem, Morris renders as a pair for each month of the year, signifying changing nature. These poems include such favourites as *Atalanta's Race*, *The Man Born to Be King*, and *The Death of Paris*. The poems are rich in images of nature and its seasonal changes: "the waste of snow", "a yearning for spring", the "mountain's brow", "misty as the autumn sea" and so forth.

Readers of *The Earthly Paradise* - if indeed there be many today - will have their own interpretation of the voyage, the quest, the difficulties and the final resigned acceptance of mortality. Morris, an ardent, proselytising, utopian socialist, poet, designer, craftsman and keen gardener spent a highly creative life in numerous and varied efforts to overcome the manifest social and cultural failures of industrial England. In his utopian novels, *The Dream of John Ball* and *News from Nowhere*, he envisaged a better world through images of beauty, and a healthy, creative, active and artistically rich earthly life. For him, paradise was a quest capable of being realised, with understanding, effort, love and communal fellowship, in the natural and social worlds.

Yet some lines at the end of Morris' poem seem to raise a question, doubts even. Can paradise ever become a reality? He apostrophises his poem,

In some old garden thou and I have wrought,
And made fresh flowers spring up from hoarded seed,
And fragrance of old days and deeds have brought
Back to folkweary; all was not for nought.
- No little part it was for me to play –
The idle singer of an empty day.

14.6 Nature and Paradise in the Mogul Empire

As both image and aspiration, "paradise" was to become a profound and enduring expressive element of the Mogul Empire, a dynasty of Mongolian-Turkic origin which came to dominate large parts of the Indian subcontinent over the course of some three centuries.[16] The Empire was founded early in the sixteenth century, in Kabul, by a Turkic prince, Babur, descended on his mother's side from the Mongolian tribal leader and conqueror, Genghis Khan and, on his father's, from the Turkic warlord, Timur (Tamburlaine), who also claimed descent from Genghis Khan.

14.6.1 Timur, Heir of the Mongol Tribal Warriors and Mogul Forebear

Although the ethnic origins of the Mogul Emperors were Mongolian and Turkic, culturally they were substantially Persian, and their religion was that of Islam, an example of military conquerors assimilating much of the governmental and administrative practice of the conquered, their religion and their literary and artistic taste. The religion of Islam, which the Arab conquerors had brought to Persia in the seventh century, was adopted by the Mongols, the Turkic people and their Mogul descendants. With some interesting variations, the Mogul emperors were Sunni Muslim, generally tolerant of Hinduism and Buddhism when they extended their imperial grasp across India, with the notorious exception of the Emperor Aurangzeb, who grew into a bitter and intolerant zealot.

At the end of the twelfth and into the thirteenth century, the Mongol tribal warrior, Genghis Khan, had led a military confederation of nomadic cattle breeders and pastoralists across the grasslands of the Asian steppes, occupying large areas of central Asia as well as parts of China. His sons, grandsons and other Mongols further enlarged these domains by conquest and vassaldom, extending their control across a vast territory, from China and Korea in the east to the Caucasus in the west.

The conquests by the Muslim Mongols and Turks, following those by the Arabs, together with the effects of Muslim missionaries, traders, and local converts ensured that, by the beginning of the sixteenth century, the Eastern Christian world and much of India and Asia had chosen or been pressured into adopting the Islamic faith. Nomadic Mongols and their descendants exchanged the round, wooden-framed and felt-covered yurtis and their rough pastures for marble palaces, hunting lodges, sumptuous gardens, orchards and cultivated fields. At least, the tribal leaders who became rulers of nations lived in the palaces while their subjects tilled the fields and served as cavalry and foot soldiers. These eastern nomadic people mingled with the Turkic nomads and, as they reached further west, with the sophisticated Persians. A new culture developed from the fourteenth century combining military power, settlement in towns, agriculture, pastoralism, industry, crafts, poetry and painting, trade with the west and the Islamic faith. It was the seedbed of the Mogul Empire, which emerged in the sixteenth and greatly expanded in the seventeenth and eighteenth centuries, then collapsed in the nineteenth.

The misleading popular image of the Mongols is of hordes of wild and bloodthirsty bandits, rampaging across the steppes, plundering settlements and murdering their inhabitants, a threat to civilisation. This is only part of the story. As well as their skill as horsemen, breeders of cattle and adaptation to demanding environments, they brought the Far East into close and creative contact with the Middle East. Moreover, they had an artistic tradition of their own, deriving from their nomadic way of life. In bold and vigorous paintings, they represented this life both materially and spiritually: encampments, family groups, hunting, fabulous creatures, demons, gods, dancing shamans, Buddhist religious symbols, wild and domesticated animals, and birds. In a two-way cultural exchange, following conquests in the Middle East they also illustrated Abraham's sacrifice, Muhammad's ascension, and Arabic and Persian literature.[17]

It was the Mongol-Turkic practice, at the demise of a ruler, to divide his domain among his sons and closest relatives. As a consequence, conflicts within the dynastic succession, and military ventures to maintain control of conquered territory, featured during the centuries of their rule. Success in one such series of conflicts resulted in Timur (1336–1405), a leading member of a Turkic-Mongol family, establishing a new (Timurid) dynasty in Transoxania (Uzbekistan). It was here that the first of the Mogul Emperors, Babur, grew to manhood.

Timur, the last of the nomadic conquerors of the Eurasian steppe, presenting himself as "the Sword of Islam", sought to restore and reinvigorate the Mongol empire of Genghis Khan and his immediate successors. If not, as he claimed, a linear descendant, he saw himself as the warrior heir of the great military leader of the Mongols. As well, he was a patron of art and architecture both within and beyond his capital, Samarkand, where he established religious and educational institutions, thereby cementing Sunni Islam as the state religion. The Turks were generally Sunni Muslims, the Persians Shi'a. Through conquest, Timur became the most powerful and feared ruler in the Muslim world. Persia was one of his conquests, as was the Delhi Sultanate, parts of Russia, India and Iraq. He developed contacts in the West, in Spain and France, which included exchange of ambassadors and trading relations. His daring exploits and ferocity in battle were legendary, both horrifying and fascinating the West, as in Christopher Marlowe's late sixteenth century play, *Tamburlaine the Great*.

Not only a ferocious military conqueror, notoriously massacring people who resisted, destroying their settlements, Timur was an empire builder in more senses than one. What he established, in blending diverse ethnic, cultural and political elements, adopting Persian court culture and fashioning a new realm, was to serve as an inspiration if not indeed a foundation of the later Mogul Empire. Gardens around palaces such as those in his capital Samarkand became public symbols of power and prestige and displays of wealth. These gardens were described by Clavigo (a celebrated Spanish envoy to his court) and, a century later by Timur's great grandson Babur (1463–1530), first of the Mogul emperors, as a mixture of orchard, vineyard and pleasure park, enclosed, quartered, irrigated, a grand setting for palace and pavilion. Clavigo, following his long journey by sea and overland from Castille, wrote enthusiastically about Samarkand as a place of beauty and distinction: magnificent buildings, long lines of trees, ornamental (plane, cypress, poplar) and fruiting (cherry, apricot mulberry), fountains, water tanks, colourful tents, fluttering banners, rich costumes and plentiful jewels worn by members of Timur's family and court.[18]

14.6.2 Babur and Humayun, the First Moguls

Babur, first of the Mogul Emperors and pioneer of its distinctive garden art, followed in Timur's footsteps as a military leader and conqueror of territory but without the fearsome image. Born in the minor kingdom of Fergana (now part of Uzbekistan), he first sought to capture Samarkand, Timur's former stronghold. Failing to hold Samarkand, with his followers he went further afield, eventually capturing and setting up his power base in Afghanistan at Kabul where in 1508 he laid down the Bagh-i-Wafa (Garden of Fidelity), best known of his gardens. From Kabul, needing more human and physical resources to feed his ambition and pay his army, Babur raided North India, eventually toppling the Sultan of Delhi. Although he never liked India, its culture or its people, from it as a base, Babur progressively extended his empire. At Agra, one of his first actions, in 1528, was to establish a riverside garden, as he had previously done at Kabul, thereby setting a trend which was to become a notable feature of Mogul rule in the plains of Northern India and the mountain valleys of Kashmir. That garden, a variation with terraces of the Persian fourfold *chahar-bagh* dating from the time of Cyrus the Great at Pasargadae, exists today, albeit altered, neglected and restored, as is the norm for old gardens.

Visiting Samarkand nearly a hundred years after Timur's death, Babur admired the city, not only as a political and military stronghold of the Timurids. In his autobiography, the *Baburnama,* he wrote of it as "a wonderfully beautiful town", its people orthodox (Sunni) in religion and pure in faith.[19] It remained in his memory as a model for the settlements he was later to establish in Afghanistan and India. The gardens, fields, orchards and natural surroundings of Samarkand were vivid in his fond recollections of the place. Several of them could well have provided inspiration for his later projects in Kabul and India. Recorded in miniature paintings, they were to remain influential throughout the Mogul dynasty.[20]

The Samarkand gardens displayed elements of the classical Persian *chahar-bagh*, the division of a walled enclosure into four symmetrical sections (squares or rectangles) with a central axis in the form of a stone-lined water channel flowing across a raised pavement and under a raised platform. This became the standard layout in a tradition extending from ancient Persia, impacted by Islam with its analogous heavenly paradise, through to the Mogul Empire and beyond. Garden historian Jane Brown called it the origin of all formal garden design.[21] Other water channels ran at right angles to the main axis suggesting the Edenic rivers of milk and honey, wine and pure water. The gardens had pavilions or some other kind of shelter raised above or overlooking the central stone platform, cut into which there might be a water trough or tank. Trees and shrubs were either scattered or in regular lines and groups and flowers bloomed in the ground beneath and around them.

The Timurid and later Mogul gardens with their origins in ancient Persia, in the imperial "paradise" or hunting park and in imperial palace gardens such as Cyrus's citadel, Pasargadae, also drew in the Q'ranic image of Paradise, the heavenly domain of the faithful with its echoes of the Garden of Eden. We should not, however, overlook, as a source, the desert oases which provided shelter from heat and dust, and cool refreshing water together with palms and grasses. That is to say, the Persian – Muslim – Mogul garden (the fourfold *chahar-bagh*) was alike a well-considered response to the environmental conditions, an expression of religious faith and a symbol of a rich artistic culture. Babur and his imperial successors appreciated these various considerations, blending them over several centuries in grand garden settings for magnificent forts, tombs, mosques, and other enclosed gardens across the imperial domains.

For Babur, more than all the emperors, vegetation both natural and cultivated was a source of great delight, as in Kabul where he planted Bagh-i-Wafa, the "Garden of Fidelity", on the *chahar-bagh* plan of a fourfold division of the land, with a central water tank and criss-crossing paths. He compared the region's fruits with those of Samarkand; gardens there, he wrote in his memoirs, "arranged symmetrically terrace above terrace", planted with poplar, elm and cypress, had few to compare for "beauty, air and view". He delighted in the crops of sour cherries, oranges, citrons, quinces and walnuts. He introduced sugarcane and plantain and arranged for the planting of trees for ornament and utility. He also remarked on the fine pasture grasses, the prevalence of wild goats, wild sheep, deer, ducks, the flight pattern of birds and ways of catching them. He thought the holmoak a most valuable tree for firewood due to its vigorous flame, aroma and deposit of fine ash. Very little in the natural environment escaped Babur's observant eye, enquiring mind and practical intelligence (Fig. 14.10).

What we know of Babur and his attitudes toward nature and gardens comes mainly from his autobiography, the *Baburnama*, based on diaries in which over years of military campaigns and imperial processions he recorded vivid impressions

Fig. 14.10 Two of the many valued trees of India, from *Baburnama*. Late 16th century. Metropolitan Museum of Art, New York. Purchase, Louis E. and Theresa S. Seley Purchase Fund for Islamic Art, Anonymous Gift, in honor of Amy Poster, and Friends of Islamic Art Gifts, 2013. Accession Number: 2013.576. Public domain

in his domains of animals, plants, gardens, agricultural production, towns, settlements and people. This autobiography, no mere catalogue of imperial possessions, was compared by British Orientalist Henry Beveridge to the greatest of the genre, notably those of St Augustine and J.J. Rousseau and the memoirs of Newton and Gibbon.[22]

Page upon page of the *Baburnama* reveal the freshness, vitality and acute interest in whatever its author experienced and heard. He was a most acute and reflective observer of the natural and social worlds. His interest in the natural world dates from childhood; he wrote of the charms of his birthplace, Fergana, in terms of the abundance of its flowing waters, blossoming tulips, violets, roses, orchards of apricots, almonds, grapes, melons and pomegranates, its productive stone quarries and turquoise mines, its abundance of hares, pheasants and other animals for hunting and fowling. Moreover, willow trees provided branches for staves and shafts for arrows. Even at an early age, Babur was demonstrating an exuberant and highly practical interest in resources of economic, military, aesthetic and recreational value. In the *Q'ran* the faithful are admonished to seek knowledge and it seems that Babur did just this in respect of plants, animals and people across his domains, as did grandson Akbar in matters of religion (below).

The person that emerges from the *Baburnama* is a remarkable individual, a successful military leader and ruler of people who was at the same time a lover of gardens, a practical gardener, and an astute and admiring observer of landscapes, mountains, valleys, rivers, pastures and cultivated fields, forests and human settlements from the smallest to the grandest. In the course of his travels, he commented on agricultural land production, orchards and well-watered and vegetated landscapes. When visiting settlements, he appraised local poets, painters, musicians, architects and builders. He wrote about palaces, shrines, tombs, mosques, temples, almshouses, hospitals, fishponds, racecourses, and the impact on conditions of life of local rainfall, wind, snow and so on. His observations had practical point; as a ruler and legislator he was concerned that his domains should be ordered, well-managed, economically productive and pleasant places in which to live. From his early education in Persian culture, he may have absorbed from the great mediaeval Sufi poets, Sadi and Hafiz, the delight of the Persian spring, of pale pink almonds flowering against the deep green of cypress trees, of apricots against bare brown hillsides, and then of the succession of bulb flowers from scilla and narcissus, thence to lilac and jasmine.

No armchair gardener, Babur is depicted in a miniature painting of one of his gardens, in conversation about planting or cultivation with one of his gardeners (Fig. 14.11). As a practical plantsman, he maintained the tradition of the Arab satraps (provincial governors) who, like Cyrus in ancient Persia, ranked husbandry with the arts of war and did some of their own fieldwork and planting. For example, he continued planting his Garden of Fidelity (Bagh-i-Wafa) in Kabul over many years, as would any serious gardener. He enjoyed being among plants and animals, with his family and subjects of similar tastes. Visiting Bagh-i-Wafa in "the days of the garden's beauty", he delighted in its verdant lawns which "were one sheet of trefoil, its pomegranate trees yellowed to autumn splendour, the fruits full red; fruit on the orange trees green to gold." Not just the vivid text, the *Baburnama* is adorned with brilliantly illustrated plants, animals, birds, landscapes and garden settings, enjoyment of which no longer requires visits to distant museums and galleries, many can be viewed online.

Through a mixture of military campaigns and imperial processions, with their encampments, and visits to gardens and places of outstanding natural beauty, Babur had many opportunities for open-air receptions, parties, feastings, poetry recitals

Fig. 14.11 The Emperor Babur overseeing his gardeners in the act of measuring flower beds at his favourite Bagh-i-Wafa (Garden of Fidelity) within today's Afghanistan, perhaps near Kabul. Mughal miniature, India. c.1590–1600. 2013GB0924. © Victoria and Albert Museum, London

and musical performances. There was also daily prayer with his Sunni congregation and reading the *Q'ran*. Garden visits and gardening were an integral part of this lifestyle.

As he extended his campaigning further into India, Babur at first found it "a wonderful country". With closer acquaintance, he modified this opinion. It lacked perennially running water on level land, irrigation was inefficient and there was an insufficiency of bullock-driven waterwheels. Concerned with quality of performance in his domains, he criticised the way waterwheels were managed, a laborious and filthy process as dung and urine from the working buffaloes were allowed to pollute the well.

But the new lands were rich in animal life including crocodile, rhinoceros, wild buffalo, hog, deer, monkey, squirrel and elephant, the latter "an immense animal and very sagacious". One parrot he described as "an excellent learner of words", whereas another had "a most unpleasant voice, like the drawing of broken china on a copper plate". His autobiography is rich in such acute observations. Descriptions are precise, often colourful, as of a jackfruit which he likens to "a sheep's stomach stuffed, and sickeningly sweet". So adhesive was the fruit that people "are said to oil mouth and hand before eating it".

Although he felt the need for imperial control, as time passed, Babur, the cultivated, fastidious connoisseur and man of action became increasingly disgusted with India and its people as a land of "few charms"; its people "have no good looks"; they lack social graces (i.e. Persian manners). The handicrafts were dismissed as deficient in form, symmetry, method and quality. At times, although the air was fine, just before the arrival of the monsoon it became unpleasantly hot and humid and fierce, dust-laden winds swept in from the north west. What then, apart from selected landscapes, fauna and flora and the gardens that he constructed and visited, was the main attraction? In a word, wealth. India was a huge land mass; it had a very large workforce; there existed "masses of gold and silver".

Despite its many defects, India was thus valued by Babur for its plentiful human and natural resources. He persisted in India, restoring and installing waterwheels, digging tanks, sinking wells, building and maintaining urban settlements, encouraging farming, mining and manufacture and maintaining order by a mix of military force, centralised administration and treaties. At Agra, disgusted by the ugly, muddy banks of the River Jumna (over which, today, presides the mighty Taj Mahal), he built the Ram Bagh, first of the Mogul gardens in India. He had stone-lined channels and tanks dug, and set up water wheels instigating garden-making with parterres and flower beds. A cool bathhouse, to whose construction he devoted close attention, was made to protect against the three oppressive elements of Hindustan, "its heat, its violent winds, and its dust". These achievements notwithstanding, Babur and his successors should not be credited with first bringing stone lining of water channels and related technologies to India. There were many local structures and ideas to build upon. For example, tanks, pools, fish ponds and even lakes, all lined with cut stone, and built prior to Mogul occupancy, were observed by the early seventeenth century Spanish traveller, Pietro della Valla, who, arriving by sea, travelled overland from Surat, reaching Calicut (Calcutta) before returning to Europe. [23]

For reasons of state that must have outweighed his personal distaste, Babur extended his Indian domain from the Indus River in the west to Bihar in the east and south from the Himalayas to Gwalior, and he made peace with the Hindu kingdoms of South India. It was left to his descendants, the emperors Akbar and Jahangir, both to stabilise the Empire and to discover the early summer delights of Kashmir, free from the dust and heat of Babur's oppressive North Indian plains. With well-watered hill slopes, there emerged for his successors an opportunity to develop another version of the fourfold *chahar-bagh* on the shores of Lake Dal and along the river valleys of the Vale of Kashmir.

Babur's spiritual home and final resting place was Kabul where in 1646 his descendant, Emperor Shah Jahan, built him a mausoleum in a long terraced garden which in Babur's lifetime had served as a place of relaxation and pleasure, as was to become the norm for tomb gardens. He had launched a great new dynasty and founded an empire. With varying degrees of success, his descendants progressively built on these beginnings, developing and strengthening governance and administration, drawing upon the indigenous culture and creating for a large part of the Indian subcontinent a unified state. At Babur's death, the empire was more incipient than extensive. It fell to his successors, and most of all to his grandson, Akbar, to consolidate and ensure durable control. Others, notably Jahangir and Shah Jahan, were to develop to the full the artistic and cultural potential that was never far below the surface of the empire and the societies it ruled (Table 14.1).

Babur's son and successor, Humayun, was something of a dilettante. Unlike his father and his son Akbar, he had no great gifts as a military leader or governor. He was at any rate a weak ruler who succumbed (like some later Moguls) to alcohol and opium. Deposed, he fled to Herat where, in return for support from the Persian, Shah Tarmas, to regain his kingdom, he was obliged to declare public allegiance to Shi'a Islam, while retaining his personal Sunni convictions. He returned from Persia, characteristically with two famous Persian artists to embellish his court. The years he and his family and followers spent in and around Herat were important, on their later return to India, in refreshing the flow of Persian art, architecture and literature into the Mogul Empire.

Humayun was a patron of the arts including Persian poetry and miniature painting. He was said to have in his possession the world's greatest collection of illuminated manuscripts. His premature death resulted from a fall on the stairway of his

Table 14.1 Reigns of the Great Moguls. Based on Gascoigne, B. (1987) *The Great Moghuls*. New Delhi. Time Books International. p.250

Reigns of the Great Moguls 1526–1707	
Mogul Emperor	**Reign**
Babur (1483–1530)	1526–1530
Humayun (1508–1556)	1530–1556
Akbar (1542–1605)	1556–1605
Jahangir (1569–1627)	1605–1627
Shah Jahan (1592–1666)	1627–1658
Aurangzeb (1618–1707)	1658–1707

library, when he turned in response to the muezzin's call to prayers. Whatever his limitations as an imperial ruler, Humayun sustained the cultural thread from the Persians and the Timurids which was to attain such fine expression under his successors. He is commemorated in a splendid tomb in Delhi, commissioned, in 1569–1570 (Fig. 14.12). Set in a large *chahar-bagh*, the first of a succession of monumental Mogul tombs in India in a grand, formal garden setting, this tomb garden cemented what was to become a widely used royal and aristocratic style of symbolic religious expression in monumental architecture.[24]

In the classic Persian four-fold, geometrical layout, Humayun's tomb garden domain spread over some 30 acres, while stone-lined water channels cut across pavements and disappeared under the tomb building. Re-appearing on the opposite side, they alluded to the Old Testament and Q'ranic verses on the four rivers flowing beneath the Garden of Paradise. Elizabeth Moynihan, American garden designer, symbolically interpreted this tomb, commissioned by Humayun's widow in association with Emperor Akbar, "The large square enclosure, divided with geometric precision, was the ordered universe; in the centre, the tomb itself rose like the cosmic mountain above the four rivers which were represented by the water channels. In the perpetual growing season of Hindustan, there were eternal flower and fruit, water and the perfume of Paradise."[25] Like the well-furnished tombs of the Egyptian pharaohs and nobility, those images of paradise, the symbolic trees of eternal life in the Mogul tomb gardens were indicative of a hoped-for after-life. Perhaps the fruit trees were to provide nourishment for the intended journey to Paradise.

Fig. 14.12 Humayun's Tomb and garden, Delhi, restored 2014. swati bhati, CC BY-SA 3.0, via Wikimedia Commons

The paradise image and design was a widely used ancient decorative motif not just in gardens and tombs but in carpets. For example, the Sasanian prince of Persia, Chosroes 1 (531–579), had an immense carpet sixty yards square, called *Spring*, and woven on the plan of a royal pleasure garden. It had streams, paths, trees and flowers and was worked in gold on an earth-coloured ground. The leaves of plants were in silk and coloured stones formed fruit. A central tank was enclosed by paths, canals and parterres. Such designs prevailed for centuries in royal circles, continuing to the *chahar-bagh* designs of Middle Eastern rugs today. Carpets designed in this fashion, together with decorative ornaments, paintings and so on, bring the beauty of the garden into the house.

14.6.3 Akbar, the Greatest Mogul

Akbar (1542 – 1606), Humayun's son, succeeded to the throne in 1556 at the age of 13. He is generally regarded as the greatest of the Mogul emperors, at least in terms of consolidating and further extending the military domain that he inherited, and establishing a more or less unified civil state with a strong system of central administration and a vigorous economy including trading relations with European powers. Thus he was effectively Babur's successor as empire builder, achieving a secure system of governance and imperial authority in a region of scattered petty kingdoms and competing chieftains.[26] Among many other achievements, he had the *Baburnama* translated from its original Turkic into Persian, and copiously illustrated by the greatest artists of the age, drawing on Babur's descriptions of humans, animals and places.[27]

Marrying a Rajput princess, who became mother of the next emperor, Jahangir (who in turn married a Rajput princess), Akbar sought to assimilate Hindu civilisation into the Mogul blend of Mongol – Persian – Turkic – Islamic culture. In its development up to and including the rule of Shah Jahan, the Empire achieved great distinction in architecture, monumental building construction and garden design. If largely Persian and Timurid in origin, these became distinctively "Mogul", with the employment of architects, designers and - especially craftsman not only from Persia and Afghanistan but from India and all parts of the expanding Empire. Significantly, Akbar, a Sunni Muslim who politically anchored Islamic state identity, had a deep personal interest in other religions, inviting Hindus and Christians, together with mystical Sufis, to meetings. These activities disturbed the orthodox Sunnis in his entourage, but he persisted in pursuits that appeared to separate his personal religious convictions from those of the state.

No scholar, even said to be illiterate, Akbar often led animal hunts as well as military campaigns, rode a fighting elephant, wrestled, and took pleasure in visiting gardens. He may have been illiterate, but he greatly enjoyed storytelling, and had his grandfather's *Baburnama* read to him. He had a large library and art collection, with a retinue of painters, translators and writers, commissioned translation of Sanskrit literature and encouraged learning. He is credited with forging, by his patronage, a distinctive Mogul style of painting, drawing on Indian as well as Persian and Timurid sources. It was naturalistic rather than the repetitive patterning of Persian painting, with manuscript illustrations, court scenes and portraits.[28] He believed that painters were divinely inspired in a unique way, "There are many that hate painting, but such men I dislike. It appears to me as if a painter has quite peculiar means of recognising God, for a painter in sketching everything that has life, and in devising its limits, one after the other, must come to feel that he cannot bestow individuality upon his work, and is then forced to think of God., the giver of life, and will thus increase his knowledge."[29]

The paintings illustrating the *Akbarnama* and translations of the Hindu epics in Sanskrit, the *Mahabharata* and the *Ramayana*, were peopled by realistic figures. Indian miniature paintings, inspired by ancient legends, philosophy, religion and poetry, came to the fore. Less refined than the Persian style, the work was vivid and colourful; subjects included landscapes, lovers in gardens, animals, birds, trees, flowers and devotional scenes. "Each miniature is an expression of the soul and its sensitivity, its admiration for natural beauty and an illustration of its most secret dreams and its highest aspirations."[30]

Akbar was said by the translator of his biography to be "both ruthless and self-indulgent"[31], but this comment must be set against a many-faceted life. Whether or not related to his idiosyncratic religious beliefs, on one occasion when in the midst of a great hunt he ordered the encircled animals and birds to be released, "not the feather of a finch was to be touched" and he distributed alms and gold to faquirs.[32] This was but one of many marks of a strong, idiosyncratic character. He ordered a garden and temple to be erected at the site of his spiritual experience. Among his religious rituals were pilgrimages to tombs of Hindu as well as Muslim saints, a testimony also to his statesmanship in a multi-faith empire.

Another curious period in Akbar's life was the construction of a new royal city, Fatehpur Sikri, on a hot, stony slope some forty kilometres west of Agra. A palace and garden complex, including a magnificent mosque and houses, it was built to celebrate the long awaited birth of an heir, Salim, the later Emperor Jahangir. Visited today, the remains of this remarkable structure of rich red sandstone surmounted by a white marble tomb, all in a predominantly Hindu style, stand empty and almost forlorn against the red soil of the semi-arid plain.

In a distant echo of Kublai Khan's interest in religious diversity at Xanadu, under Akbar's leadership, assemblies took place at Fatehpur Sikri, according to his biographer Abu-L-Fazl, of "the wise of every religion and sect." In exposing the "bigoted Ulama and the routine lawyers" and choosing the course of enquiry and reflection in preference to accepting revealed truths, Akbar might seem to have rejected the religion of the book (*Q'ran*). At any rate, he abandoned orthodox belief in favour of "insight" or "that which without schooling illuminates the pure temple of the heart. The inner soul receives rays from heaven."

The mystery of the building and very short occupancy of Fathepur Sikri as a capital (ten years) is not adequately explained by his biographer and has puzzled historians. When he was already at work on the massive Red Fort at Agra, why did Akbar launch out into the semi-arid, fiercely hot and dry plain for another lavish, costly structure? Perhaps he wanted seclusion from prying orthodox eyes to pursue his heterodox religious interests – or just a "country retreat"? Again, why leave Fathepur Sikri deserted after less than two decades? Was it because of lack of water? Or the political pressure to be seen more at Agra? Since the city was established on a site where a saint had forecast that a son and heir would be born to Akbar – as he duly was – Fatehpur Sikri may have been a grand gesture of humble thanksgiving, not envisaged as a permanent royal power base.

For whatever reasons, Akbar abandoned Fatehpur Sikri, resettled his court and, with his son Salim, instituted the practice of spending the summer months far from the blistering plains, in the well-watered vale of Kashmir, following its conquest in 1586.

In the words of Abu-L-Fazl, capturing the spirit of Akbar's journey into Kashmir: "The world's lord keeps his eyes open for wonder and regards the old world as a fresh ornament of the Creator... His heart is especially drawn to any place where there is wondrous work of destiny. Hence he always has Kashmir in mind." He looked upon plants, his biographer wrote, as "one of the greatest gifts of the Creator." Akbar was charmed by the entry to the mountain province, "From Bhimbar to Hirapur there is a continuous range of hills which for narrowness and difficulty, and for ascents and descents, is unrivalled. The groves, the blossoming flowers, the glorious air, the melody of the waterfalls, increased astonishment ... and removed from the heart the troubles of the journey. But when we on the day passed from the hills to the plain, there was a splendid spectacle. A new world appeared, and a new paradise withdrew the veil from her countenance."[33]

That Akbar was able to undertake six week-long journeys there and back with his court and followers and to spend lengthy periods on reaching remote Kashmir is a tribute to the military authority, relatively settled governance and the organised administration established under his reign. Equally, it is a tribute to his determination to cement his hold on this beautiful province, as to his desire for a safe and comfortable journey to and from it, that he employed literally thousands of rock splitters, stone cutters, diggers and others to smooth and level the precipitous tracks through the mountains. Armies as well as his family and court traversed this route.

In Kashmir, Akbar indulged his love of hunting and water fowling and visited the region's valuable saffron fields around Lake Dal. It was in the surrounds of this lake that Akbar, the court and followers, planted or encouraged the planting of hundreds of gardens. Set against ochre, pink and purple mountains, they extended down the slopes from the foothills to the lakes and river banks. Many of them resulted from the efforts of his female relatives – a family affair.

By contrast with the smaller, enclosed *chahar-baghs* of the North Indian plains, which cosseted meagre flows of water in narrow stone channels, the Kashmir gardens commonly had access to flowing mountain streams which were harnessed to feed fountains, water stairs and falls. These produced dazzling displays of tumbling water, catching the light and filling the air with the welcome sounds of a plentiful and continuous flow. Those remaining today, which include the splendid Shalimar Bagh, give the impression of large, orderly spaces, cut by long regular lines of water channels, pathways and planting – faint echoes of distant Pasargadae and Cyrus' "paradise." The imperial gardens contained pavilions, some still standing, with viewing platforms behind which were enclosed, private areas with provision for the harem and the emperor's private entertainment. Figures of birds and animals were carved as ornamentation of the fountains, much as in Renaissance Italian gardens.

On the palace wall of Verinang Bagh in Kashmir, the favourite garden of Nur Jahan, the powerful consort of Akbar's son, Emperor Jahangir ("World Seizer"), a religious memorial was inscribed, "The king raised this building to the skies: the angel Gabriel suggested its date – 1609." Seven years later the mason's tablet read: "God be praised! What a canal and what a waterfall! Constructed by Haidor, by order of the King of the World, the Paramount Lord of his Age, this canal is a type of the canal in Paradise, this water fall is the glory of Kashmir."[34]

Inscribed on one of the pavilions is a famous Sufi Persian text: "If there is paradise on earth, it is here, it is here, it is here." It may be read either as a pious utterance of one of the faithful envisaging the heavenly paradise of the *Q'ran* or as an expression of delight in the earth and its wonders, or as both. Paradise in the heavens may be hoped for, but for the Moguls it certainly existed among superb mountains, gardens of flowing water and beautiful plants around Kashmir's Lake Dal.

This feeling of an earthly paradise in gardens and among nature seems like an echo of those mediaeval Sufi poets, Hafiz and Sa'ad, whose work similarly lends itself to both literal and metaphorical interpretation:

The forest, letting me walk amongst its naked limbs, had me on my knees again, in silence shouting – yes, yes, my holy friend, let your splendor devour me.[35]

Such metaphorical if rather less poetical uses of plants, gardens and the natural environment are found in the numerous flower (not to say flowery) references that are scattered throughout the *Akbarnama*. Thus Akbar's reign was "the period of decking the garden of dominion and prestige" and "the parterres of the Caliphate are watered by the stream of his scimitar and by the hexagonal earth irradiated by his banner's Crescent", while (in an allusion to north India's challenging environment) "the gleaming ensign of his greatness lightens up the dark, dust-stained world." Then there was the meeting, called for "the decorating of the young plant of the Divine rose garden, to wit, for the circumcision of his Majesty." And "Thanksgiving to Providence who had made fortune's rose bloom in the thorn-brake of misfortune." Exaggerated and overblown as this language (in translation) is to readers today, it was doubtless part of a long established rhetorical convention which authors prudently observed when writing about the ruling emperors and their families.

14.6.4 Jahangir, Master Garden Maker

Akbar's tomb was erected by his son, Jahangir (1569–1627), between 1606 and 1617. Of red sandstone and white marble, it was set in a *chahar-bagh* at Sikandra a few kilometers from Agra (now in the suburbs).

In a 1632 sketch by English traveller Peter Mundy, the tomb is set in a walled garden thickly vegetated with trees, shrubs and what look like flower beds

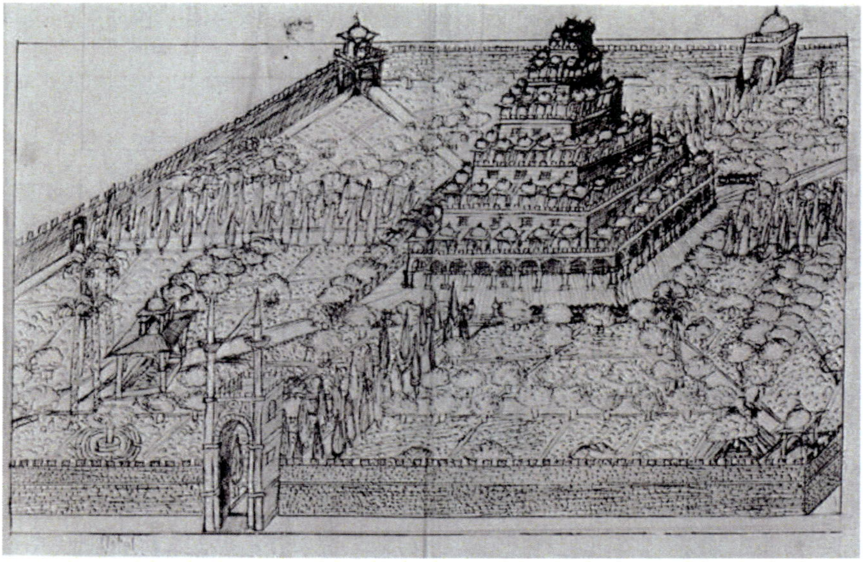

Fig. 14.13 Akbar's Tomb, Sikandra, showing layout of classical Mogul garden. This sketch by Peter Mundy from around 1632 indicates the richness of the original planting. Courtesy Hakluyt Society

(Fig. 14.13). During Aurangzeb's reign the tomb was plundered and even Akbar's bones were scattered. In the colonial era, under Viceroy Lord Curzon, it was restored and is today one of the tourist sites of Agra. [36] Mrs Villiers Stuart, in her pioneering study of Mogul gardens, drew out Indian as well as Persian sources, suggesting an association in the garden layout with an element of Hindu mythological geography of the world - Mt Meru in the middle, from which waters of a hidden spring flow, in all four directions of the compass. A sacred tree grows (of knowledge or good and evil) with Naga, the holy water snake, coiled at its roots. The magic, sacred, mountain, holy tree, sacred spring, guardian snake, she wrote, all featured in ancient conceptions of Paradise – not only those of the *Book of Genesis*. In her interpretation, the tree on the mount was replaced by a temple, then the Muslim tomb. Whatever place such an interpretation may have occupied in the mind of Jahangir and his designers, it evokes the fusion of cultures which emerged during the Mogul dynasty. It was on a visit to Kashmir that he ordered a sacred spring to be edged with stone and enveloped in a garden, with a canal and places of worship for recluses (the Vir-nag i.e. Snake Spring).[37]

During the reign of Jahangir, architecture, landscape design and gardening took their place among the pre-eminent arts and forms of religious expression, and as settings for pleasurable events of the Mogul courts. Jahangir is said to have commissioned more gardens than any other Mogul, mainly for sport and pleasure. Aided, if not dominated, by his wife, Nur Jahan, he devoted much of his life as emperor to the arts. Jahangir also, in Akbar's footsteps, succeeded in bringing together the Muslim overlords and their Hindu (and other) subjects in a generally harmonious mix. Jahangir also adopted the Hindu royal custom of having himself weighed in the scales against gold, silver, jewels, cloth, rice, bulbs and so on which, as observed by Sir Thomas Roe, the (unsuccessful) British envoy to the court, he then distributed among the Brahmins.[38]

Through his father Akbar, Jahangir developed a keen appreciation of the Kashmir landscape which in his memoir he called "a garden of eternal spring … a delightful flower bed, and a heart-expanding heritage for dervishes." Together with his son, Khurram, (later the Emperor Shah Jahan), he developed the splendid Kashmir garden, Shalimar Bagh, which survives to this day, although in an altered state with a road passing through, on the north eastern shore of Lake Dal. Nur Jahan, who was a powerful figure in imperial politics, travelled with Jahangir to Kashmir where she designed gardens including her own, Jarogha Bagh, on Lake Manasbal. She also rebuilt Babur's Ram Bagh at Agra.

Francis Bernier, the French court physician who accompanied Emperor Aurangzeb on his only trip to Kashmir, wrote "It is not without reason that the Mogols call Kachemire the terrestrial paradise of the Indies."[39] Bernier observed that the, "whole kingdom wears the appearance of a fertile and highly cultivated garden … an enchanting scene." Even so, he found the quality of the fruit inferior to that of his native France!

Jahangir's greatest garden, Shalimar Bagh in Kashmir, was described by English landscape architect, Sylvia Crowe, as "one of the loveliest gardens in the world"[40] (Fig. 14.14). It rises, she wrote, above Lake Dal, from which in a boat the Emperor

Fig. 14.14 Shalimar Bagh, Kashmir, creation of Akbar and son Jahangir. A "new paradise." Pradeepkjoshi, CC BY-SA 4.0, via Wikimedia Commons

is said to have admired his masterpiece. The deep, dark waters of the lake with their floating islands of vegetation and flower-bespattered surface were a kind of overture. Above them, through rippling stone watercourses, stone ramps (chadhars) and falls, the broad stream of water passed under the emperor's Hall of Public Audience, his Hall of Private Audience, his private chamber and the zenana, or women's quarters. The zenana, of black marble, was built later by Jahangir's son, Shah Jahan. The great canal of glittering tumbling water with its fountains and falls, passed through an avenue of stately chenars (plane trees) on its way to the lake.

The Kashmir Shalimar Garden has some correspondence with the Q'ranic description of paradise – a series of pavilions of ease and delight in a beautiful, well-watered garden. But whatever prayers may have been intoned and pious sentiments voiced, Shalimar and the other gardens and their pavilions in Kashmir reveal a deep affection for the earth and its beauties of flowing water, luxuriant vegetation, majestic trees, mountains and colourful flowers. The diaries, biographies, written accounts, paintings and other artefacts may suggest divine grace to the faithful but unquestionably they revel in the beauties of nature.

Writing of the Kashmir gardens collectively, Sylvia Crowe emphasised that: "Site and level, prospect and orientation, planting, and above all water in all its forms, from the reflecting pool to the spectacular waterfall – these were the real constituents of the Kashmir garden, to which the buildings were incidental."[41] Shalimar Bagh was said to have had seven terraces, down which the water tumbled, corresponding to the seven degrees of Paradise in the *Q'ran*.

Jahangir's memoirs record his close interest in the natural world, similar to that of his forebear, Babur. From an early age, he became an excellent calligrapher and took a close interest in botany, history, geography and art. In Kashmir, he studied saffron plants, flowering shrubs, trees, animals and landscapes. A thoroughgoing naturalist, Jahangir dissected the wind-pipes of birds to investigate their singing and cut open a snake that had swallowed a hare. He discoursed on animals with albino variation, on the mating of pet cranes and on the hatching of chicks. He commissioned paintings of birds and other creatures. In Kabul, he enjoyed delicious and plentiful peaches; even so, none of its fruits could compare with the Hindustani mango. His memoirs yielded nothing to the *Baburnama* in the close observation of the products of nature, both wild and cultivated and in sensual pleasures, whether sniffing attar of roses, tasting pineapples, or delighting in a landscape of meadow, river and mountain – and discoursing on the breeding of goats.

14.6.5 Shah Jahan – Creator of the Taj Mahal

Jahangir was succeeded by Shah Jahan (1592–1666), creator of the Taj Mahal. From his mother and grandmother, Rajput blood flowed into his veins. The impact of Hinduism on Mogul culture is shown in elements of the design and finish of the Taj whose masons and craftsmen included numerous Indians. As a builder and garden maker, Shah Jahan's greatest triumph was the Taj, erected on the banks of the

Fig. 14.15 Oblique view of the Taj Mahal and gardens at Agra, set alongside the Yamuna River. Ujjwal goel2005, CC0, via Wikimedia Commons

River Jumna (Fig. 14.15). As well, he was responsible for work on the Red Fort in Agra and the Shalimar Bagh at Lahore, the best surviving Mogul garden outside Kashmir.

Now a UNESCO World Heritage site, the Lahore garden was built on the model of his father's, Shalimar Bagh, in Kashmir but, on the Afghan plains not the slopes of Kashmir; irrigation was by flooding of the sunken beds, not the tumble of mountain streams. This method of heavy watering of flat land just a few times a year is used by horticulturists today, in semi-arid lands. Enclosed in high ornamental walls, the Lahore garden is large, spacious, terraced, with pools, tanks, fountains, carved water chutes, a broad canal, pavilions, trees, shrubs and flowers and extensive lawns. Built between 1637 and 1641, this splendid garden has since passed through challenging times, like practically all of the Mogul gardens, but it is now restored as a monumental pleasure park for the enjoyment and relaxation of the public.

Shah Jahan's connoisseurship extended from magnificent buildings and spacious formal gardens to precious stones, of which he had a remarkable collection. He built a "country house" near Delhi but according to that rather acerbic critic, the French traveller and court physician, Francois Bernier, it did not reach the standard of Versailles or Fontainebleau. Bernier praised the Mogul practice of planting trees in their gardens but in his estimation their landscapes did not equal the peerless view from Pont-Neuf in Paris!

In common with his predecessors and successors, Shah Jahan was a military campaigner. In his last years, he was a prisoner of his murderous son, Aurangzeb, in the Agra Red Fort from where he was able to gaze at but never again to visit the Taj Mahal, the tomb-garden he had erected as a mausoleum for his beloved wife, Mumtaz Mahal. Like earlier emperors, Shah Jahan had passionate artistic interests. These are no more evident than in the ethereal beauty of the Taj, its exquisite balance, symmetry and sense of suspended time, its superb dome, seeming to float above a delicately moulded cube, its lustrous inlaid white marble and its elegant long rectangular garden layout. The architect is unknown and may even have been Shah Jahan himself. Certainly he took a close interest in its construction and would have approved the final design. He patronised artists who created brilliant, sharp images in miniature paintings of flowers birds and animals. Some of these feature in the coloured stone inlays of the Taj Mahal, in striking contrast to the glowing marble.

On tiles, as on carpets, in miniature paintings, on vases, plates, embroidery, pavements and platforms, and around the dado of the Taj, the idea of paradise is expressed in Persian and Mogul art through images of angels, houris, land animals, birds, fish, trees, fruit, flowers and running water, all conjoined in harmony. The entwined cypress (eternity) and fruit tree (beauty, sustenance) symbolise life, hope. The exquisite patterning of inlaid semi-precious stones on the white marble surfaces of the Taj speaks eloquently of the artistic tastes of the Emperor as of the Indian and other craftsmen, whose descendants still work on the streets around the Taj carving and inlaying pieces of marble for sale to tourists. Bernier, despite his unfavourable comparisons of Mogul artistic achievements with those of Paris and Versailles, regarded the Taj as "an astonishing work", more worthy as a wonder of the world than the pyramids, "the unshapely masses and heaps of stone in Egypt"[42].

There is a story that Shah Jahan had planned a black marble tomb to himself paralleling the Taj Mahal in an extension across the river. Further research, by Elizabeth Moynihan among others, on foundations in the riverbank suggests that what was planned was not another tomb but an enclosed garden corresponding to the one fronting the Taj Mahal, the Jumna River (presumably with cut stone banks) providing the right-angled crossover of the traditional, religiously inspired *chahar-bagh*.

Completed in 1654, the design, layout and construction of the Taj Mahal derived from several sources; the dome, also used by Nur Jahan in the design of her father's tomb (*I'timad-ud-Daulah*) at Agra, was Persian. Some commentators have observed Indian influences and they are certainly apparent in the stone carving and ornamentation of the building, including the lotus flower bud that atops the dome. The garden design is in the Persian *chahar-bagh* tradition as influenced by Islam and modified over the centuries (Fig. 14.16). The total composition is, however, uniquely Mogul, a monument as much to Shah Jahan himself as to his beloved wife. Over the entry arch, inlaid in black marble, is a message for "the pure of heart" who are invited to enter "the Gardens of Paradise". The Taj Mahal, a munificent gift to posterity, is unquestionably one of the world's greatest architectural and landscape masterpieces. While 22 years and 20,000 workers were required for its completion and its cost contributed to a crisis in the Empire's finances, it continues to cast glory - reflected from the shimmering marble - on the artistic and nature loving dynasty which it so well symbolizes.

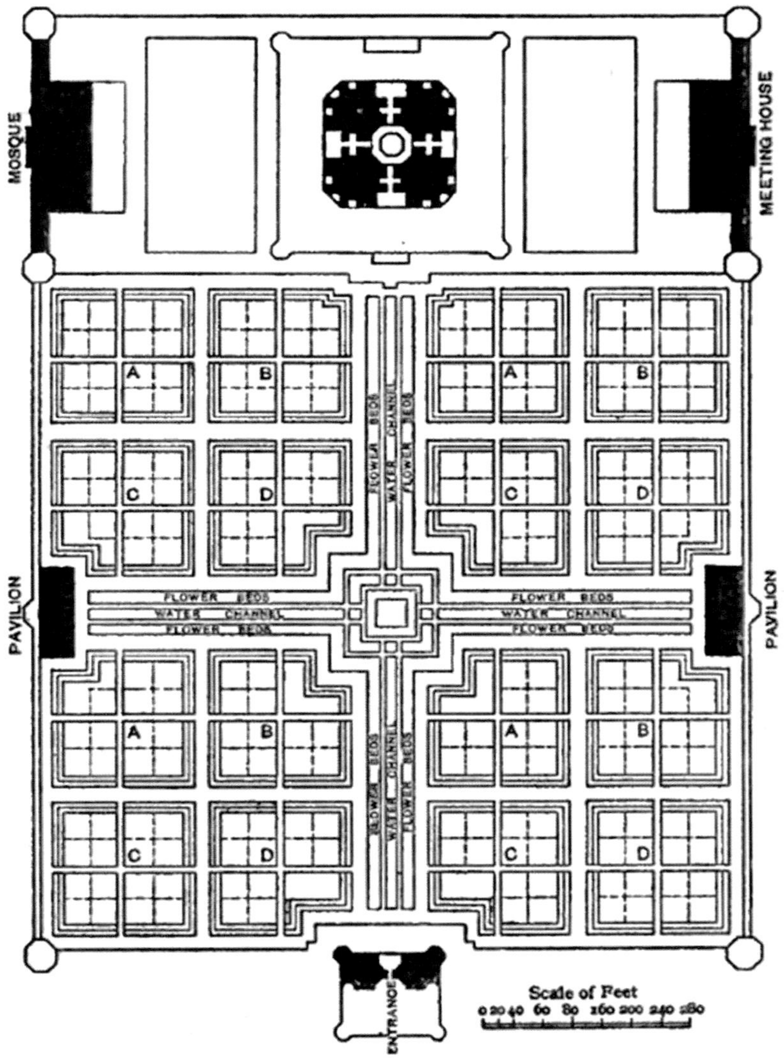

Taj Garden.

(Colonel Hodgson's Plan.)

Fig. 14.16 Plan of the fourfold Mogul Garden at Taj Mahal, prepared in 1878 by the Surveyor-General of India, Colonel Hodgson. Villiers Stuart, Public domain, via Wikimedia Commons

14.6.6 Aurangzeb, the End of Empire and the Aftermath: The Last "Mogul" Garden

With the advent to the throne of Shah Jahan's son, Aurangzeb, through murder of his brother and imprisonment of his father, the tradition of political expansion by military force continued. But the new emperor was religiously intolerant, imposing Sharia law across the whole Empire. Orthodoxy had been prevailing under Shah Jahan but not the harsh intolerance that was to come. Aurangzeb, who destroyed Hindu temples while erecting mosques, is credited with few if any significant contributions to the rich, Persian-inspired cultural tradition which his predecessors had preserved and enriched. Unlike them, who had been brought up in the tradition of classical Persian culture, Aurangzeb experienced in his early education a mix of practical studies and modern subjects. Potentially useful, for Aurangzeb they failed to develop the personality needed to sustain either the riches of his cultural inheritance or a stable, prosperous Empire. Becoming increasingly morose and tyrannical, unable to share power, he brought the golden age to an end and was succeeded by minor figures under whom the Empire fragmented, coming to an end in the nineteenth century. A devout, orthodox Muslim, Aurangzeb persecuted Hindus as infidels and had none of the interest in religious diversity of Akbar. In old age, anticipating the crumbling of Mogul order in Hindustan, Aurangzeb wrote to one of his sons, "I myself am forlorn and destitute, misery is my ultimate lot."[43] As described by one historian he was "a Puritan of overweening ambition".[44] Harking back to the ruthless and murderous actions of his ancestors, Genghis Khan and Timur – but not to the Mongol-Turkic love of the arts - he was unable to sustain the civil and cultural values that his Mongol-Turkic forebears had absorbed from the societies they ruled. The last of the Mogul emperors, Bahadur Shah I, a puppet of the British in the colonial era, was propelled to reluctant leadership of the Indian Mutiny (the first War of Independence). As a result, in 1863 he was banished from the Red Fort in Agra to Rangoon.

There was to be one last outpouring of the great Mogul tradition of garden-making. This was the early twentieth century garden designed for the new Viceroy's Palace in New Delhi not by a Persian, Mogul, Turk or Indian, but by the distinguished British architect and garden designer, Edwin Lutyens. Together with another prominent British architect, Herbert Baker, Lutyens was commissioned by the British government in 1913 to design and oversee the construction of a new centre of imperial power in India, New Delhi. Baker designed the administrative buildings and was involved in the layout of the new city while Lutyens, who was married to the daughter of a former viceroy, had responsibility for the Viceroy's Palace (now the Presidency of the Republic of India) and its extensive formal gardens and lawns.[45] The grand Presidency architecture is not without wit as Lutyens included features in carved stone which echoed a signature drawing of him by William Nicholson wearing a large pair of spectacles. He knew that some of the visitors would recognise the architect in pairs of prominent stone circles around entrances and exits.

An aerial view of the vast open space (some 139 ha.) including the buildings, lawns, trees, water courses, pavements, gravel areas and flower and shrub beds, shows how Lutyens in the overall Presidency project managed an impressive blend of classical western and Indian architecture with a Mogul-style garden of some 6 ha., unifying palace and its garden setting into a single harmonious entity (Fig. 14.17). While it is not obviously a *chahar-bagh* and is named on his plan *Imperial Delhi Viceroy's House and Formal Indian Garden*, there were no suitable Indian garden models to draw on, whereas Lutyens visited Mogul gardens and sites in Kashmir, Lahore and Agra and consulted an authority on Mogul gardens, Mrs Villiers Stuart, who had rushed to complete her book, *Gardens of the Great Mughals*, in order to influence decisions about the design and construction of New Delhi.

From early in his career, Lutyens' ideals were Palladio, Wren and Renaissance Classicism. Like Herbert Baker, as a young man he had been an assistant in the office of architect, Harold Peto, who became well known for his Italian-revival style of garden design. This was a time of burgeoning Italian neo-classicism in England, when the super rich likened themselves to the merchant princes of the Italian Renaissance and there was a vogue for the neo-classical paintings of Lawrence Alma-Tadema and Lord Leighton. In Lutyens' years of working in England, in collaboration with the famous English garden designer, Gertrude Jekyll, a unique landscape-garden style combining formal layouts and cottage style planting had

Fig. 14.17 Modern "Mughal" Garden of Rashtrapati Bhavan, official residence of the President of India in New Delhi, India. Likhith N.P, CC BY-SA 3.0, via Wikimedia Commons

emerged, combining formal landscaping with romantic planting.[46] These influences are apparent in the New Delhi Presidential Garden. The building has the feeling of a blend of classical Indian and western architecture whereas in the garden Lutyens produced what we might think of as a lushly planted latter day grand, formal "Mogul" garden.

Designed as an interlocking series of long canals, two running north-south and two east-west and a grid of squares with tanks, fountains, pathways, tree avenues, hedges, shrub borders, rose garden, stone pergola and flower beds, the garden is a ceremonial space where, as once in Mogul times, official events, receptions and parties are held. The garden also provides a private retreat for president, family and entourage. It still performs many of the functions of the imperial Mogul gardens while capturing the spirit of great Indian and Mogul traditions which some at least of the emperors had managed to bring together in harmony and mutual enrichment.

14.7 Paradise Regained?

Unlike the heavenly paradise of Islam and Judaeo-Christianity, the Mogul imperium was hardly peaceful for any great length of time. Warfare and poverty left numerous scars on the rural landscape and in the towns and cities. Religious tolerance and scepticism prevailed at times but so also did Sunni religious orthodoxy and intolerance. The earthly paradise of the Moguls was, like all empires, sordid and cruel in parts. Over time a cohesive Empire did, however, emerge, with substantial progress in governance, public administration, economic development and the arts. The standard histories of the Empire and of the great gardens and monumental buildings tell us little about the mass of people outside the court. The emperor and his entourage practised the charity enjoined on them by their religion, similarly they cherished the army, as Akbar's biographer indicated they must in his "eight essentials of Empire". Furthermore, they acknowledged other imperial "essentials" in their great designs and their political administrative functions.

According to Akbar's biographer, Abu-L-Fazl, when writing of the birth of the future Emperor Humayun, there were "eight essentials of empire". These were mostly of a political character:

1. High fortune
2. Great designs
3. Conquering power
4. Administrative capacity
5. Civilising facility
6. Devotion to the welfare of God's servants
7. The cherishing of the army
8. The restraining it from evil

Humayun was deficient in several of these, whereas his son, Akbar, excelled in most if not all of them. Babur made a strong start. Succeeding emperors enlarged and consolidated the empire until the British instituted another, albeit comparatively short-lived, Indian Empire under Queen Victoria.

Abu-L-Fazl's catalogue of essentials of empire, does not identify other great attributes of several of the emperors: the capacity to find contentment and beauty in the earth and to enjoy the natural world; and the aesthetic sensibility that finds expression in superb artistic creations ranging from exquisite miniature paintings and mystical poetry to brilliant architectural designs and noble structures, gardens among them. These were not seen as optional extras by the succession of emperors and their families; they expressed the essence of a good life under sound imperial rule. They owe much to the Persian inheritance dating from the second millennium BC which included, in the decorative and literary arts, representations and images of animals, birds, plants, idealized landscapes and enjoyable human life in beautiful surroundings. Without military conquest and effective government, sustained over the centuries, there would have been no Mogul Empire and few of the cultural achievements. Likewise, there had to be, and there was, political and administrative capacity of a high order, with respect for the conquered people. Indeed, tolerance of diversity including religious belief and practice was a hallmark of the most notable of the emperors. Although Sunni Muslims, with the exception of Aurangzeb they were able, on the whole, to avoid the extremism and intolerance that Islam (like Christianity) has over the centuries fostered in not a few of its adherents. If they saw a heavenly paradise as their ultimate destination – perhaps their share of "good fortune" (first of the essentials of Empire) - this was a remote goal, to be reached, if at all, not only through religious piety but a life of pleasure, enjoyment and achievement on earth.

We don't know just how observant in practice the emperors, their families and entourages were in respect of the virtues and requirements proclaimed by the founder of their religion, Muhammad, but we do know that they did pay homage to human pleasures and a fulfilling, happy life among the gifts of nature. As military conquerors, at times they plundered settlements and cities and massacred people, but they did not plunder the earth (on the evidence we have) and they found in nature the deepest satisfaction. Their hunting enthusiasm does not seem to have resulted in the widespread destruction of species and great forests were not destroyed for economic gain as use was made of their products. In the world today, due to the actions of some violent extremists, the religion of Islam has been exposed to virulent criticism and intolerance. One of the values of studying and reflecting on the Mogul Empire is that it provides another perspective on Islam and underlines some of the world's greatest cultural achievements which occurred under its aegis.

The image of paradise has been used as a lodestone in this chapter and a wide variety of metaphorical, poetical and secular uses of the term has emerged. What this demonstrates is that whereas, in traditional religions, paradise is commonly seen as a distant and uncertain goal, it is also a rewarding earthly quest. Indeed, as many of the metaphysical beliefs dating from the time when the Judaeo-Christian and Islamic religions were first formulated have, in modern times, dissipated or been wholly rejected, the quest for an earthly paradise, the restoration of a garden in Eden, becomes all the more pressing. At their best, the Moguls had a firm grasp of what this might entail, if not for all then at least for an enlightened minority with the necessary means.

We have only glimpsed some of the achievements in the civilization created in the centuries of the Mogul empire and its antecedents. While to a considerable extent inspired by religion, the Moguls gave another twist to religion by focusing the heavenly image of paradise on its concrete manifestation on earth. In this way, they perhaps quite inadvertently contributed to a belief that continues to gain ground in religious circles, that the modern discovery of God lies in seeking the divine in nature and that "man" is not the master of nature but a partner in its care and protection. This is an outlook that the religious and non-religious, gardener and non-gardener alike, can celebrate together.

Notes

[1] Prest, J. (1981) *The Garden of Eden.* London and New Haven: Yale University Press. ch. VIII Nature.

[2] King, R. (1979) *The Quest for Paradise. A History of the World's Gardens.* New York: Mayflower Books. p. 279.

[3] In an "Essay on the rise and progress of gardening in Ireland." cited in Walsh, W., Ross, R. and Nelson, C. (1985) *An Irish Florilegium. Wild and Garden Plants in Ireland.* London: Thames and Hudson.

[4] Sedding, J.D. (1891) *Garden Craft Old and New.* London: K. Paul, Trench, Trubner and Co. p. 3.

[5] Hobhouse, P. (2003) *Gardens of Persia.* Glebe, New South Wales: Florilegium. p. 8.

[6] Xenophon (c430 – 354 BC) *The Oeconomicus. A Discussion on Estate Management IV.* The Loeb Classical Library. London: William Heinemann Ltd; Cambridge, Mass.: Harvard University Press.; Hobhouse, P. (2003) *Gardens of Persia.* op.cit. ch.1 "The Elements of the Persian Garden"; King, R. (1979) *The Quest for Paradise. A History of the World's Gardens.* op.cit.. ch.2 "The Four Rivers".; Lehrman, J. (1986) "Islam, Gardens of" in Jellicoe, G., Jellicoe, S., Good, P., Lancaster, M. (eds) *The Oxford Companion to Gardens.* Oxford: Oxford University Press; Prest, J. (1981) *op.cit.*

[7] Boucharlat, R. (2018) "Cyrus and Pasargadae: Forging an Empire – Fashioning "Paradise"" in Shayegan, M. R. (ed) *Cyrus the Great – Life and Lore.* Boston Mass: Ilex Foundation. pp. 131–149.

[8] Stronach, D. (1994) "Parterres and Stone Watercourses in Pasargadae. Notes on the Iranian Contributions to the Evolution of the Garden Designs." *Journal of Garden History.* Vol.14, 1994. pp. 3–12; Boucharlat, R. (2011) "Gardens and parks at Pasargadae: two 'paradises'? in Rollinger, R., Truschnegg, B. and Bichler, R. (eds) *Herodotus and the Persian Empire.* Wiesbaden: Harrassowitz Verlag. pp. 557–574; Boucharlat, R. (2009) "The "Paradise" of Cyrus at Pasargadae, the core of the Royal ostentation" in Ganzert, J. , Wolschke-Bulmahn, J. (eds) *Bau und Gartenkultur zwischen "Orient" und "Okzident".* Munich: Martin Meidenbauer Verlag. pp47–64. Findings are summarised by Andrew Selkirk (2019) "Pasargadae. Seeking the Origins of a Garden Palace." *Current World Archaeology* Issue 97, pp. 30–35.

[9] Boucharlat, R. (2016) "*Qanat* and *Falaj*: Polycentric and Multi-Period Innovations" in Angelakis, A.N. et al. (eds) *Underground Aqueducts Handbook.* Boca Raton/ London/ New York: CRC Press Taylor & Francis Group. Chapter 17, pp. 279–301.

[10] Burckhardt, T. (1976) *Art of Islam: Language and Meaning.* World of Islam Festival Publishing Company. p. 191.

[11] Musée de Louvre (1989) *Arabesques et jardins de paradis: collection francais d'art Islamique.* Paris: Ministere de la Culture, de la Communication, des Grands Travaux et des Bicentennaire.

[12] Burckhardt *op.cit.* p. 189.

[13] Gul-Baden Begam (Princess Rose-Body) *The History of Humayun (Humayun-nama)* translated with Introduction by A. S. Beveridge (1902) London: Royal Asiatic Society.

[14] Quest-Ritson, C. (1996) *The English Garden Abroad.* London: Penguin Books, p. 223; The inspiration for the restoration of Arcadia in the "Elysian Fields" of the great English gardens of Stourhead and Stowe in the eighteenth century was Virgil; Chambers, D.C.C. (1993) *The Planters of the English Landscape Garden. Botany, Trees and the Georgics.* New Haven and London: Yale University Press. ch. 2 "The Translation of Antiquity: Pliny and Virgil."

[15] Nettleship, H. and Sandys, J.F. (1891) (eds) *Classical Antiquities, Mythology, Religion, Literature and Art.* London: William Glaisher Ltd.

[16] Majumdar, R.C. (1984) (ed) *The Mughul Empire.* Bombay: Bharatiya Vidya Bhavan. *Passim.*

[17] Ipsuroglu, M.S. (1969) *Paintings and Culture of the Mongols.* Trans. With Introduction by D.S. Phillips. London: Thames and Hudson.

[18] Ruy Gonzalez de Clavijo (1859 English trans.) *Narrative of the Embassy of Ruy Gonzalez de Clavijo to the Court of Timu at Samarkand AD 1403–06.* (by Timour Beg) London: Hakluyt Society.

[19] Beveridge, A. S. (1922) *The Babur-nama in English (Memoirs of Babur)* Vols 1 and 2. Translated from the original Turki text of Zahiru'd-din Muhammad Babur Padshah Ghazi. Available on the Internet Archive 2007.

[20] Wilbers, D.C. (1962) *Persian Gardens and Garden Pavilions.* Rutland, Vermont and Tokyo: Charles E. Tuttle Company. ch 2 "Timurid Gardens: from Tamerlane to Babur."

[21] Brown, J. (1999) *The Pursuit of Paradise – A Social History of Gardens and Gardening.* London: Harper Collins. p. 320.

[22] Beveridge, H. Introduction to *The Akbar Nama of Abu-L-Fazl History of the Reign of Akbar Including an Account of his Predecessors* Trans. From the Persian by H. Beveridge. Vol.III. New Delhi: Ess Ess Publications. p.xiii.

[23] Grey, E. B. (1892) (ed) *The Travels of Pietro della Valle in India.* From the Old English Translation of 1644, by G. Havers. In two volumes. London: Printed for the Hakluyt Society.

[24] Saraswati, S.K. (1984) "Mughul Architecture" in Majumdar, R.C. (ed) *op.cit.*

[25] Moynihan, E. (1982) *Paradise as Garden in Persia and Mughal India.* London: Scholar Press. pp. 111–112.

[26] Roy, S. (1984) "Akbar" in R.C. Majumdar, R.C. (ed) *op.cit.*

[27] Akbar's life is told in a most fulsome biography: *The Akbar Nama of Abu-L-Fazl. History of the Reign of Akbar Including an Account of his Predecessors.* Translated from the Persian by H. Beveridge. Vols 1–111. New Delhi. Ess Publications (1973 reprint).

[28] Anand Krishna (1981) *Early Mughal Painting.* New Delhi. Lalit Kala Akademi; Losly, J.P. (1986) *Indian Paintings in the British Library.* New Delhi. Lalit Kala Akademi.

[29] Quoted in Welch, Stuart Cary (1975) *Imperial Mughal Painting.* New York. George Brazziler, p. 19

[30] Brunel, Francis (n.d.) *The Splendour of Indian Miniatures.* Delhi. Clarion Book/ Paris. Editions Delroisse. p. 4.

[31] Henry Beveridge. Introduction to Volume 3 of the Akbar Nama.

[32] S. Roy (1984) *op.cit.* p. 135.

[33] Akbar-Nama Vol.3(2) p. 824

[34] Villiers Stuart, C.M. (1913) *The Gardens of the Great Mughals.* London: A & C Black. p. 188

[35] Interpretation of Hafiz's poem "The Mountain Face Lifted" by Daniel Ladinsky, quoted in Ladinsky, D. (2017) "The mystical poet who can help you lead a better life". 9 January, 2017. http://www.bbc.com/culture/story/20170109-the-mystical-poet-who-can-help-you-lead-a-better-life (Accessed Feb 18, 2020).

[36] Thacker, C. (1979) *The History of Gardens*. Sydney: A.H. and A.W. Reed. p. 26.

[37] Villiers Stuart C.M. (1913) *op.cit.*

[38] *The Embassy of Sir Thomas Roe to India, 1613–1619*. London: The Hakluyt Society.

[39] Constable, A. (1891/1968) 2nd ed. *Travels in the Mogul Empire by Francois Bernier AD 1656 – 1668*. A revised and improved edition based upon Irving Brock's translation. Delhi: S. Chard and Company. p. 401.

[40] Crowe, S. and Haywood, S. (1973) *The Gardens of Mughul India – a history and a guide* Delhi: Vikas Publishing House. p. 95.

[41] Crowe, S. and Haywood, S. *op.cit.* p. 135.

[42] Constable, A. (1891/1968) *op.cit.* p. 5.

[43] Aurangzeb, *Rukaat-i-Alamgiri*, tr. Joseph Earles, Calcutta, 1788. pp. 119–120.

[44] Gascoigne, B. (1987) *The Great Moghuls*. New Delhi. Time Books International. p. 243.

[45] Nath, A. (2002) *Rashtrapati Bhavan Dome Over India*. Mumbai: India Book House.

[46] Brown, J. (1982) *Gardens of a Golden Afternoon. The Story of a Partnership: Edwin Lutyens and Gertrude Jekyll*. Harmondsworth, Middlesex: Penguin Books.

Complete Listing for Wikimedia Commons Images

14.1 Bosch – Haywain Triptych. https://commons.wikimedia.org/wiki/File:Bosch_-_Haywain_Triptych.jpg Source: subscription art database. Attribution: Hieronymus Bosch or workshop, Public domain, via Wikimedia Commons.

14.2 Samuel Taylor Coleridge at age 42. Mezzotint print by Samuel Cousins after Washington Allston. 1854. https://commons.wikimedia.org/wiki/File:Samuel_Taylor_Coleridge_at_age_42.jpg Source: Library of Congress. Attribution: Samuel Cousins, Public domain, via Wikimedia Commons.

14.3 Yuan Shangdu. https://commons.wikimedia.org/wiki/File:Yuan_Shangdu.jpg Attribution: Flaumfeder, CC BY-SA 4.0 https://creativecommons.org/licenses/by-sa/4.0, via Wikimedia Commons.

14.4 Marco Polo – costume tartare. https://commons.wikimedia.org/wiki/File:Marco_Polo_-_costume_tartare.jpg Author: Grevembrock. 18th century. Source: Scanné de *Coureurs des mers*, Poivre d'Arvor. Attribution: Grevembrock, Public domain, via Wikimedia Commons.

14.12 Humayustomb. https://commons.wikimedia.org/wiki/File:Humayustomb.JPG Source: swati bhati, own work. Attribution: swati bhati, CC BY-SA 3.0 https://creativecommons.org/licenses/by-sa/3.0, via Wikimedia Commons.

14.14 Shalimar Bagh, Srinagar. https://commons.wikimedia.org/wiki/File:Shalimar_Bagh,_Srinagar.jpg Author: Pradeepkjoshi, own work. Attribution: Pradeepkjoshi, CC BY-SA 4.0 https://creativecommons.org/licenses/by-sa/4.0, via Wikimedia Commons.

14.15 Taj Mahal from the sky. Author: Ujjwal goel. Source: own work. https://commons.wikimedia.org/wiki/File:Taj_Mahal_from_the_sky.jpg Attribution: Ujjwal goel2005, CC0, via Wikimedia Commons.

14.16 TajPlanMughalGardens. Source: *Gardens of the Great Mughals*. By C.M. Villiers Stuart.1913. https://commons.wikimedia.org/wiki/File:TajPlanMughalGardens.jpg Attribution: Villiers Stuart, Public domain, via Wikimedia Commons.

14.17 Mughal Garden, Rashtrapathi Bhavan – panoramio (5). Author: Likhith N.P. https://commons.wikimedia.org/wiki/File:Mughal_Garden,_Rashtrapathi_Bhavan_-_panoramio_(5).jpg Attribution: Likhith N.P, CC BY-SA 3.0 https://creativecommons.org/licenses/by-sa/3.0, via Wikimedia Commons.

Chapter 15
Voyagers, Explorers, Travellers and Collectors

For many centuries before the first of Captain James Cook's famous voyages to the South Seas (1769–1771), European geographers and cartographers among others had speculated about the existence and nature of a great south land – Terra Incognita, Terra Australis. According to the tradition of global mapping established in the second century by Greek astronomer and geographer, Claudius Ptolemy, spatial symmetry and global stability required a southern counterbalance to the known northern hemisphere land mass. Speculation had theoretical foundations grounded in the science of the time. They were to be tested by a succession of great navigators who extended the dominion of the West to the farthest reaches of the globe.

15.1 Mapping the Globe

Resident in Alexandria, Ptolemy (c.100–c.170 AD) had access to its great library, thus to Classical Greek learning including astronomical and mathematical accounts of the spherical form of the earth and its distances. He also had the itinerary maps or descriptions of the Mediterranean trading routes (*portalanos*) identifying coastal features and staging routes. Drawing on these sources and on travellers' tales, Ptolemy was the first to develop a systematic geography of the world. His *Geographia* provided projections and geographical co-ordinates of thousands of known locations using itineraries, sailing directions and topographical descriptions. Whether he produced actual maps as distinct from directions on making maps is debated.

Like many classical texts, *Geographia* was lost to Western Europe until the twelfth century although it was known among Arab scholars. Since no text of the *Geographia* as he wrote it has survived, *Ptolemy's map*, however authentic, is a creation of mediaeval and post mediaeval cartographers and geographers.

M. Skilbeck, *Loving and Studying Nature*, International Explorations in Outdoor and Environmental Education 10, https://doi.org/10.1007/978-3-030-80751-1_15

Versions of *Geographia* were edited, updated, illustrated and translated over many centuries, starting from mid fifteenth century Italian prints, with illuminations, improved projections and more detailed outlines of manuscript maps. By the end of the sixteenth century the locus of Ptolemaic map-making had shifted north. In 1513 an edition of the *Geographia* attributed to Martin Waldseemüller was published in Strasbourg: "The map's form, with his other works – the *Cosmographiae introductio*, a globe, and two world maps of 1507 and 1516 – a related body of old and new geography, anticipating the scheme of Gerardus Mercator."[1] Regarded as the peak of Ptolemy's influence on topography, these maps were to be superseded by those incorporating the discoveries of the Portuguese and Spanish then the Dutch, English and French voyagers.

The image of Ptolemy's maps did not fade and several ideas that he expressed were reiterated, extended and elaborated by his post-mediaeval editors. They continued to influence beliefs about the southern oceans long past the Ptolemaic heyday early in the sixteenth century. To complete the world map and the European view of the world that it modeled, the oceans of the southern hemisphere had to be crossed, land masses identified, explored, mapped – and occupied (Fig. 15.1).

Ptolemy's Map was thus a process, not a single static document. And it was not only an essay in cartography, but just as much a history of voyages, discovery and exploration, a revelation of the full extent of the physical world. The succession of

Fig. 15.1 Abraham Ortelius' World Map, 1570, including both Ptolemy's postulated Great South Land (here – Terra Australis nondum cognita) and Marco Polo's kingdoms of Beach, Lucach and Maletur. Ortelius utilised the projection of Mercator's 1569 map. Abraham Ortelius, Public domain, via Wikimedia Commons

world maps, progressively updated, are a cartographic and cultural saga of immense influence. In the early sixteenth century for the first time the Americas were included, incorporating the charts and maps of the European voyagers who crossed the Atlantic oceans to the "New World". Later in that century, the projection bearing the name of the Flemish cartographer, Gerardus Mercator, was incorporated into the *Geographia*, greatly strengthening its scientific character and its utility for voyagers.

There were two keys to these developments: first, written and oral accounts of discoveries and the ocean and land charts made by the European ocean voyagers from the middle of the fifteenth century; second, developments in scientific knowledge and instrument making, navigation and shipbuilding.

Cartography drew upon progress being made in astronomy, mathematics and geography and these in turn involved theorising, surmise, observation and measurement. Optical instruments, including the compass, astrolabes, sextants, crude devices for measuring time (later the chronometer) and distance on sea and land were invented and improved. Navigation across the seemingly boundless southern oceans necessitated sturdy ships able to carry substantial supplies of water and food stores, capable of withstanding huge seas and fierce storms, to say nothing of commanders and crews possessed of fortitude and courage. The knowledge they contributed to an understanding of the natural world was dependent on a mix of material and psychological factors. It was only achieved at great human cost.

15.2 Sailing the Oceans, Colonising the Land

The era of long ocean voyages, commencing in the middle of the fifteenth century, yielded empirical data which led to ever greater accuracy and sophistication in mapmaking and theorising about what lay beyond the known world whose parameters were steadily expanding. At the same time, interest in the advancement of scientific knowledge in the form of natural history gradually became a major factor in the voyages of discovery late in the eighteenth century, when naval ships carried teams of naturalists. But national power and commerce were probably of greater moment.

When King Henry the Navigator sent ships south from Lisbon along the coast of West Africa in the mid fifteenth century, Bartolomeu Dias rounded the Cape of Storms (renamed the Cape of Good Hope by the Portuguese) in 1487 and Vasco da Gama opened the sea route from Europe to India in 1522, the primary interests were trade and competitive national advantage. In 1492 Christopher Columbus, sailing on behalf of European royal and mercantile interests, brought back tobacco plants, turkeys, pineapples and reports of chili pepper. Failing to find the China he sought for in the West Indies, he reported on commodities of use and value to Spanish traders.

Spain was beaten to the wealth of the Spice Islands (East Indies) by Portugal but was compensated by the spoils of Central and South America. Rivalry between these two southern European powers was to an extent regulated by Papal intervention in the form of the Treaty of Tordesillas whereby in 1494 the (newly

Western-discovered) world was divided between Spain and Portugal, control of Africa and Asia being vested in Portugal (plus Brazil – the line of division was the meridian 370 – to the west Spain, to the east Portugal) while Spain took the Americas and the ocean and lands of the Pacific – still to be observed by Balboa from his "peak in Darien" (1511) and entered by Magellan through the Straits bearing his name (1522).

While trade and accruing wealth were primary motives for undertaking what were often years' long and always hazardous voyages, they were not the only ones. As already mentioned, scientific inquiry only became of major importance in the eighteenth century. Nevertheless, discoveries of value to natural history were frequently made. Mariners' logs reported on new species of sea and land birds, fish and other forms of sea life previously unknown to western science, on climatic conditions, wind and water currents, tidal movements and ocean floors, coastal topography and southern skies. On land they found new and potentially useful plants and animals and observed but often did not respect, the customs and habits of indigenous people. The discovery of gold and silver, precious stones and timber trees added to the store of knowledge. Of practical use to the voyagers and the settlers who followed were water sources, provisions, navigable rivers, flat lands, mountains, havens for the repair of ships and protection against storms and, for commerce, trading posts. Not least, mariners' reports and those of early settlers enabled the testing and correcting of established views about the natural world and its plant, animal and mineral resources.

As well, the great ocean voyages and landfalls and mariners' tales of sea folk, monsters and mysterious lights and other atmospheric phenomena kept alive ancient superstition and folklore while providing rich material for poets, playwrights, chroniclers and novelists – in England, Shakespeare, Daniel Defoe and Coleridge among them.

Economic benefit and the enhancement of royal authority and national power and prestige were evidently of primary importance. But other reasons to endure the hardships and take the risk of shipwreck, loss of life, piracy, disease and prolonged anxiety and fear were the spur of fame, curiosity and a sense of adventure. If planet earth truly was a Ptolemaic and Copernican globe and not a flattish sphere as believed by many of the ancients and some of the moderns, what lay beyond the known lands of Europe and west and central Asia? If the hemispherical balance theory was true, what did the counterbalance look like and in what ways might the lands, their inhabitants and resources differ from those of Europe?

And what uses might Europeans make of them? In what ways and to what degree did the plants and animals of distant lands and in far-off oceans resemble those of myths, legends and artistic imagination, as projected in classical and mediaeval texts and maps – and travellers' tales? The questing European mind needed answers.

15.3 Gateways to the Pacific

While curiosity and the economic and imperial power motives were from the outset strong and enduring, and natural history was to become so from the eighteenth century, there was yet another consideration – religion. Christianity in various forms played a significant role in the settlement of the Americas. It was to serve as a continuing undercurrent in nineteenth and early twentieth century Australia.

In the founding of the New England colony by the Puritans early in the seventeenth century, in the Portuguese drive into India, in the Spanish conquest of Central and South America in the sixteenth century, and in the opening of the Pacific and the settlement of Australia religious beliefs about the natural world – its control, direction and destiny – were to remain a significant and often understated factor.

As for the European move into the Pacific, the Spanish King Phillip III was urged in a memorandum by one Dr Julian Arias to counter the irreligious threat, as he saw it, from the northern nations of Holland and Great Britain,

> For the English and Dutch heretics, whom the devil unites for this purpose by every means in his power, most diligently continue the exploration, discovery, and colonisation of the principal ports of the large part of the world in the Pacific Ocean, and sow in it the most pernicious poisons of their apostasy, which they put forth with the most pressing anxiety in advance of us, who should put forth the sovereign light of the gospel.[2]

On the long established Ptolemaic surmise that there must be a continental land mass in the great stretch of uncharted ocean between Peru in the east and the Cape of Good Hope, Arias called on Phillip III to discover and settle it for Spain. The motives he adduced were sovereign power, extension of the realm of the Catholic church and "the conversion of the Gentiles of the southern land". For "sovereign power", the Spanish conquistadors of Central and South America had already substituted gold and silver, or economic benefit. Christianity was being enforced on the indigenous people, a "boon" in return for their slavish labour and for occupancy of their land and its resources. The attention the European powers gave to the south-bound voyages, the interests they served and the benefits accruing, entailed a variety of interpretations and justifications. But in the "natural order of things", the imposition of Christianity was high in the political ordering and rationalisation of conquest and occupancy. Notable in this regard was the voyage into the Pacific from Peru in 1605 led by Pedro Fernandez de Quiros, with six Franciscan friars on board, and in search of the great south land. Quiros abandoned the voyage in the New Hebrides and command was taken by Luis Vaz de Torres who sailed through the island-studded strait to the north of Australia that now bears his name. The Dutch followed the Spanish, describing the north coast and its Aboriginal inhabitants in the most unflattering terms.

European attitudes were shaped by the biblical doctrine that God's making of the world was for the use and benefit of mankind and this was further elaborated by the conviction of the supremacy of European culture, hence of particular forms of Christian belief and prevailing socio-economic values of the European nations.

Consequently it was by "nature" that the resources of the newly discovered lands were thought to be of value for, and to be opened to exploitation by, the voyagers and settlers.

The first known sighting by a European of the Pacific Ocean was, as mentioned above, in 1511, when Vasco Núñez de Balboa waded into the incoming tide, formally taking possession for Spain of this Mar del Sur (South Sea) and its lands (Fig. 15.2). Magellan's passage in 1522 through the straits at the southern tip of the Americas combined with expansion by the conquistadors and missionaries along the west coast of the Americas, led first to coastal exploration. Then the voyages were extended across the Pacific in a quest for the wealth of the East Indies and east Asia. Together with their occupancy and control of the western seaboard of the south, central and parts of North America, this gave Spain dominance in the Pacific throughout the sixteenth century and a strategic foothold that lasted until defeat in the Philippines in the Spanish American War in 1896. For the sixteenth century, and despite the maraudings of Francis Drake and others, the Pacific was, in the words of Oskar Hermann Christian Spate, *The Spanish Lake*.[3]

Fig. 15.2 Spain, in the figure of Vasco de Balboa, claiming the South Sea. Nineteenth century engraving by unknown artist. Unknown author, Public domain, via Wikimedia Commons

Dr Arias' memorial, while ineffectual in persuading the Spanish monarchy to fulfill its proselytising mission, throws light on the struggle among the European powers for sovereignty, influence, economic gain and cultural hegemony – globally in general and specifically in the Pacific. It was the piratical ventures of Francis Drake eying Spanish treasure ships (1577–1578) and of Thomas Cavendish (1586–1588) that made the South Seas known to the English. From late in the sixteenth century, the balance of power began to shift, from the southern to the northern countries. Accordingly, the image of the Pacific shifted from a Spanish lake to an Anglo-Dutch sphere of influence. It was mainly the voyages, starting in the seventeenth century, of the Dutch and English navigators and then the French that resulted in the redrawing and completion of the charts of the southern oceans, the dispelling of the belief in a great southern continent , the delineation of Terra Australis by the establishment of straits and sea passages to the north, east and south, the mapping and ultimate European settlement of the Australian and New Zealand coastlines and the islands of the South Pacific.

15.4 Landfalls: From Terra Australis to Australia via New Holland

Terra Australis first became New Holland then the term Australia was adopted. Advocated by explorer and navigator, Matthew Flinders, at the beginning of the nineteenth century and recommended some two decades later by Lachlan Macquarie, Governor of New South Wales (which then included modern Queensland and Victoria), the naming of "Australia" signified both an end to many centuries of speculative invention and partial delineation and the beginning of a century of land exploration when the agenda of natural history was turned inland from the oceans and greatly intensified.

Whether Dirk Hartog in 1616 and Abel Tasman in 1642 and the Englishman, William Dampier, in 1688 were preceded by the Portuguese along the south western and western coasts of Terra Australis has long been a matter for speculation and search for buried ship remnants. Whether the Dutch ship *Duyfken*, whose crew were the first Europeans known to have sighted and charted the Australian continent (the west coast of Cape York Peninsula) – or Dirk Hartog who is the first European known to have landed – in 1616 at Shark Bay, Western Australia, were preceded by Chinese, Malaccans or others are questions that are still debated. What is clear is that none of them, apart from the Aboriginal inhabitants, permanently settled until the English arrived on the east coast in 1788.

Remains of a supposed Portuguese ship have not been found in the coastal sands around the western Victorian seaside town of Warrnambool. Certainly much of the northern and western coastline of Australia was accurately mapped by the end of the seventeenth century. Abel Tasman's discovery of the present-day Tasmania (but not the strait separating it from the mainland) first named after Governor General of the

East Indies, Anthony Van Diemen ("Anthony van Diemenslandt"), the identification and naming of several landmarks on the west coast and proclamation of the ownership rights on behalf of the Dutch East India Company, further extended the Dutch interest in the south seas. Bowen's map of Tasman's discoveries show a complete absence of surveys along the east Australian coast, while the two thirds of Australia to the west and north are recognisably mapped. This pre-dated Cook's discoveries by more than a century (Fig. 15.3).

It was the voyages, discoveries and map making of the Dutch navigators, especially in the western and northern coastal areas, that resulted by the end of the seventeenth century in the adoption of the name New Holland. It persisted until well after the English had established their colonies. Ernest Scott has underlined the importance of the Dutch: "The whole of the important discoveries in the Australian region made between 1606 and the time of James Cook were Dutch. We may dismiss as fabulous the supposition that either the Portuguese or the Spanish discovered any part of these coasts."[4]

Fig. 15.3 Emanuel Bowen's map "A complete map of the Southern Continent survey'd by Capt Abel Tasman depicted by order of the East India Company of Holland in the Stadt House of Amsterdam." This map shows Tasman's 1642 partial mapping of Tasmania and New Zealand along with a complete absence of surveys of the east Australian coastline. This was the most accurate data available to Europeans until the voyage of James Cook in 1770. Historic Maps Collection. Princeton University Library. Public domain

Tasman's journal records his party's New Holland landing, following several attempts in dangerous weather conditions. He reported edible wild vegetation "fit to be used as pot-herbs", "abundance of excellent timber", of great size and notched trees, fresh albeit shallow water streams, foot prints of animals resembling tiger claws, tree gum. He reported, too, on the large number of gulls, wild ducks and geese, on mussels – but no fish. Like many observers a century and more later, Tasman commented favourably on the open forests, "unhindered by dense shrubbery and underwood, which would prove a great advantage in exploring the country". Whether this was a result of burning by Aborigines (a claim discussed in Chapter 16), at any rate Tasman observed clouds of dense smoke. He appears to have seen no Aborigines but suspected their presence. After erecting a pole declaring possession, the two ships sailed east and Tasman reached New Zealand's south island where he again saw smoke. Tasman "discovered" New Zealand but did not establish that it comprised two islands, separate from the great south land. Boats manned by Maoris approached and in an attack several of Tasman's crew were killed or injured. Tasman condemned "this outrageous and detestable crime" and in Governor General Van Diemen's Report it was further castigated in that the natives were found to be "of a malignant and murderous nature". By contrast, lack of contact with the native inhabitants of New Holland was an intimation of the shyness or timidity often attributed to Aborigines by later explorers.[5]

On return to Batavia, Tasman was censured for failing to explore Tasmania and New Zealand beyond the coastal fringe, for failing to follow instructions which would have led to mapping details of Australia's east coast and for bringing no reports of commodities to the potential benefit of the sponsoring Dutch East India Company. While his encounter with Maoris discouraged further investigation, in Tasman's account of landing in Tasmania, several interesting gleanings of natural history and cultural interest are mentioned. Neither Dirk Hartog, nor other Dutch, English and French expeditions in the seventeenth century added much to an understanding of what might be of European interest in Australia except for the emerging picture of its human inhabitants. Dampier, however, as discussed below, made a contribution as a naturalist which was appreciated only much later.

Some accounts of Aboriginal people drew out shyness and timidity; others mentioned unwelcoming and unfriendly behaviour. Terms such as "foul and treacherous behaviour" showed little understanding of a people who quite justifiably felt that their livelihood might be threatened. It was the adventurer and former buccaneer, William Dampier, who, in the three months he spent on and around the coast of present Western Australia, gave the most extended and insightful account prior to the eighteenth century of the land and its inhabitants (Fig. 15.4). From East Timor late in 1688 or early 1689, he headed south in the *Cygnet*, aptly named given the prevalence of black swans in the northwest of New Holland.

Dampier was confident that New Holland was of great size, either an island or a continent and not joined to Asia, Africa or America. He found the littoral sandy, dry and waterless on a surface that was thinly grassed, supporting a sparse, low tree cover. Animals, birds and fish were observed but not in great number. In some infamous comments he found that "The Inhabitants of this Country have no Houses and

Fig. 15.4 William Dampier (1651–1715), author of the first European study of nature in Australia (New Holland). portrait by Thomas Murray c 1697. National Portrait Gallery, London. NPG 538. Thomas Murray, Public domain, via Wikimedia Commons

skin Garments, Sheep, Poultry, and Fruits of the Earth. Ostrich Eggs etc. … And setting aside their Humane Shape, they differ but little from Brutes."

Continuing, he found no favour in either the persons of the Aboriginals, or their impoverished food gathering, lodgings, tools or weapons. Tellingly, Dampier found that the Aboriginals had no admiration for "anything that we had"[6]. A century later, Cook too found that the Aborigines appeared to be disdainful of the material culture of the European intruders.

In short, Dampier initially made no attempt to understand the life and culture of the native inhabitants, ridiculing and despising them. Yet in his second voyage in 1698/9 (the Roebuck), Dampier showed himself an appreciative, close and perceptive observer of the birds, sea creatures, vegetation and animals en route to and on the shores of New Holland. He found beauty in the speckled shells of Shark's Bay – which he named for the abundance of the creatures when caught which he measured and whose flesh he and his crew savoured – over that of the huge turtles. Judgmental, ethno-centric attitudes prevented Dampier, as it did many others, from valuing the indigenous people amidst the beautiful creatures of the new-found land.

Of the land, as he cruised the coast, Dampier's journal is replete with vivid descriptions, expressing delight in the animal and plant life. It seems that among the numerous plants observed, he may have been the first European to extol the brilliantly flowered creeping plant named much later as Sturt's Desert Pea (Fig. 15.5), "There grew here 2 or 3 Sorts of Shrubs, one just like Rosemary; and therefore I call'd this Rosemary Island. It grew in great Plenty here but had no Smell. Some of

Fig. 15.5 Sturt's Desert Pea (*Swainsonia formosa*), a celebrated Australian wildflower, by artist Ellis Rowan, William Dampier the first European to record it. Watercolour on paper. c1900. P2549. Collection: Museum of Applied Arts and Sciences. Gift of Sydney Technical College, 1922. Photographer unknown

the other Shrubs had blue and yellow Flowers; and we found 2 Sorts of Grain like Beans: the one grew on Bushes; the other a Sort of a creeping Vine that runs along the Ground, having very thick broad Leaves and the Blossom like a Bean Blossom but much larger, and a deep red Colour, looking very beautiful."[7] Not only plants, sea creatures and birds also attracted Dampier. He admired the form and colours of the rocks and cliffs. Seeds of the "very beautiful" plant (Sturt's Desert Pea) were among the specimens he collected. Stored in lengths of bamboo cane, together with his notes they were preserved when the Roebuck was later wrecked in the Atlantic and are now in a herbarium at Oxford University.[8]

Reflecting on the time he spent, "ranging off and on the Coast of New Holland" late in the seventeenth century, Dampier qualified his general sense that it "was not very inviting": He commented favourably on its situation, "so very advantageously in the richest Climates of the World"[9], its potential of fruitful land. He anticipated inhabitants with whom there might be useful trade in commodities (this despite his dismissive account of those inhabitants he did encounter). Finally, he rather fulsomely alluded to the fruits of his own surveys as being of value and use to future explorers while regretting his inability to extend them to the as yet uncharted eastern seaboard of New Holland. We might add that through his plant collecting, he wrote the first European chapter of nature study in Australia.

15.5 Cook and Banks

Lieutenant James Cook R.N. received a commission from the British Admiralty in 1766 (Fig. 15.6). Its famously declared purpose was to observe the transit of the planet Venus across the surface of the sun from sites in and around Tahiti. This was a relatively rare phenomenon which, carefully timed, would be of value in developing more precise longitudinal measures, hence of use to the Royal Navy and the country's navigators and explorers. For its scientific significance, the Royal Society joined the government in sponsoring Cook's voyage to the South Seas. For the opportunities it provided to greatly extend the scope of his botanical and wider natural history interests, the young and wealthy Joseph Banks contracted with the Admiralty to assemble a team of botanists and artists to join him on the voyage (Fig. 15.7). Not publicly known, not even to Cook until he opened a set of instructions on leaving Tahiti, was that for the British government there was a second purpose, which was to prove of greater import, valuable as the astronomical observations were thought to be. This was to confirm the existence or otherwise of the mysterious south land, puzzled over and projected by cartographers in various forms and dimensions in the long succession of Ptolemy's world maps from the twelfth century onwards.

Fig. 15.6 James Cook R.N. (1728–1779) Lionised in the eighteenth and nineteenth centuries as one of the greatest of navigators, in the twentieth century Australian Aboriginals demonized him as the forerunner of British settlement of their land. Portrait by Sir Nathaniel Dance-Holland, c. 1775, National Maritime Museum, Greenwich. Nathaniel Dance-Holland, Public domain, via Wikimedia Commons

Fig. 15.7 Sir Joseph Banks, botanist on Captain Cook's voyage to Australia, and later President of the Royal Society. Engraved by A. Cardon after a drawing by W. Evans from the picture by T. Lawrence. Wellcome Trust. CC BY 4.0 via Wikimedia Commons

Cook's "discovery" and mapping of the east coast of the Australian continent and his prowess as a navigator and leader are so well known through numerous historical and nautical studies as to require no more than a brief reference. It is in the pages of his *Journal* and the work undertaken by Banks and his team that we can pick up the two trails started a century earlier by Dampier in the west of the continent: perception of the indigenous people; the study and appreciation of the plant and animal kingdom.

The relationship between Aboriginal people and the country's flora was of scant interest to Cook and Banks, only becoming a subject of profound interest a century and a half later, as we see in the following chapter. What does emerge, however, from Cook's *Journal* is an observation both dramatically different from Dampier's and quite subversive of prevailing doctrines concerning human happiness, the good life and the benefits of civilisation as understood by European Enlightenment rationalists – of whom Banks was one.

Cook, who was familiar with routes taken by earlier voyages, Quiros and Torres among them was, in the *Journal* of his first voyage, dismissive of the belief in a great southern continent at least north of latitude 40 degrees south. From his second voyage, with a passage far to the south, he firmly concluded that there was no such continent although with characteristic caution, he allowed that some future voyage might conceivably disprove his own firm conviction.

Cook's narrative of his first voyage, charting the eastern coast of Australia, his disproof of the long held mariners' and cartographers' conviction that New Guinea and Australia were a continuous land mass and his remarkable passage through the reefs, shoals and channels of the north Queensland waters, constitute one of the great stories of maritime discovery and survival. Although his *Journal* is primarily a detailed, dispassionate record of compass bearings, winds, tides, fathom soundings, islands, reefs, channels and sailing manoeuvres required to ensure safe and continuous progress up thousands of kilometres of coast, there are passages of incisive cultural observation, high emotion and great humanity. In the mariner's log book are notes on bird life, indigenous food sources in the sea and on land and quite clear pictures of the 'kangooroo'. Although lightly sketched, this great length of coastline emerges in European consciousness from vague obscurity into a definiteness and clarity that was almost inevitably to lead to settlement.

From time to time the botanising activities of Banks and his assistant, Daniel Solander, emerge from a mariner's text. Thus the landing in a bay in the central coast of the present New South Wales leads to the naming of Botanists' or Botany Bay in recognition of the prolific and unfamiliar flora that Banks and Solander studied and collected. Cook, however, noted the presence of stingrays in the shallow water, had them caught and eaten and thought first to name the landing Stingray Bay. Here, as elsewhere, he recognised that the seas and the land have an abundance of food or potential for food production. Soils are often sandy and dry, as Dampier found them on the northwestern coast, but in places they are also moist and loamy. Grassy meadows suggest scope for pasturing animals while the well spaced trees betoken relative ease of cultivation. Supplies of fresh water may not be bountiful, but are found to be adequate. Timber in the hardwoods is dense and "ponderous" but could be useful and in any case Cook makes no claim to expertise in these matters which are, "the province of Mr Banks and Dr Solander in their botanical forays".

The high drama of Cook's *Journal* is the near disaster of the hole ripped in the hull of the *Endeavour* when she struck a coral reef in the shallows of the northern coastal waters (Fig. 15.8). Inching along one of the world's most treacherous coasts, in uncharted waters, "intangled among Islands and Shoals", confronted with "vast foaming breakers", "in the very Jaws of distruction (sic)", Cook's seamanship and leadership of men met perhaps its greatest challenge when he was faced with the distinct possibility of shipwreck and marooning on this remote coast between Cape Tribulation, so named "because here began all our Troubles" and Cape Providence named to express gratitude for disaster averted, thanks, according to his account, to the heroic and skillful efforts of all members of the ship's crew. After recovering from this "alarming and, I may say, terrible circumstance", Cook's equanimity was quickly restored and the *Journal* entries focus on the gathering of edible plants and the hauls of fish from the sein nets.

What Cook demonstrated in the remarkably successful passage along the east coast of Australia, was that an order by the Admiralty, which enabled completing of the work he and his crew had come to do, also sustained the pioneering botanical field work of Banks and his team. With the aid of the campaign for colonisation that Banks was subsequently to lead, Cook had provided clear and definite answers to

Fig. 15.8 *A View of Endeavour River on the coast of New Holland*, by Ignaz Sebastian Klauber, 1795, based on a drawing by Sydney Parkinson. Public domain. National Library of Australia. nla.obj-135961675

centuries of questions and uncertainty about the mysterious great south land. If there was no great south land encircling the southern oceans, there was the prospect of yet another extension of the European diaspora to an island continent of considerable proportions, intriguing natural features and life forms, and of potential use and value.

Cook's *Journal*, in a passage of great humanity and perspicacity, offered something more. By contrast with Dampier's derogatory observations, Cook found in the Aboriginal people with whom he had contact in his coastal sightings and landings, a force of nature against which to appraise his own culture, and Europe's claims to a superior civilization. Cook could almost be writing a coda to Rousseau's 1750 prize winning essay for the Academy of Dijon, *A Discourse on the Sciences and the Arts*, which called into question the Enlightenment claim for the superiority of the arts and sciences of civilisation over a more "natural" existence, respectful of "feelings and desire and free of unattainable wants".

Starting with a matter of fact description of the physical features, adornment, tools and equipment, habitation, social grouping, food and technology of the Aboriginal people, Cook respectfully drew up, "a small Vocabulary of a few words in the New Holland Language which we learnt when in Endeavour River" (where the damaged ship was, over several weeks, careened and repaired). Then, quite suddenly he delivers his remarkable encomium of a life so different from his own: "From what I have said of the Natives of New Holland they may appear to some [Dampier, doubtless] to be the most wretched People upon Earth; but in reality they are far more happier than we Europeans, being wholly unacquainted not only with the Superfluous, but with the necessary Conveniences so much sought after in Europe; they are happy in not knowing the use of them. They live in a Tranquility which is not disturbed by the Inequality of Condition. The earth and Sea of their own accord furnishes them with all things necessary for Life. They covet not Magnificent Houses, Household stuff etc; they live in a warm and fine climate, and enjoy every wholesome Air, so that they have very little need of Cloathing … in short, they seem'd to set no Value upon anything we gave them, nor would they ever part with anything of their own for any one Article we could offer them. This, in my opinion, Argues that they think themselves provided with all the necessary's of Life, and that they have no superfluities."[10]

One wonders whether Cook discussed these reflections with Banks, the wealthy landowning paragon of Enlightenment values, or indeed whether together Cook and Banks drew any parallels between two secular versions of Paradise, which they had experienced, the one with Aboriginal people on the banks of the Endeavour River, the other under the palms and among the bougainvilleas and hospitable natives in Tahiti.

On return to England, Cook and Banks were celebrated, Banks even more than Cook by being invited to meet King George III. Cook prepared for further voyages whereas Banks launched into a public life of scientific patronage and leadership including, ultimately, Presidency of the Royal Society. For reasons never explained, the superb collection of drawings and paintings prepared by the artists in his team was not published in his lifetime, although copper plates were etched. Some have become available to the public since, testimony to one of the finest periods in the history of botanical art, and of the hardships and even the mortality entailed in gathering specimens to draw and paint.[11] Artist, Sydney Parkinson, was among several crew members who did not survive the voyage home, while his sketches and paintings did (Fig 15.9).

15.6 Naturalists, Collectors and Botanical Artists

With two ships, *Le Geographe* and *Le Naturaliste*, Captain Baudin set sail from Le Havre in 1800 for the Southern Ocean on a voyage whose purpose was primarily scientific. "He had twenty-four scientists aboard, who gave the crew no end of trouble," the captain said. "They clutter up the deck, they suffer from seasickness, they do nothing useful and they panic at the slightest gust of wind."[12] The crowding of naval ships with naturalists and their collecting, preserving and recording equipment did not always lead to harmonious relations at sea. Nor were they better on land when the naturalists wished to linger while the captain had coastlines ahead to chart.

Following the new world views opened by travelers on land, such as Marco Polo who, late in the thirteenth century, left Italy for China where, as we saw in the previous chapter, he spent seventeen years in the service of Kublai Khan's court, the great

Fig. 15.9 Sydney Parkinson (1745–1771), botanical artist on *HMS Endeavour* who did not survive the voyage. Self portrait. Sydney Parkinson, Public domain, via Wikimedia Commons

ocean voyagers between the fifteenth and eighteenth centuries became the source of a vast increase in knowledge and unprecedented wealth for the European powers, with enormous scope for political expansion. This led to empire building on an unprecedented scale. The huge growth of private and public wealth concentrated in Europe, facilitated a great surge in scientific inquiry and in research, notably in zoology, botany and biological science more generally, extending over several centuries. Darwin's theory of evolution by natural selection had as its inspiration observations and inquiries in the Galapagos Islands during his time as a naturalist on board *HMS Beagle* in its circumnavigation of the globe over the course of five years from 1831 (Fig. 15.10).

Collectors of plants had been active for centuries before the era of the great European voyagers. Cultivation of economic plants for consumption and use both depended on and generated the study of plants and the introduction to cultivation of new species. But the interest in importing species from distant places – as Egyptian Queen-Pharoah Hatshepsut is supposed to have imported palms from Sudan – began in earnest when in the eighteenth century not only economic plants – such as quinine bark, rubber, tea, breadfruit – were transported across oceans and continents, but plants of interest to private gardeners, nursery proprietors and Botanical Gardens were collected in expeditions which criss-crossed the globe.

In order to explain this phenomenon, we must turn briefly to the European history of gardening, with Great Britain, notable for its gardening culture, as an illustration. It seems that ornamental gardening – as distinct from agriculture and the cultivation of vegetables – was introduced to the British people from 43 AD onwards, by the Romans who in Italy had used plants for ornamental and health-giving purposes around their villas. There is archaeological evidence of similar practices in

Fig. 15.10 Charles Darwin 1840. The young Charles Darwin following his circumnavigation of the globe in *HMS Beagle.* Portrait by George Richmond, 1840. George Richmond, Public domain, via Wikimedia Commons

Roman villas in the English countryside, suggesting that the occupying army brought plants from the continent of Europe. However, with the departure of the Romans in 410 AD the practice of household gardening died out and, with but few exceptions, did not recur on any scale until Tudor times and into the seventeenth century. Gardening as an art and craft alike became steadily more important for all classes of society. Today the "English garden" connotes a style, or a genus with many species, which is world renowned – even if rather vaguely defined, other than as referring to ornamental gardens commonly seen in the English countryside and villages, whether around cottages, villas or mansions. At any rate, England has been for centuries a nation of notable gardeners, garden visitors, writers and broadcasters.

It is of some interest that the English garden and so-called English flowering plants and trees are practically none of them indigenous. With a few exceptions, "English" plants are exotic, except in the sense, that after centuries of growth – reproduction and successful adaptation to soil, climate and cultivation many have become virtually indigenous – hollyhocks, violas, oaks, planes, mulberries for example. Viewed historically, the English garden is a product of socio-economic and cultural conditions on the one hand and, on the other, generations of voyagers, travelers, diplomats, specialist collectors and traders who imported seeds and plants from all the continents except Antarctica and numerous islands scattered across the world's oceans. Plant collecting for the purposes of advancing the science of botany was one purpose for collectors as is shown by the encouragement given by the Swedish botanist, Carl Linnaeus (1707–1778) (Fig. 15.11), the great taxonomist. To

Fig. 15.11 Carl von Linne (Linnaeus) in the traditional dress of the Sami people of Lapland he visited early in his scientific career. Linnaeus holds flowers of *Linnaea borealis* in his right hand. Portrait by Martin Hoffman, 1737. Museum Boerhaave, Leiden. Naturalis Biodiversity Center, Public domain, via Wikimedia Commons

assemble specimens for classification, Linnaeus, a professor of botany and medicine at Uppsala, encouraged his students and assistants to set forth on plant hunting expeditions. As we saw above, one of those assistants, Daniel Solander, accompanied Banks on Cook's *Endeavour* and together with Banks botanised along Australia's east coast. Linnaeus was responsible for the now universal bi-nomial system (for each species) whereby all living things, not only plants, are classified for simplicity into genus and species. Actually his system, now much modified, was elaborated into six categories – genus and species being just the end point of the classification which comprises: kingdom, class, order, family, genus and species.

Another scientific collector of note was Joseph Dalton Hooker who became, like his father before him, director of London's Kew Gardens of botanical fame. Hooker undertook several excursions to the Californian Rockies, to Antarctica and to the Indian Himalayas, making many discoveries of value to science but also to gardening. Most famous is his collection of species of rhododendrons in the mountains of Sikkim and India in the years 1847 to 1851. This species of flowering plant became a great favourite of large landowners and smallholders alike with many new varieties bred from specimens imported by numerous collectors.

By the nineteenth century, plant collecting had become a large scale, serious business, literally so in that many collectors for lengthy periods, extending perhaps over most of an adult lifetime, were paid professionals. They were recruited by commercial plant nurseries, professional botanists and directors of botanic gardens. Kew and the Paris *Jardin des Plantes* were leaders but botanic gardens everywhere had their own paid collectors or relied on distribution of specimens and scientific data from the major centres. Collectors were also patronised by wealthy or simply enthusiastic individuals, as were the Bartrams, father and son, who as we saw in Chapter 12.7, for decades sent specimens to private garden owners in England in return for payment in cash or kind.

The great majority of nineteenth and early twentieth century collectors were men, many from humble beginnings – commonly as Scottish under-gardeners. But there were women who took up the challenge of collecting, at times experiencing the hardships and setbacks of their male counterparts. In the young Swan River colony (of now Western Australia), Georgiana Molloy (1805–1843), wife of Captain John Molloy, a government official, lived in the sparsely settled south west corner of the colony (Fig. 15.12). Often her husband travelled on public business around the territory and in his absence, Georgiana took her small children into the bush on collecting expeditions. At first she found the indigenous flowers less interesting than those "back home". But that changed with closer observation. She sent numerous specimens of the colony's splendid flowering plants to the naturalist Captain James Mangles RN FRS FRGS in England with whom, to alleviate her lonely condition, she also corresponded.[13]

Collecting did not always lead to the dispatch of plants and seeds. Two of the prominent members of the profession were painters: from England, Marianne North (1830–1890)[14] (Fig. 15.13); from Australia, Ellis Rowan (1848–1922) (Fig. 15.14). Both travelled extensively, especially in tropical regions. Among the thousands of their painted flowers, shrubs, trees, vines, insects and birds are species scarcely known or unknown at the time to western science. A selection of North's paintings is on permanent display at Kew in a building evoking an Indian hill station dwelling, which she financed. Many of Rowan's works are in the Australian National Library in Canberra and in galleries across the nation.[15]

In Australia, Baron Ferdinand von Mueller, Government Botanist in Victoria and Director of the Melbourne Botanical Gardens had a numerous group of collectors,

Fig. 15.12 Georgiana Molloy c.1829, Western Australian botanical collector. Original artwork believed to be a miniature painted before she left England in 1829, artist unknown. (Hasluck, 1955. p.6), Public domain, via Wikimedia Commons. Photographic plate attribution reads "Reproduced by permission of Mrs V.M.R. Bunbury, 'Marybrook'. Public domain, via Wikimedia Commons

several of whom were women, who either on their own initiative or at his frequent request sent him material for identification or for propagation and wider use, whether in gardening or agriculture.[16] Mueller himself collected, at times travelling alone in remote areas of the Australian "bush".

Some of the collectors have become familiar names in horticulture and gardening by virtue of their at times perilous exploits and popular books and magazine articles. From England, Reginald Farrer (1880–1920) ventured into alpine regions in East Asia, collecting specimens still prized by gardeners, and writing about the best ways to grow them in English conditions; *The Rock Garden* is a first-rate book on growing alpine plants. Apart from a single guide and not always then, Farrer travelled alone, reportedly dying at an early age from alcohol poisoning. Another who came to a tragic end was the greatest of the many Scottish gardeners who became collectors, David Douglas (1799–1834), "arguably the greatest botanical explorer collector of them all ..."[17] Douglas travelled extensively in the marvelous conifer forests of northwest America and Canada by foot, horseback and canoe, often through unexplored territory, for the Royal Horticultural Society of London, thereby serving both scientific and commercial interests. He popularised the widely planted Douglas fir which was first identified for European botany in 1792 by another Scottish collector, Archibald Menzies. Douglas was responsible for the popularisation of many economically important trees into forestry, as well as for their ornamental uses. They include the Sitka spruce and *Pinus radiata*, now planted worldwide for timber products. Still popular garden plants that he collected include lupins, Californian poppy, penstemons and evening primrose, orchids, oak and the Oregon grape (*Mahonia aquifolium*). Douglas sought to extend his botanical and geographical interests by travelling to Hawaii. Making his way along a forest track, he fell into a pit dug for trapping wild cattle and was reportedly gored to death by a

Fig. 15.13 Marianne North (1830–1890) in front of her home, Mount House, Alderley, taken by Susan Hodgson c.1869. Source: *Recollections of a happy life, being the autobiography of Marianne North.* Vol. 2. 1894. Public domain, via Wikimedia Commons

maddened animal. Collecting plants for the enrichment of gardens back home could be a most hazardous occupation abroad.

Perhaps most famous in the peak period of collecting was Robert Fortune (1812–1880). (Fig. 15.15) From a gardening background in Scotland and supported by the London (later Royal) Horticultural Society, Fortune travelled to China

Fig. 15.14 Ellis Rowan's painting of Rothschild's Bird-of-Paradise (*Astrapia rothschildi*) Papua New Guinea c.1917. National Library of Australia. nla.obj-138748166

following the Treaty of Nanking (1843) whereby England and other European powers forced their way into the country. He shipped back, in protective (Wardian) cases, what soon became popular garden plants, abelias and campanulas among many others, and learnt about bonsai and fishing with cormorants. But Fortune's real importance to horticulture arose when the East India Company recruited him to organize the transplantation of a large quantity of tea seedlings and seeds (*Camellia sinensis*) from China to India. The success of this daring operation led to the establishment of the Indian tea industry which in the course of the century outstripped that of China. This outcome meant employment and wealth for one country, at the expense of another.

Fortune is usually presented admiringly by horticultural and other authors but not in the recasting of colonial history by critics of European imperialism. Martin Hoyles writes, "His discovery of new plants is inextricably linked to the opening up of China for economic exploitation and to the colonisation of India. The manner in which many of the plants were taken is indicative of the developing power relations between Britain and its empire.[18]

Fig. 15.15 Robert Fortune (1812–1880), doyen of nineteenth century plant collectors. "A Chinese Bird's-eye View of the Stream of "Nine Windings" and strange Rocks." This image by W. Jackson pictures an area near the Woo-e-shan limestone hills (today in Fujian province) where China's finest souchong and pekoe teas grew at the time of Fortune's visit. Source: Fortune, R. (1852) *Two visits to the tea countries of China and the British tea plantations in the Himalaya*. London: John Murray. Biodiversity Heritage Library. 17207596603. Public domain. CC BY 2.0 via Wikimedia Commons

In their authoritative history of plant collecting, Mary and John Gribbin detail the value to European gardening of Chinese plants. A most versatile plant was (and remains) the bamboo. Philip Short enumerates its many and varied uses in nineteenth century Chinese society.[19] Until Fortune's expedition to China, much plant collecting was for scientific purposes and the pleasure in gardening of patrons whether individual or institutional. But the economic potential of certain species resulted in a growing commercial and political interest. The establishment of the rubber industry in the imperial domains of the East Indies depended on the transportation of specimens from South and Central America and resulted in the decline of the native industry (or of its potential for growth), a direct consequence of its growth in Malaysia and Sumatra under the auspices of the imperial powers. Similarly, discovery of the anti-malarial qualities of the powdered bark of the South American chinchona tree (quinine) by Westerners (but long known to indigenous people) and shipment of plants to India and establishment of a successful industry there, was another imperial benefit.

In 1787 Captain William Bligh, commissioned by the British Admiralty, sailed to Tahiti to collect plants of the bread fruit tree to transport to the West Indies where they were expected to yield a nutritious and tasty food source for black slaves working in the sugar cane plantations. With him was another of Banks's proteges, Scottish collector, Francis Masson. Bligh and his crew were at loggerheads, resulting in a mutiny. Bligh and loyal crew members, including Masson, were bundled into a long boat in which they traversed 6,000 kms of ocean, landing in Indonesia. Meanwhile, the mutineers sailed back to Tahiti, thence to Pitcairn Island. Their descendants were later relocated to Norfolk Island where some have remained until today. They settled amidst groves of another splendid botanical specimen, new to science and now widely planted around coastal resorts in Australia – the Norfolk Island Pine (*Araucaria heterophylla*). This tree belongs to the same family as the South American monkey puzzle (*Araucaria araucana*) whose seeds were encountered by plant collector, Archibald Menzies, who at an official dinner with the Viceroy of Chile, noted its nuts in a dessert. He pocketed some and brought the seedings they produced back to Banks in London.[20] The monkey puzzle tree became a favourite ornament of British gardens in the Victorian times and has remained so. The handsome Bunya pine (*Araucaria bidwillii*), native to Queensland, is widely planted in Australia, including in my country garden.

Collecting had other outcomes than the dispatch of plants and seeds to European and American plant nurseries, scientific centres and private patrons. The voyages of discovery and the growth of interest in horticulture, especially in flowering plants, was a stimulus to botanical and other artists. What has been called the golden age of botanical art extended over more than two centuries from late in the seventeenth century. It continues to thrive today – as for example Philippa Nikulinsky's frontispiece of this volume and in the celebrated series of paintings of all known banksia species by fellow Australian, Celia Rosser, commissioned by Monash University. Their work ranks with that of such notable figures as Georg Dionysius Ehret who collaborated with Linnaeus and Pierre-Joseph Redouté, of rose painting fame.

Fig. 15.16 Albrecht Dürer, *Great Piece of Turf*. 1503. Watercolour, pen and ink. Albertina Museum, Vienna. Albrecht Dürer, Public domain, via Wikimedia Commons

The forerunner of this movement is Albrecht Dürer (1471–1528), the Nuremberg painter, engraver, print-maker and art theorist. In addition to the portraits and altar pieces for which he is famous, Dürer produced superb pieces of nature art, among them a Young Hare and a Great Piece of Turf (Fig. 15.16). These works, in meticulous detail, show great empathy with the subject, its vitality and beauty. In addition to their scientific purpose and value, it is these qualities that distinguish the botanical and zoological art of the age of the great voyages, when artists came to be included in a ship's complement – such as Sydney Parkinson and Ferdinand Bauer respectively in James Cook's first voyage to the east coast of Australia and Matthew Flinders' circumnavigation of the continent of Australia.

Notes

[1] Crone, G.R. (1958) *Maps and their Makers*. London: Hutchinson's University Library. pp. 75–76.

[2] Major, R.H. (1859) *Early Voyages to Australia*. London: Hakluyt Society. Cited in Scott, E. *Australian Discovery. Vol. 1. By Sea*. London and Toronto: J.M. Dent and Sons. Ltd. 1929.

[3] Spate, O.H.K. (1979) *The Spanish Lake*. Canberra: Australian National University Press.

[4] Scott, E. (1929) *Australian Discovery* Vol. 1 By Sea. London and Toronto: J.M. Dent and Sons Ltd. p xix.

[5] Scott, E. *op cit*. "Tasman's Discovery of Van Diemen's Land and New Zealand" pp. 21–24; 35.

[6] Scott *op.cit*. p. 63.

[7] Scott *op.cit*. p. 99.

[8] Seddon, G. (2005) *The Old Country. Australian Landscapes, Plants and People*. Port Melbourne. Victoria: Cambridge University Press. ch.1 First Encounters.

[9] Scott *op.cit*. p. 107.

[10] Scott *op.cit*. p. 226.

[11] Gooding, M., Mabberley, D., Studholme, J. (2017) *Joseph Banks' Florilegium. Botanical Treasures from Cook's First Voyage*. New York: Thames and Hudson.

[12] Lacour-Gayet, R. (1976) *A Concise History of Australia*. Ringwood, Victoria: Penguin Books Australia. p. 125.

[13] Hasluck, A. (2002) *Georgiana Molloy. Portrait with Background*. Fremantle, W.A.: Fremantle Arts Press.

[14] Vellacott, H. (ed) (1986) *Some Recollections of a Happy Life – Marianne North in Australia and New Zealand*. Caulfield East, Vic.: Edward Arnold Australia; North, M. (1980) *A Vision of Eden – the Life and work of Marianne North*. Exeter, England: Webb & Bower.

[15] Ponsonby, L. (2000) *Marianne North at Kew Garden* London: Royal Botanic Gardens, Kew; Hazzard, M. and Hewson, H. (1982) *Flower Paintings of Ellis Rowan*. Canberra: National Library of Australia.

[16] Olsen, P. (2013) *Collecting Ladies – Ferdinand von Mueller and Women Botanical Artists*. Canberra: NLA Publishing.

[17] Gribbin, M. and Gribbin, J. (2008) *Flower Hunters*. Oxford: Oxford University Press p. 161.

[18] Hoyles, M. (1991) *The Story of Gardening*. London: Journeyman Press. p. 109.

[19] Short, P. (2012) *In Pursuit of Plants*. Crawley, Western Australia: University of Western Australia Press.

[20] Gribbin and Gribbin *op.cit*. p. 96.

Complete Listing for Wikimedia Commons Images

15.1 OrteliusWorldMap1570. Source: Library of Congress. https://commons.wikimedia.org/wiki/File:OrteliusWorldMap1570.jpg Attribution: Abraham Ortelius, Public domain, via Wikimedia Commons.

15.2 Balboa südsee.19th century engraving by unknown artist. https://commons.wikimedia.org/wiki/File:Balboa_s%C3%BCdsee.jpg Attribution: Unknown authorUnknown author, Public domain, via Wikimedia Commons. Source: http://www.trainweb.org/panama/history.html

15.4 Dampier-portrait. https://commons.wikimedia.org/wiki/File:Dampier-portrait.jpg Attribution: Thomas Murray, Public domain, via Wikimedia Commons. National Portrait Gallery. Source: http://www.npg.org.uk/collections/search/portrait/mw01706/William-Dampier?LinkID=mp01176&role=sit&rNo=0.

15.6 Captainjamescookportrait. National Maritime Museum, Greenwich. 1776. https://commons.wikimedia.org/wiki/File:Captainjamescookportrait.jpg Attribution: Nathaniel Dance-Holland, Public domain, via Wikimedia Commons.

15.7 Engraving of Sir Joseph Banks Wellcome M0011324. https://commons.wikimedia.org/wiki/File:Engraving_of_Sir_Joseph_Banks_Wellcome_M0011324.jpg Attribution: See page for author, CC BY 4.0 <https://creativecommons.org/licenses/by/4.0>, via Wikimedia Commons.

15.9 Sydney Parkinson. https://commons.wikimedia.org/wiki/File:Sydney_parkinson.jpg Source: Natural History Museum. Attribution: Sydney Parkinson, Public domain, via Wikimedia Commons.

15.10 Charles Darwin by G. Richmond. Collection: Darwin House. https://commons.wikimedia.org/wiki/File:Charles_Darwin_by_G._Richmond.jpg
Attribution: George Richmond, Public domain, via Wikimedia Commons.

15.11 Naturalis Biodiversity Center – Martin Hoffman – Carl von Linne (Linnaeus) in Lapland costume – painting. https://commons.wikimedia.org/wiki/File:Naturalis_Biodiversity_Center_-_Martin_Hoffman_-_Carl_von_Linn%C3%A9_(Linnaeus)_in_his_Lapland_costume_-_painting.jpg Attribution: Naturalis Biodiversity Center, Public domain, via Wikimedia Commons.

15.12 Georgiana Molloy 2. Source: Scanned from *Portrait With a Background* by Alexandra Hasluck (1955). https://commons.wikimedia.org/wiki/File:Georgiana_Molloy_2.jpg Attribution: Original artwork believed to a miniature painted before she left England in 1829, artist unknown. (Hasluck, 1955. p.6), Public domain, via Wikimedia Commons. Although copyright has expired, photographic plate attribution reads "Reproduced by permission of Mrs V.M.R. Bunbury, 'Marybrook'.

15.13 Marianne North Hodgson. Photo: Susan Hodgson c.1869. Source: *Recollections of a happy life, being the autobiography of Marianne North.* Vol. 2. 1894. https://commons.wikimedia.org/wiki/File:Marianne_North_Hodgson.jpg Attribution: Photo taken by Susan Hodgson, wife of Brian Houghton Hodgson, Public domain, via Wikimedia Commons.

15.15 Two visits to the tea countries of China and the British tea plantations in the Himalaya (17207596603). By Robert Fortune. p.203. Biodiversity Heritage Library. Source : https://www.flickr.com/photos/biodivlibrary/17207596603 https://commons.wikimedia.org/wiki/File:Two_visits_to_the_tea_countries_of_China_and_the_British_tea_plantations_in_the_Himalaya_(17207596603).jpg Attribution: Fortune, Robert, Public domain, via Wikimedia Commons. CC BY 2.0.

15.16 Albrecht Dürer – The Large Piece of Turf, 1503 – Google Art Project. Source: NgELdACk3I8Jkg at Google Cultural Institute https://commons.wikimedia.org/wiki/File:Albrecht_D%C3%BCrer_-_The_Large_Piece_of_Turf,_1503_-_Google_Art_Project.jpg Attribution: Albrecht Dürer, Public domain, via Wikimedia Commons.

Chapter 16
Ways of Valuing and Appreciating Nature

This penultimate chapter explores a wide and diverse array of attitudes, values and practices whereby people have attached themselves very firmly and happily to the earth and to nature. As we have seen throughout the book, attitudes toward the living earth have vacillated between two poles, one a spiritual journey toward a hope for eternal heaven, the other acceptance of the challenges and rewards of earthly existence. Over time there has been a growing realisation that addressing the challenges and taking up the opportunities of present life in the natural world can also be a fulfilling spiritual life for which no supernatural framework is required. From a religious standpoint, this allows for a sense of God firmly in and of nature.

16.1 The English Vicarage Garden

16.1.1 The Ideal English Garden

The close attention to his melon patch, flower beds, birds and other creatures paid by The Revd Gilbert White at Selborne (Chap. 13) was becoming typical of a feeling for nature among an expanding middle class in eighteenth century England. Clergymen had long been conspicuous among naturalists, botanists, gardeners and garden writers but as a wholesome and thoroughly respectable – and respectiful – occupation, gardening was also to become a favourite pursuit of other middling members of polite society, as it had been since Tudor times a pastime of the aristocracy, wealthy merchants and manufacturers. Generally denied to the urban poor who were beginning to crowd into back-to-back dwellings in manufacturing cities and towns, care of a small plot of land from the earliest times had been a necessary means of succour for rural workers with their patches of cabbages, potatoes, herbs and "sallets".

Gardens in England, whether around village cottages, suburban villas or on grand estates have often been looked upon as a distinctive expression of the "Englishness" of everyday English life and of national character. These rather shadowy terms have been reduced to an idea and an entity, "the English garden", a form

M. Skilbeck, *Loving and Studying Nature*, International Explorations in Outdoor and Environmental Education 10, https://doi.org/10.1007/978-3-030-80751-1_16

and style which, whatever may be its Roman origins and antecedents in an island off the coast of the European continent, refers now to designs, layouts, choice of planting, an attitude toward the activity of gardening and to ways of writing about it. Thus the term, "the English garden", evokes more than a style and an activity, it conjures up a culture, a way of life now experienced in different countries around the world much as does "Mediterranean" or "prairie" gardening. Like them, it takes many different forms, according to the tastes of the gardener and the scale and location of the garden.

Many books with the words "English garden" in their title trace the historical development and the character of this gardening culture, so it is necessary here to identify only its leading features, at least as I understand them,

- "To garden", regardless of scale and location, means expressing positive, often loving, attitudes toward the natural world. Rosarian Samuel Reynolds Hole, Dean of Rochester Cathedral, wrote, "He who would have beautiful Roses in his garden must have beautiful Roses in his heart. He must love them well and always."[1] Gardeners in the English tradition seek to capture an image of nature in shaping the space, and the placing and cultivation of plants, together with envisaging their actual or imagined relation to the world of nature beyond the garden wall. The garden is a place, however circumscribed, in which to experience nature directly. Hence the English country cottage is, if not always in reality then in fashionable paintings of Helen Allingham and Alfred Parsons, something that is simple, apparently rooted in the earth, sustained by a bed of common garden flowers and vegetables and fruit trees.[2] By contrast, the artifice of Marie Antoinette's *Ferme Ornée* at Versailles where she and her attendants dressed and played as dairy maids, is looked upon as a falsification of a simple, natural, virtuous rustic life.
- While regularity, balance, symmetry and other formal elements found in other traditions (Italian, French, Dutch) are not excluded, as for instance in the use of cut stone, clipped hedges, regular terracing, statuary and fountains, the form and planting of the English garden are not dominated by these features but provide a modest setting for them. As expressed by seventeenth century poet, Andrew Marvell, "a green thought in a green shade".
- Plantsmanship is of the essence – the knowledge and skill required to grow every individual plant as best it may, in accordance with its "natural" tendency, is more a mark of pride than is the design of an elaborate array of paths or parterres.

The style of "the English garden" is often characterised in terms such as elegant, restrained, simple, modest, painterly, conservative, features of a supposed, attributed or idealised "English character". Clearly only one of several possible characterisations (consider, for example, behaviour of football fans on Britain's overcrowded stadium benches!), this is an attempt to define a style of gardening akin to the search for a polite and prideful "Englishness" in general. In the 1955 BBC Reith Lectures, the distinguished architectural and cultural critic Nikolaus Pevsner, identified as masters of English landscape art and design Hogarth, Reynolds, Constable, Adam, Wren – and the celebrated eighteenth century landscaper, Lancelot (Capability) Brown, so named because he found that large country estates showed "capability" when he accepted commissions to reshape them. Brown was a controversial figure. His landscape style, on which more below, became influential among the well-to-do whom he persuaded that, under his guidance, their land was "capable" of being remade (Figs. 16.1 and 16.2).

It was followed by Humphrey Repton (1752–1814) who designed for the middle as well as the upper class, and was satirised by Jane Austen in a mischievous passage in her novel *Mansfield Park*. In more recent times, these "improvers" have been roundly criticised for sweeping away the formality and detail of ancient gardens, thereby destroying prized heritage. Repton famously produced "Red Books" for his clients, in which an illustration of the garden as it existed was overlaid by an enticing illustration of what it might become, under his designing hand (Figs. 16.3, 16.4 and 16.5).

In 1806, Repton laid down principles which have become standard tenets of landscape gardening in the "natural" English style, "The perfection of landscape gardening consists in the four following requisites: First, it must display the natural beauties and hide the natural defects of every situation. Secondly, it should give the appearance of extent and freedom, by carefully hiding or disguising the boundary. Thirdly, it must studiously conceal every interference of art, however expensive, by

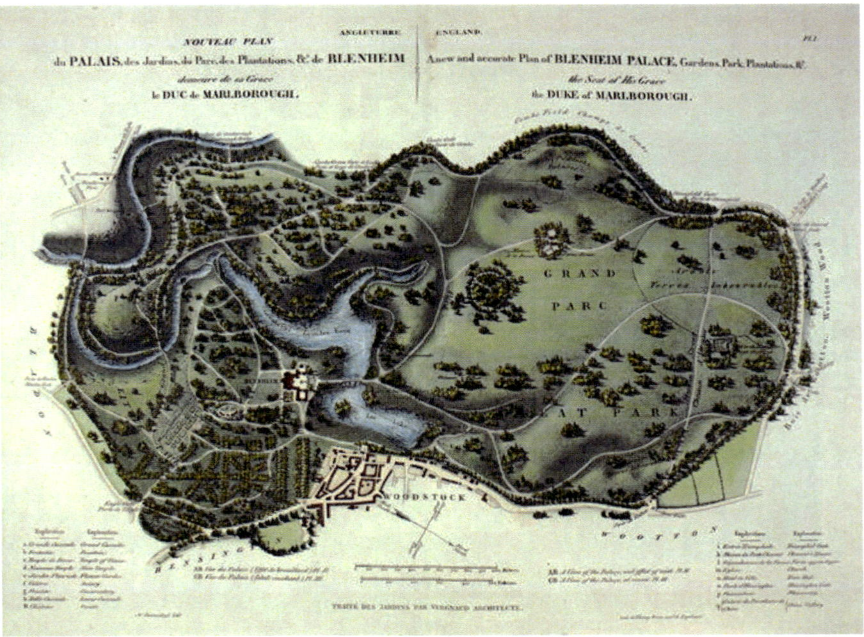

Fig. 16.1 Plan of Blenheim Palace, Gardens, Park and Plantation, 1835, shows distinctive features of Capability Brown's landscaping. *L'Art de Créer les Jardins*. Nicolas Vergnaud, CC0, via Wikimedia Commons

Fig. 16.2 View to Blenheim Palace across the Lake. Simon Q from United Kingdom, CC BY 2.0, via Wikimedia Commons

Fig. 16.3 Before. Humphry Repton's image of the estate to be landscaped at Wentworth, South Yorkshire. From his Red Books. Humphry Repton (1752–1818), Public domain, via Wikimedia Commons

Fig. 16.4 After. Humphry Repton's image of proposed landscaping at Wentworth, South Yorkshire. From his Red Books. Humphry Repton (1752–1818), Public domain, via Wikimedia Commons

Fig. 16.5 Humphrey Repton, Landscape Gardener. Northmetpit at the English Wikipedia, CC BY-SA 3.0, via Wikimedia Commons

which the natural scenery is improved; making the whole appear the production of nature only; and fourthly, all objects of mere convenience or comfort, if incapable of being made ornamental, or become proper parts of the general scenery, must be removed or concealed."[3]

An example of what part of that heritage looked like before Brown and Repton "improved" it is to be found in Colonial Williamsburg (Fig. 16.6). Its lovingly restored and reconstructed mansions and cottages, neatly laid out gardens, with closely scythed lawns, straight gravel paths, white painted picket fences, geometrical beds, clipped box hedges, topiary, flower beds of foxgloves, asters and tulips and scattered native trees, may be reached, as I recall, in a drive from Washington DC through the beautiful countryside of North Virginia or viewed on the computer screen in a generous display of photographs and text.[4] Either way, it is an open-air museum to enjoy and to study when reflecting on what Lancelot Brown and Humphrey Repton erased in order to "naturalise" the English landscape.

In destroying terraces and Italianate vistas through great avenues of exotic trees and ornamental structures, Brown sought to simplify, to eliminate fussiness. He aimed to ensure that noble architecture was set in gentle, grassy slopes and valleys, streams dammed into sinuous lakes and trees planted in apparently natural clumps (as they now appear in his surviving landscapes). He recognised that his wealthy

Fig. 16.6 The formal layout of the garden surrounding the Governor's Palace, Williamsburg, Virginia. Reconstruction during the 1930s drew on many original documents including, from Oxford's Bodleian Library, an early copper plate engraving of the principal façade together with a portion of its gardens. Photo: Jim Weiss LLC

clients were paying for magnificence and grandeur but provided it by the simplest, least assertive means – the very antithesis of Le Nôtre's Versailles and Vaux-le-Vicomte. A fine example is at Petworth, in the Sussex countryside. This attitude of hostility to French formalism became political; as against the rigidities of the French autocracy of Louis XIV at Versailles, English ideas of liberalism and democracy were to find expression in the flexibility and freedom of a manner of gardening which embodied the beauty and the "naturalness" of a gently orchestrated landscape of pasture, lakes, and "native" trees (in fact Brown planted naturalised exotics such as the "English" oak). The landscape whose "naturalness" was being recaptured had in reality been created over centuries of cultivation, pasturage and forestry, by canals, by climatic change and geological forces, and by the shaping of ports, rivers and roadways. "Nature" was an image, conceived at a certain period of time, of an Edenic countryside whose carefully selected virtues were expressed by a long line of pastoral poets in the tradition of Virgil's *Ecologues* and *Georgics* and the dream-like settings of Shakespeare's pastoral dramas.

Of course, only a relatively small minority of gardens styled "English" ever were on a grand scale, encompassing large tracts of prized rural land which might have been forcefully enclosed. Amongst others who drew attention to the social injustice entailed by the making of the great country estates was Oliver Goldsmith, in *The Deserted Village* and the ditty,

> The law locks up the man or woman
> Who steals the goose from off the common
> But leaves the greater villain loose
> Who steals the common from off the goose.
> (*The Goose and the Common*)

Mention was made above of the fondness for country cottages on the scale of small rural villages surrounded by compact fields. As towns and cities rapidly expanded in the course of the nineteenth century, the typical "English garden" had to make its way through ever smaller urban plots of expensive land. Even so, the image of a space in which nature flourished (but not so much nature's unwanted weeds!) against the inroads of asphalt, brick and concrete was cultivated in the mind as much as in the soil itself.

16.1.2 The Vicarage and Its Garden

A species of "the English garden" which has become as legendary as the (human) species itself, and now driven toward near extinction, is "the English vicarage garden". Flourishing as did the Church of England in the centuries following Henry VIII's breach with the Church of Rome, the vicarage became the protected domain of the parson in the midst of a community whose souls he was charged to preserve from evil and steer heavenwards. The parish, to whose spiritual needs the vicar tended, became much more than a spiritual space in its potential as a branch of secular governance, developed by the law-makers and civic authorities. Thus tithes payable to the church by local landowners also contributed to the funding of roads, hospitals and the implementation of the Poor Laws as well as the household of the vicar, the curates and their families. The story is too long and complicated to embark upon here. Suffice to say that, after several centuries vicars in the crowded industrial cities had their hands full dealing with urban poverty, crime, avarice and social and economic inequality on an unprecedented scale. By contrast, while these features of nineteenth century life were common across the country, vicarages in rural areas often provided opportunities for their incumbents, still attending to their flocks, to enjoy rural pursuits. Socially subservient to the landowner on whom they often depended for their livings, they nevertheless were, with other professionals, medical practitioners and lawyers for instance, part of the rising middle class who expected to enjoy gentlemanly pursuits. Conspicuous among these were the outdoor activities of the naturalists, amateur geologists, fossil and shell collectors, bird watchers, antiquarians – and gardeners.

The nature-loving rural vicars were already a declining species by the end of the nineteenth century, and vicarages were becoming redundant. Their gardens fast became a target for real estate agents and property developers. Their heyday ended with the Edwardian age and when so much of Western civilisation was brought to its knees in 1914. Nevertheless, remnants of the vicarage garden remain, as I was able personally to experience when residing in the United Kingdom in the early 1980s.

My wife and I spent a summer weekend with friends, Wendy and Frank, in a quiet Hampshire village well away from the noisy capital. For several hours on the Friday night of our arrival, over dinner and late into the night, we discussed a common topic of the time, the uses to which nuclear energy was being put, both for generating electric power and arming weapons of mass destruction. The devastation of the Japanese cities of Hiroshima and Nagasaki and the surrounding natural environment by the first atomic bombs, which brought the second World War to a sudden end, still seared the consciousness and conscience of the victorious allied nations. The "Cold War" between the West and the Soviets was then raging and the threat of "Nuclear Winter" was a fearsome reminder that not only human life on a massive scale but much of the natural world was threatened with annihilation. The combination of scientific artifice and the so-called "military- Industrial complex" seemed to many of us less a tribute to the intelligence and ingenuity of an advanced civilisation than to a disordered creation of a Frankenstein-like deformed monster.

Wendy, a science journalist, was a former Australian diplomat who had met Frank, an English physicist, in Sweden where he was director of the Stockholm International Peace Research Institute (SIPRI), a leading global agency researching and campaigning against the proliferation of nuclear-armed weapons. Frank had also been Executive Secretary of the Nobel Peace Prize winning Pugwash, the group of eminent scientists who, across the barriers of the Cold War, sought to diminish the threat of the Soviet-West nuclear confrontation. Frank was a leader among those people who not only perceived and understood the looming threat to the earth and all of its animal and plant inhabitants of the arms race, but worked tirelessly to persuade the public and political leaders of both sides to care for the earth and to find ways of peaceful co-habitation, avoiding military confrontation and devastation. Their ideas and activities remain as important today as they were during the decades of the Cold War. Nuclear devastation is not only an ever-present military threat, but as meltdown at Chernobyl and Fukushima nuclear reactors remind us, peacetime uses of nuclear energy can result in huge destruction of the earth and shocking loss and poisoning of so many forms of life.

During the years of intense hostility between the Western and Soviet camps, fear and anxiety formed a dark cloud over what should have been a world rejoicing in the advent of peace and rising prosperity following half a decade of armed conflict as German Nazism and Japanese Imperialism were subdued and reduced to pale shadows of their former hideous selves. Fear and anxiety subsided but, as that era of threat to the world has passed, it has been succeeded by a regime of consumer capitalism, materialism and complacency whose ever swelling disproportions bring no less dangerous threats to the life of humanity and the very survival of the natural world. Moreover, armed conflict seems to be ever-present.

In that Hampshire village, on the Saturday afternoon following the gloomy and worrying night-time conversations about possible global scale disaster, our hosts invited us to a calming stroll around the village. In the warm sunshine, all was still and quiet. The clouds of the previous night lifted and in soft light we passed by pleasant cottages and modest villas with their flower gardens and occasional trees in full deep green leaf. This once ancient hamlet, with surrounding fields of corn and arable land, in the eighteenth century became a village which served as a staging post on a well-travelled route from London to the West Country. Now it seemed to be so settled, rooted to the land as if forever intact and immovable.

At the end of a winding street, we came to a long paling fence, enclosing a low hedge of neatly clipped conifers. Pushing open a barred wooden gate, we made our way across a wide space of gently sloping, close-mown grass. Our hosts explained that we were joining a fund-raising afternoon tea party in the local vicarage. As we reached the top of the rise a large, rambling house of uncertain vintage rose above a line of mixed shrubs, several in pink or white flower. Behind it, to one side, was the top of a square Gothic tower in grey stone; a familiar sight in the southwest of England, it anchored house and garden to the soil and told us what the house was and why it was there. A group of some 15 or 20 people in summery garb had gathered loosely around a middle-aged man in a black satin front, white dog collar and brown tweed coat, the vicar. As he caught sight of us we were beckoned to his side. We made our way past a massive copper beech (*Fagus sylvatica*) and a monkey puzzle tree (*Araucaria araucana*), through flower beds cut into the grass: tall deep blue delphiniums, alongside red paeonies, pinks and carnations, white roses, crowds of blue forget-me-nots and many more. We were welcomed most warmly and handed cups of hot, milky tea.

Ever since this long ago visit to a peaceful, beautiful English country vicarage garden, and the convivial gathering of local folk, I have reflected on the contrast between the bitter, destructive politico-military conflicts that dominated much of the twentieth century and a calm, quiet, peaceful undercurrent of everyday life which continued unabated, a way of living in harmony with one another and with the natural world. That vicarage garden has served me as a symbol of the love and study of nature as exhibited in "the English vicarage garden". I recall this scene, in vivid detail, after the lapse of more than three decades. It left me with an image - the delightful, companionable setting of a modest event in a garden, as if that were the purpose of the garden. A once only but quite memorable visit to an English vicarage garden.

The English vicarage, house and garden has stirred depths of emotion in many English hearts, at home and abroad. Miss Read, popular author of novels and stories of the countryside, when introducing a collection of essays by Piers Dudgeon on notable country gardens and their vicar-gardeners, wrote nostalgically of, "the happiness I feel when the subject of vicarage gardens crops up, that vision of a blessed plot where it is always a summer afternoon."[5] That vision seems to be widely shared. Is it simply because so many summer functions were held there, in the local out-of-doors parish room? Or is it because the English vicarage garden embodies a wider vision of England's "green and pleasant" land?

Whatever may be the nostalgic memories stirred by Piers Dudgeon's essays, they provide insight into the horticultural skills and aesthetic values of generations of gardener-vicars, most of whom took a very practical interest in the craft of gardening. Such were the Hon. Revd. John Townshend Boscawen and his son, The Revd. Canon Arthur Townshend Boscawen, who famously gardened in Cornwall over several decades from late in the nineteenth century.

Following a visit in the middle of the nineteenth century to John Townshend Boscawen's garden at Lameiran, a tiny village west of Truro in Cornwall, the distinguished rosarian, The Revd. Samuel Reynolds Hole wrote, "Believing in Eden as thoroughly as though I had seen it ... believing that our love of horticulture and our happiness in a garden are reminiscences of our first glorious home, and longings to reproduce it ... I shall never forget that "goodly place and goodly time ... He has established, and long may he enjoy, the most perfect example of a wild garden, which, as I believe, is to be found in England."[6] This garden anticipated by decades William Robinson's *The Wild Garden*, widely heralded as the foundation of that now popular style of gardening not only in Britain. The Boscawens, father and son, were pioneers in the introduction to the south-west of England of such Australian and New Zealand trees and shrubs as hoherias, olearis and senecios. Arthur Townshend Boscawen won the Royal Horticultural Society Gold Medal and The Gardeners Chronicle Cup for new plants in 1912 for six crimson manukas (*Leptospermum scoparium Nichollsii*) raised in the rectory garden from seed sent from New Zealand. In 1922 he was awarded the ultimate accolade of the R.H.S., the Victoria Medal of Honour.

Other vicar-gardeners achieved distinction amongst the not so wild rose-growing community. The Revd. Samuel Reynolds Hole, mentioned above, cultivated, wrote about and popularised roses from his Caunton Manor near Newark, built early in the nineteenth century from proceeds of cotton manufacture by his grandfather, where he was successively curate, vicar, cathedral canon and dean. The aristocratic rose seemed a fitting interest for the owner, not of an old-style country vicarage but an industrialist's mansion. In 1869, Hole published *A Book About Roses* which went through many editions; in 1876, together with The Revd. Henry Honywood D'Ombrain, he was instrumental in founding the National Rose Society.

A highly effective preacher, Dean Hole found the seeds of moral reform in gardening which he expressed in rotund Victorian prose in *Our Gardens*: "The enjoyments of gardening being so manifold and continuous, bringing brightness to the home, health to the body, and happiness to the mind, it is for us, who have proved them, whose daily lives are made more cheerful by their influence, out of our gratitude and goodwill, to invite and to instruct others, that they may share our joy. There is great need of such persuasion."[7]

Not burying himself in the bosom of nature, like a country vicar, Hole carried out inner city missionary work in London's impoverished and squalid East End for which he was elevated by the Bishop of Lincoln, becoming a canon of Lincoln Cathedral. Such was his persuasiveness, as a horticulturalist, when Hole declared the heavily scented, fully double, coppery coloured rose *Gloire de Dijon* his favourite, it was widely planted on rectory walls around the country – and is cultivated by numerous rose enthusiasts today.

Another noted rosarian was The Revd. Joseph Pemberton, remembered for breeding a most gardenworthy strain of roses: hybrid musks. This strain was singled out for warm praise by the Oxford ancient historian and Master Gardener, Robin Lane Fox of Pembroke College.[8] As I write this, I glance through my study window at one of Pemberton's best hybrids, *Rosa Penelope*, a dense, medium sized bush holding aloft great sprays of fragrant creamy pale pink flowers with prominent yellow stamens (Fig. 16.7).

A notable horticulturist, The Revd. William Wilks, Vicar of Shirley, near Croydon in Surrey, a much-admired parish priest with a strong social conscience, became Secretary of the Royal Horticultural Society, restoring its flagging fortunes. In his large informal garden of spreading lawns, trees, shrubs and flowers, Wilks developed the famous Shirley poppy by numerous crossings and re-crossings of wild specimens collected on country walks. But perhaps best known among many well-known vicarage gardeners is Canon Henry Nicholson Ellacombe (1822–1916) of Bitton in Gloucestershire, son of another notable vicarage gardener, The Revd. Henry Thomas Ellacombe. In a 1½ acre garden, Henry Nicholson Ellacombe cultivated a huge variety of plants from common snowdrops to exotic rarities, corresponded with Kew and wrote small stylish classics of garden literature, *In a Gloucestershire Garden* (1895) and *In My Vicarage Garden and Elsewhere* (1902) (Fig. 16.8). He did not hesitate from spicing his formidable horticultural understanding with plant lore as found in the works of Chaucer, Gower, Shakespeare and Milton.[9]

Fig. 16.7 *Rosa Penelope* – gardens at Lark Edge. Photo: Helen Connell 2014

Fig. 16.8 "The Canon and Ferula glauca". Sepia photograph of Canon Henry Nicholson Ellacombe in his garden at Bitton. Public domain. Courtesy Jim Heavens, St Mary's Bitton.

Finally, although there are many more gardener-vicars worthy of mention who reached toward Eden by burying their hands in the earth and raising emblems of its reputed beauty, the famous author, Christian socialist, Cambridge Professor of History, serious naturalist and vehement social critic, The Revd. Canon Charles Kingsley, must be singled out. Remembered today as the author of the boys' adventure story of the piratical exploits of Francis Drake, *Westward Ho* (with brilliant accounts of natural scenery), and the delightful young children's book *The Water Babies*, full of nature observation, Kingsley as curate then vicar at Eversley, Hampshire threw himself into country life and crafts. In the middle of the nineteenth century, his vicarage was surrounded by farmland, forests and moors. He wrote invitingly to his fiancée Frances, enclosing a drawing of their future domain, "The view is beautiful. The ground slopes upward from the windows (of a large three storied red brick vicarage with grey slate roof) to a sunk fence and road, without banks or hedges, then rises in the furze hill … which hill is perfectly beautiful in light and shade, and colour ... Behind the acacia on the lawn you get the first glimpse of the fir-forests and moors, of which the five-sixths of my parish consist. Those delicious self-sown firs! Every step I wander they whisper to me of you, the delicious past melting into the most delicious future."[10]

Kingsley continued writing to Frances, as much a naturalist as a curate and lover. His book *The Water Babies* is not only a tale of chimney sweep Tom's escape from cruel master and urban misery, fellowship with the "water babies" and an evocation of unpolluted nature as governed by a divine redemptive spirit, it is a biologically knowledgeable account of the plants and creatures inhabiting banks and streams in the English countryside. It was written at a time when industrialism, population growth and agricultural innovation were flexing themselves for the great modern assault on nature.

Today the vicarage garden still exists here and there. It spread to parts of the erstwhile empire; for example, to a leading orchid specialist in Sydney in the first half of the twentieth century, The Rev. H.M.R. Rupp. Nephew and protégé of a "celebrated naturalist" (headmaster of Geelong Grammar School where he was a high performing pupil), Ruff showed in his care of plants the same devotion as to his clerical calling.[11]

However, the glory days of gardening vicars are past and the firm conviction of a God of biblical revelation has been loosened as has the presence of devout gardeners across rural landscapes. Yes, there are remnants and religious conviction in some quarters is as strong as ever, even if biblical revelation for theologians and religious scholars now appears to be less engrossing than deep seated concerns about deprivation, social justice and the spiritual malaise of secular life. Some of them find more conviction in a deity immanent in the natural world than as portrayed in the *Bible* and other sacred texts. The fate of the natural world, as the waves of species extinction continue and the threats to those that remain increase by the year, leads many other religiously-minded people to search for a redeeming God in nature, bracketing when not altogether dismissing holy literature. But that God, if present, is in those parts of nature that remain in pristine condition, those tracts of the earth that are protected by restorative and constructive human action from the destructive forces. If there is an enduring message to be taken from the memory of the English vicarage garden, it is that individually and together we can create havens for all creatures, ourselves included, whether in the countryside or in the canyons of huge cities, by patiently and lovingly studying nature, working to sustain it and enjoying the fruits of our labours. In so doing, we need to maintain a spirit more jocose than mournful, as did Rupert Brooke when ill and homesick in Berlin in 1912 he penned *The Old Vicarage, Grantchester*,

> … And in that garden, black and white,
> Creep whispers through the grass all night;
> And spectral dance, before the dawn,
> A hundred Vicars down the lawn;
> Curates, long dust, will come and go
> On lissom, clerical, printless toe;

And oft between the boughs is seen
The sly shades of a Rural Dean ...[12]

Those pirouetting vicars are no more, but gardens today can provide rich habitat for birds, insects, small reptiles and mammals. Gardening, avoiding poisonous herbicides and insecticides, is a form of healthful physical activity and mental solace. Gardeners necessarily interact with many aspects of the natural world in ways that readily foster sympathetic understanding of what is required to protect rather than destroy that world.

16.2 Neot Kedumim

We continue the rooting of beliefs and values in the earth and only distantly in the heavens, with a modern episode celebrating nature and culture. The place is contemporary Israel, its present environment and its ancient traditions. Some readers, appalled by the continuing hostility involving the state of Israel, the West Bank occupancy, and the Gaza Strip, may find it difficult to recognise that a project right alongside Palestine, in Israeli territory, has value in a study such as this, but the following pages may suggest another conclusion.

Since the 1948 Arab-Israel War, high tension has prevailed in the eastern land mass of the Mediterranean, with sporadic outbursts of armed conflict. The occupancy of, or the return by, Jews to Palestine, part of the *Old Testament* "Promised Land," and the expulsion of large numbers of Arab Palestinians have been vehemently resisted by much of the Islamic Arab world. They continue as a constant vexation in global international relations and have for long been a running sore in the UN General Assembly and Security Council.

It is not necessary to reach further into the circumstances of this huge, unresolved set of deep seated problems to see that religious issues from the ancient past play a role among the political, economic and social factors that are generally to the fore in the current disputations. In the ancient biblical texts, this land was a sacred promise relayed through the medium of the prophets to the tribes of Israel, in return for the tribes' suffering and their obeisance to Yaweh. During the expansion of Islam from the seventh century, the promised land became part of the Arab world. It has profound emotional, symbolic and experiential meanings for the diverse peoples of the region and far beyond.

For traditionally-minded Jews, the land and the identity of the Jewish nation are one, incorporating the far flung diaspora. In the *Book of Genesis*, God observed the Jews of the Exodus building a city east of Egypt. He said, "Behold, the people is one, and they have all one language; and this they begin to do; and now nothing will be restrained from them, which they have imagined to do." As the *Old Testament* story unfolds, the Jewish imagination beckoned its people far beyond that early settlement. To humble their pride, God scattered them, breaking the unity of the culture and confounding their language in the tower they had erected, Babel.

God had promised Abram (Abraham) "to make thee a great nation" (*Genesis* 12:1). The bounty of land stretching from the Nile to the Euphrates was given, in the first covenant, to the Hebrew people. As to why only one of the many peoples in the region was chosen is a puzzle unless it was that, as a literate people, Hebrew scribes could record the numerous messages received by the prophets from God. Other people, lacking the scribes and the zealous prophets, never produced anything comparable to the *Old Testament*, a history which, deeply etched into their hearts, gave the tribes of Israel their sense of righteous occupation of "all the land of Canaan for an everlasting possession". (*Genesis* 17:8).

Land, its control and care, thus came to have an abiding and tenacious religious significance for the Jewish people, just as it did for Australian Aboriginal people. (Sect. 16.4) However, and awkwardly for their claims, both of the other two great monotheistic religions of the region also have claims: the Muslim Arabs to a Palestine, if not in whole then in large part, and the Christians to a presence in the holy sites of Christianity.

It is impossible to acknowledge the impressive nation-building and nature-respecting achievements of the modern Jewish state without setting them in this context of conflicting claims. In the Arab-Israel War of 1948, some 70,000 Palestinians were expelled from lands which in the UN Partition Plan of 1947 were outside the area designated for the Jewish state. The Arab nations, or prominent and powerful militant forces within them, have in turn consistently denied land rights to the Jews, even conceding them no land rights at all, i.e. total expulsion. Military force and fractured diplomacy have for well over half a century prevailed over the *obiter dicta* of the *Old Testament* and the resolutions of the United Nations.

Yet, it is against this background of injustice, armed conflict, and hatred that the modern state of Israel has flourished. One of its most impressive achievements has been in arid land management. The Kibbutzim movement has been a leader in dry-land farming, pasture improvement, agronomy and drip irrigation systems. Conservation and restoration of devastated lands, xero-scaping gardening (minimum water use) have been among the many practices pioneered, adopted and developed in Israel. They have resulted in some of the world's leading achievements in dry-land farming and horticulture.

In occupying disputed land, the Jewish people have generally made productive and environmentally intelligent use of it. While this does not justify occupation where it has been, and continues to be, illegal (as in the West Bank), as a vast program of sustainable land use, its value is undeniable.

One project of considerable environmental, cultural and historical interest has been established in one of the politically sensitive areas of Israel, right on the border with Palestine. This is *Neot Kedumim*, an extensive park-like dry land garden designed to bring the life of the Jewish people as portrayed in the *Old Testament* into the imagination and understanding of people today, Jews and non-Jews alike. Through a variety of precincts spread over some 550 acres (222 hectares) the project retells ancient stories and events through individual plants mentioned in the *Bible*, plantations, land use and water features. Historical, religious and environmental values are combined in a unique, inter-disciplinary public education facility.[13] (Fig. 16.9).

Fig. 16.9 Young almond trees blossom in the Garden of Choice Products at Neot Kedumim. c.1980. Image: Hareuveni, Nogah (1980) *Nature in Our Biblical Heritage*. Translated from Hebrew and adapted by Helen Frenkley. p.4. Courtesy Neot Kedumim Park

The origins of this remarkable project lie in the interdisciplinary scientific, ethnographic, historical and biblical research of the parents of its founding director, Nogah Hareuveni. Ephraim and Hannah Hareuveni were field naturalists and laboratory scientists whose field trips early in the twentieth century took them and their children across the country's valleys, mountains, semi-arid lands and deserts, observing and collecting plant and other specimens. In the Hebrew University on Mount Scopus, they established a Museum of Biblical and Talmudic Botany.

Destroyed in the Arab-Israel War, this museum has been re-conceptualised at Neot Kedumim, as an open air museum and demonstration site on the rugged and virtually barren hill country between Jerusalem and Tel Aviv. In creating, now over several decades, a set of Biblical and Talmudic garden sites, Nogah Hareuveni, colleagues and volunteers envisaged an environmental-historical-religious study centre. A combination of opportunities for varied sensory experiences, active labour on environmental restoration projects, engagement with museum displays and textual study connect project workers and visitors alike with natural and cultural worlds, as seen through the lenses of the Biblical scribes and the Talmudic sages and as handed down in the Jewish Oral Law.

Unsurprisingly, Neot Kedumim has become a popular tourist venue while continuing to provide systematic study opportunities for young and old. The term "Neot Kedumim" translates into "pleasant pastures" or" habitations of old". The collection of theme-based sites recalls the physical and cultural settings of up to 3000 years ago, the period during which early books of the *Old Testament* were being written. Clearly, there is a strong effort at proselytising in these displays. They also suggest the retracing of virtuous footsteps along several pathways, back to Eden, when the world was young, happy and contented, fully in tune with nature and nature's mysterious God.

The natural features, pastoral and agricultural landscape in this Biblical and Talmudic garden have been appositely named: the Vineyard of Isaiah, the Field of the seven Variations at the Hill of the Menora, the Forest of Milk and Honey, the Dale of the Song of Songs, and others. They are signposted, with quotations from relevant religious texts.

Early land restoration at Neot Kedumim depended in part on imported soil, a practice that might be frowned on nowadays. Reservoirs were constructed; ancient terraces remade; wine presses, ritual baths, pathways and gathering places were established to facilitate understanding and enjoyment. These, together with newsletters, pamphlets, books and films, have been designed to serve pedagogical purposes rather than light entertainment.

A central concept is that of the Israelites as a pious God-fearing, not so much a stern but loving people of the land and the book. Grain-threshing floors, water wheels, olive presses, fragrant plant species, Abraham's tent, date palms and the Valley of Jericho symbolise a people with the *Old Testament* in one hand and a spade or scythe in the other. All is in honour of the one God who in their texts controls all nature and the destiny of nations.

Individual and clumped plants recall childhood images, the burning bush for example. A common misconception is exploded; the land of milk and honey was, according to prophet Isaiah, no place of gently flowing streams and verdant pasture but one of tangled briars and thorns. The ancient Israelites were pastoralists, herding sheep and goats, in need not of arable land but pasture for grazing. Briars and thorns, which were fodder for milk-producing goats, grew along with the vineyards destroyed by invading Syrians. Their flowers yielded nectar and pollen for honeybees. The land flowing with milk and honey symbolised not flourishing agriculture, but wild vegetation which was nonetheless productive.

It is refreshing to move from the ancient religious hierarchy of being and the complexities of the soul to the efforts that Moses and his followers made to wrest a living from the lands they were to occupy. At first, the forested hillsides and rugged escarpments seemed menacing to Moses' scouts, accustomed to the sandy flatlands of Egypt and the vast open spaces of the Sinai. Animals favoured by the Israelites were those trained by order of *Genesis* to serve them; sheep, goats and bullocks. Wild animals such as lions and wolves were threatening, to be eliminated where possible. It was a land for the chosen people and their herds.

As so often, land prospectuses offering milk and honey can disappoint. As doomsayers point out, Isaiah made gloomy prognostications of barren hillsides, desolate villages, rocks and thorns, to be restored but fit only for oxen "and trading of lesser cattle" (*Isaiah* 7:19–25). Goats and bees appear to have been pushed aside. In *Deuteronomy*, God offered to the humbled and suffering people in their forty years in the wilderness, "For the Lord thy God bringeth thee into a good land; a land of brooks of water, of fountains and depths that spring out of valleys and hills; A land of wheat, and barley, and vines, and fig trees, and pomegranates; a land of oil olive, and honey; A land wherein thou shalt eat bread without scarceness, thou shalt not lack any thing in it; a land whose stones are iron, and out of whose hills thou mayest dig brass." (*Deuteronomy* 8:7–9).

This might seem a land for the taking and easeful life, but the beguiling vision was conditional, first on a strict obedience to God. Then there were battles to fight and unity to establish among the Israelites themselves. Finally, it was only by intense, persistent and back-breaking labour that the land could be brought to yield its future bounty, through the rhythms of the seasons and the vicissitudes of the natural cycle.

Unlike their polytheistic neighbours (and even when in their own wavering, Baalist phase), the Israelites accommodated all of their experiences, their aspirations and beliefs, to one God, to a single universal order of nature. When sowing and harvesting occurred at different times in different parts of the country, single dates were kept to for ceremonial purposes. When a sheaf of barley was brought to the Temple, "One people brought the same one crop on one date to one Temple, an offering to One God in one city, Jerusalem."

Although Neot Kedumim, as both a physical obtrusion and a symbol of Jewish triumphalism, would be abhorrent to many Palestinians and to some modern Jews, it is an impressive and unique attempt to establish complex interrelationships between the historical development of an ancient culture and its understanding and use of land, its flora, fauna, climate, agriculture and ecology. Its purposes include scientific and historical research, exegesis of obscure texts and public education. A central theme or message is that when land is used productively it can and should be sustained over the generations, a permanent resource and, in this instance, an expression of how best to interpret the *Book of Genesis*. This is not as man's dominion, but as humanity's stewardship of nature together with reasoned understanding and enjoyment of its fertility and beauty.

16.3 Children Study Nature

When asked to reflect on some pleasant event or activity in their early school days, adults will, regardless of their age, often talk about nature study. They might refer to some classroom activity, like painting a flower, composing an essay on bees or birds, or observing seeds sprouting on damp blotting paper. Perhaps an outdoor event comes to mind – a ramble through a park or woodlot, planting lettuce in the school garden, a tree on Arbor or Environment day, or a visit to an aquarium. The activities vary but have in common direct engagement with plants and animals, handling and in some way manipulating them and interpreting them. As recalled, these are usually simple activities involving our senses – observing, handling, listening, feeling and close observation. As such they survive in memory, sometimes for decades. They tend to be agreeable memories, the happier school times. By contrast, earlier school days as portrayed in the novels of Charles Dickens, *Hard Times* for example, were far from agreeable as children experienced forced and mechanical regimes of adult discipline and mechanical instruction which dampened the spirits by eschewing the imagination. In this section, we trace some of the ideas and movements that saw the end of one regime and the dawn of more personally engaging and enriching school life. Nature study played a major part in this change.

16.3.1 Antecedents

In his writings on education, notably *Emile*, the Swiss philosophical thinker and essayist, Jean-Jacques Rousseau (1712–1778), presented to his readers and followers a strikingly novel idea about childhood. Instead of treating the child as a miniature, immature adult to be drilled into submission, Rousseau found intrinsic value in this stage of life and laid out an ideal educational regime based on deep understanding of the inborn nature and developing powers of the child - emotional, intellectual and physical and of the need for children to grow into a virtuous self-sustaining adulthood. There were antecedents in educational thought and the practice of some past teachers for these radical views, but the prevailing model was preparation for adult social roles and the life of the child was generally cast into this mould: boys as future agricultural workers, businessmen, tradesmen, professionals or military and political leaders; girls as their wives and as mothers. In the famous opening words of his *Social Contract*, Rousseau proclaimed that, "man is born free but is everywhere in chains". His notion of inherent freedom contained the idea that man is also born good, a rejection of the ancient belief in original, enduring sinfulness (Fig. 16.10). Many other writers built on the idea that human character is shaped by circumstance and will benefit from improvements in them.[14]

Taken up in the course of the nineteenth century by the Continental educators, Heinrich Pestalozzi (1748–1827) and Friedrich Froebel (1782–1852) (Fig. 16.11), founder of the kindergarten, among others, this Rousseauist developmental conception of childhood gradually led to new teaching practices in the emerging public school systems whereby to blackboards for teachers and slates for children were added classroom games and activities in order to stimulate discovery, problem solving and the ability to design and construct using child-sized materials. These served as a prelude to activities involving observation and inquiry in the wider environment including the natural world. However, traditional learning by rote continued, with large classes of children of different ages spread along benches (the "forms"), and became more common in an age of mass schooling which in England and the Australian colonies, for example, included the system whereby payment to teachers varied with the testable performance of their pupils in reading, writing and

Fig. 16.10 Jean-Jacques Rousseau (1712–1778), who persuaded philosopher Immanuel Kant and many others of the importance of feelings in human life and thought. Portrait by Maurice Quentin de La Tour. Public domain, via Wikimedia Commons

arithmetic ("payment by results"). Schools might be in operation for only a few months in the year, were often poorly attended and staffed by inadequately trained teachers who had started at age 13 or 14 as pupil teachers, teaching younger children and being instructed by teachers not much older than themselves.

Underlying Rousseau's educational ideas was a theory concerning the nature of human understanding, particularly the part played by emotions and sentiments in our perceptions and relationships with the world, both of other humans and that of nature. For this, he was admired by such luminaries as the Scottish historian and philosopher, David Hume, and the austere German philosopher, Immanuel Kant. The understanding Rousseau had of how human life is shaped and enriched by the cultivation of feeling, including feeling for the beauty of nature, became one of the (many) driving forces of the Romantic movement in literature, poetry, art and music.[15] This movement influenced developments in education late in the nineteenth century, to which we now turn.

Mass elementary schooling, enforced by legislation, was one of the great achievements of the nineteenth century, commencing in eighteenth century Europe and America and reaching fulfilment early in the twentieth century. It had many bright moments but was increasingly criticised for its dullness, routine, conformity and failure to create the morally cultivated, socially competent and responsible citizenry its advocates envisaged. English politician, author of the 1861 Revised Code (payment by results in the spirit of public charity), Robert Lowe, supposedly remarked on the passing of the Third Reform Bill of 1884 when the franchise was further extended: "We must educate our masters." Schooled they were, up to a point but in ways that fell far short of what the idealistic reformers and even the political realists expected.[16]

Fig. 16.11 Friedrich Froebel (1782–1852), pioneer of child-centred schooling and founder of the kindergarten movement. Lithograph published Syracuse, N.Y. by C.W. Bardeen, publisher, c 1897. Source: Library of Congress, Prints and Photographs Divn. C.W. Bardeen, publisher, Public domain, via Wikimedia Commons

Under the banners of "the New Education" or "Progressive Education", a fresh reforming spirit arose, late in the nineteenth century and early in the twentieth century, notably in the USA, but also in Great Britain, Continental Europe, Australia and other colonies of the British Empire. As industrialisation increased in pace and spread, growing populations crowded into urban slums, canal, rail and road networks spread across the land, forests fell under axe and saw, mines and then slag heaps, mills and factory chimneys chequered the landscape, it was increasingly realised that economic progress imposed a heavy burden on other aspects of life. Elementary schooling, and it was not more than that for the masses, was seen by its critics to mirror in its operations the drudgery of work in the factory, on the land and in the counting house. The gospel of universal happiness preached by the Utilitarians, Jeremy Bentham, and James and John Stuart Mill, did not seem to reach through the school gates into the gaunt, sooty brick buildings and the grim, grey classrooms.[17]

However overstated their concern may have been, the cultural critics, John Ruskin and William Morris, were by no means lone voices in the debates and discussions about the values of Victorian society. Socialism of a reformist, not a revolutionary Marxist kind attracted not only the working classes but many intellectuals. In Great Britain, a leading Fabian socialist, Sidney Webb, as a member of the London School Board, campaigned for system-wide reform of mass schooling and technical education as necessary for amelioration of social conditions: "I want no class of hewers of wood and drawers of water; no class destined to remain there and forbidden from rising … I cannot believe that we are only to provide the means of instruction for a certain limited number of people who we think will rise while the rest are to toil for our convenience."[18] In the United States, social, economic and political reform were widely sought as the country moved from the destruction and heartbreak of the Civil War into a confederal belief that its "manifest destiny" was one of constant advance and achievement. Education was seen as one of the domains in which the country would - and should - achieve greatness in the world.

16.3.2 Scope of the New Education

But what were the reforms whereby educational greatness was to be achieved? There were, naturally, diverse views ranging from a strengthening of the status quo, more effective disciplinary measures, greater submission to schoolmasterly and religious authority, concentrated effort by teachers and pupils alike, a closer alignment of the school curriculum to the requirements of a burgeoning, capitalist economy, to greater overall efficiency. Supporters of such approaches ranged up in criticism of the advocates of what was termed the New or Progressive Education. The latter wanted the forms and screwed-down desks of the conventional classrooms to be replaced by moveable tables and chairs, the teacher's raised dais to be lowered to the floor, the rigid syllabus to be loosened in favour of extending children's interests and experience, the method of learning by rote and the routines of instruction replaced by freer child activity, engagement more with everyday life and nature than with text books and a generally brighter, happier atmosphere in place of dull drudgery, strict discipline and harsh punishment.

In the United States of America, philosopher and educator, John Dewey, articulated a liberal democratic alternative to the perceived authoritarianism of the old order (Fig. 16.12). In England, Edmund Holmes, the retired head of Her (His) Majesty's (school education) Inspectorate, attacked the regime of payment by results and presented an image of the child-centred school of the future.[19]

Among the critics of the Revised Code and advocates of the "new education" the Scot, Simon Somerville Laurie, is of particular interest for his impact on teacher education in the colonies. Appointed to the University of Edinburgh as the first Professor of Education (the Bell Chair of the Theory, History and Art of Education) in Great Britain, he delivered his inaugural lecture in 1876, "The teaching profession and chairs of education". Laurie campaigned in books, articles and addresses for both a humanistic primary school (secondary and university too) curriculum and a highly educated teaching profession for which he believed Scotland had long been a world leader, ahead of Continental Europe and America. As if in support of this claim, there is a story told by Scottish geologist, Hugh Miller, in his *The Old Red Sandstone,* of a Scottish student who at university had topped his final year. He became a primary school teacher, whereas the third placegetter became a professor. Laurie wrote that, for 200 years, "The leading teachers of the parochial schools of Scotland have been supplied by the universities." Another example of the Scottish valuing of education is Andrew Bell who when working as an army chaplain in Madras, where he superintended an orphanage, introduced tuition of younger by older students under the overall direction of a single teacher. This was the so-called monitorial system which evolved in England and its colonies into the widely practised (and subsequently reviled) pupil-teacher system.

Fig. 16.12 John Dewey (1859–1952), leader of modern progressive education, as a young philosophy professor in the 1890s. Portrait. Artist unknown. Collection Malcolm Skilbeck

Mistrustful of what he saw as the encroachments of science and technical education into the upper years of the elementary school, Laurie came close to, but did not fully embrace, the idea of study of nature as a new element in schooling. His views derived in part from his sense-realist idealist theories – that schooling of children should cultivate sensory development and draw on everyday objects and surroundings, from a reverential attitude toward Nature and from his belief that literature would foster children's imaginative and emotional life. There are echoes of all of these in the various formulations of school nature study in Great Britain, the USA and Australia.

In his views of science, including his criticism of botany as an appropriate part of the primary curriculum, Laurie, showing touches of hostility to Darwin, Spencer and Huxley, would have been at odds with a strongly emerging current which treated nature study as a branch of elementary science – a prelude to systematic study of the sciences. To the extent that he conceptualized nature study at all, it was humanistic, not even a "subject" but part of the child's everyday world to be taken up in the study of objects, or entered through reading, drawing and the teacher's exploration of sensory experience. This is how many of its advocates and practitioners saw nature study, more a *way* of educating children than a conventional subject of instruction.[20]

For all this to be accomplished in a way that respected the primary school child's overall development as a person and future citizen, it was of paramount importance that teachers should themselves be highly cultivated, intellectually and morally. Accordingly, Laurie was tireless in his demands for teacher education to leave behind the trappings of the monitorial and pupil teacher system. Ultimately it should become a university programme of study or at any rate avenues should be open – as they were in his day in Scotland – for high performing normal school or training college students to move from the methods-practice foundations of those institutions to the proclaimed intellectual and high moral foundations of academic-professional study in the university.

In his time, Laurie had an international reputation, attracting to Edinburgh University, among others, his fellow Scot, Alexander Mackie, and Scottish-New Zealander, John Smyth. The subsequent achievements of these two, respectively as Directors of Sydney and Melbourne Teachers' Colleges and inaugural Professors of Education respectively at Sydney and Melbourne Universities, show the strong if indirect influence Laurie exerted. While there were antecedent figures in New South Wales and Victoria, it was the Mackie-Smyth regime, in tandem with Public Education Directors, Peter Board in New South Wales, and his opposite number, Frank Tate in Victoria, that marked the transformation of Australian teacher education in the twentieth century from something much less than sound craft training into the beginnings of an intellectually grounded profession (Figs. 16.13 and 16.14). Laurie's dream of a fully university-based profession was not realised in England and Australia until decades later, but the groundwork was being undertaken at the very time that the new education (seen as a transformation of schools) was getting under way at the beginning of the century. For nature study, as for the other innovations, this proved to be a serious dilemma: there were syllabuses and teachers' guides and texts; there was leadership and there were ideas. Lacking, however, was the highly educated teaching force, with a clear understanding of the aims and values, the underlying philosophy and ideas, and the general culture to which nature study was to be a point of entry and a continuing source of inspiration. Also, many other developments required the attention of administrators and teacher educators – the overall expansion of public schooling, financial challenges, other areas of the curriculum and, as the twentieth century progressed, the impact of the First World War, post-war changes and the onset of economic depression.

For both Mackie and Smyth, there were to be major frustrations and disappointments as their appointments extended into the second and third decades of the century. In the first decade, however, the light of optimism burnt brightly. With the powerful collaboration of teacher educators and state school leaders, it seemed that the reach of Laurie, the American Colonel Francis Parker at Cook County Normal School, and other highly esteemed educational luminaries in Great Britain and the USA was extending into the heartland of the emerging state systems in Australia.[21]

Fig. 16.13 Frank Tate (1864–1939), educational reformer and Director of Education in Victoria in the early twentieth century. Photographic portrait c1910. 00287. Federation University Australia Historical Collection

Late in the nineteenth century, Australia was moving from an aggregation of colonial states to an independent, unified federation, achieved in the Constitution of 1901. This alone led to optimistic forecasts of a prosperous future, periodic economic depressions, droughts and floods notwithstanding. Education was, and (except for the tertiary sector) remains, a State/ Territory responsibility with substantial Commonwealth funding. At the beginning of the twentieth century, Australia was fortunate in that in its most populous states, Victoria and New South Wales, the state Departments of Public Instruction were directed, as noted above, by two notable educational leaders, Director Frank Tate in Victoria, and Director Peter Board in NSW. What is more, and as also noted above, the Victorian (later Melbourne) and Sydney state teachers' colleges were headed by S.S. Laurie's former students, respectively John Smyth and Alexander Mackie, both capable and highly regarded leaders of their profession.

Tate and Board, Smyth and Mackie, were knowledgeable about and enthused by the New or Progressive Education movement, reading about and travelling abroad to study the ideas and practical work in schools and colleges whether in England, America or Continental Europe. They also had in many of their senior colleagues, inspectors, administrators, head teachers, lecturers and others, people who shared their enthusiasm for educational reform. In her study of the reform movement, which has New South Wales as its focus, Dorothy Kass presents strong evidence that Australia was an equal force in a trans-continental movement to change the character of elementary schooling, with nature study ideas to the fore.[22]

Fig. 16.14 Peter Board, educational reformer and Director of Education in New South Wales in the early twentieth century. Photographic portrait 1923. Source: NSW State Archives: NRS 4481, glass negatives. [7/15982] ST8607. Image: State Library of New South Wales (Mitchell Library) GPO1 17,173

South Australian school teacher, journalist and author, William Catton Grasby, was a significant figure in the Australian nature study movement, acknowledged as such by one of the American leaders, Liberty Hyde Bailey of Cornell University (author of a much cited book *The Nature Study Idea).* (Fig. 16.15) Growing up on a modest farm, Grasby had risen from a pupil teacher to senior positions including Headmaster of Roseworthy Agricultural College. He first taught nature study in a South Australian primary school and later became a leading agriculturist and journalist in Western Australia. In between, he was briefly a businessman and travelled extensively in England, Continental Europe and America studying educational changes and meeting leading educators. A resulting book, *Teaching in Three Continents* (1891), calling for reform of schooling and teacher education, had a considerable impact including on Victorian Director of Education, Frank Tate. Grasby had six major proposals for the reform of public schooling in Australia and over the following decades all were realised:

1. Free public schools
2. Abolition of Result Examinations [and payment by results]
3. Abolition of Pupil Teacher System and the Employment of Adult Trained Teachers
4. Practical Education – Kindergarten, Manual, Agricultural and Horticultural Training
5. Reform in School Buildings

Fig. 16.15 William Catton Grasby, pioneer of school nature study in Australia. Photographic portrait. c1890. State Library of South Australia. B 37455

6. Public Secondary Schools and Agricultural Colleges.[23]

As a South Australian school teacher, Grasby early in his career had introduced nature study to his pupils.[24]

Given the growing interest throughout the nineteenth century in the natural world, its animals, plants, geology and processes, among amateurs and then professional researchers, it is not surprising that educators responsible for the expanding system of public schooling should turn to their ideas and findings as sources for school curricula and ways of teaching children, who were obliged to attend schools regardless of their interest in the traditional regime of the 3Rs and strict institutional rules of conduct.

The eminent Harvard scientist, Louis Agassiz, had taught his students by "natural methods" that included direct experience of the open air natural world alongside textbook and laboratory. Oxford Professor of Botany, J.S Henslow, took his students on nature walks and worked with labourers' children to cultivate their powers of observation. Charles Darwin, as a theology student, had joined Henslow's walks in the countryside. His theory of evolution by natural selection had grown out of his global travels and decades of direct observation and reflection on diversity and change in nature. Amateur naturalists abounded, studying birds and animals in the wild, rock formations, fossils, aquatic environments and atmospheric conditions. Alexander von Humboldt (Fig. 16.16) attracted global interest by his voluminous accounts of travels in remote regions of Central and South America.[25] It was as if, in the nineteenth century, the European mind was expanding to embrace the whole natural world, not only the earth but the skies above and the whole cosmos.

Fig. 16.16 Alexander von Humboldt (1769–1859) in South America. The breadth of his studies of the natural world were described by biographer, Andrea Wulf, as "the invention of nature." Portrait by Friedrich Georg Weitsch, 1806. Alte Nationalgalerie. Public domain, via Wikimedia Commons

Nature study was a field of universal interest, literary as much as scientific. A notable example is nineteenth century French novelist, George Sand, who, in the spirit of Rousseau, accorded supreme value to simple, unaffected rural life. In her rustic or pastoral novels, letters and autobiographical writings, Sand wrote most fondly and knowledgeably of rambles in the French countryside, visits to beauty spots, wildflowers (garden flowers too), trees, streams, weather conditions, birds, insects and the customs and manners of peasants in her neighbourhood of Nohant, central France. "Nature," she wrote in *The Devil's Pond*, "is eternally young, beautiful and generous."[26]

Translated into schooling what may be fairly called a Romantic as well as a scientific attitude encouraged, nay depended on, children's observation of natural phenomena and events, the use of all their senses (imitating bird calls and counting eggs in nests), excursions into the bush, parks, zoos, ponds and gardens, pleasure in drawing, modelling, recording and conversing about encounters with nature in its many forms and manifestations. Activity learning was a watchword of progressive education and this meant a full and active engagement with the sensory and social worlds of children's individual and shared experiences. A nature study ramble, followed by classroom discussion, drawing, writing, reading, including poetry, singing and model-making, was the perfect exemplar of, "activity method". While it was, by its advocates, distinguished from the study of science, in the hands of some teachers it would have fostered a scientific outlook or frame of mind: systematic inquiry, gathering data, experimentation, the formation and testing of general ideas. At the same time, it fostered subjectivity, aesthetic feelings and reverence. It framed "nature" in terms of personal experience, not of the disciplinary domains of objective "scientific knowledge".

16.3.3 Nature Study in the Primary School

At the beginning of the twentieth century, the public authorities in Victoria and New South Wales introduced Nature Study into the primary school syllabuses and into the programme of study in the teacher training colleges. Guidelines, suggestions on what and how to teach were published in Departmental journals, together with pamphlets, magazines and books. Lectures and summer schools were instituted to support the sought for innovations. School inspectors were required to report on how well teachers were managing the new requirements, both their own nature knowledge and their readiness and ability to engage pupils in the lively, active, immediate study of plants, animals and the wider natural environment - its geology, atmosphere and responsiveness to human presence and intervention. For some leaders of the movement, politicians as well as educators, one of the objects of the new syllabus and methods of teaching was to encourage rural students to live on the land and to become knowledgeable farmers, conversant with natural processes and forces. Hence, in practice, school nature study was often linked to agricultural education. Agricultural colleges provided residential summer courses for school teachers and their lecturers wrote pamphlets and books and gave informative and inspirational talks on nature study. But overall, Nature Study was directed towards a more general, sympathetic understanding and appreciation of the natural world, to see the value of conserving and not carelessly exploiting the resources of the earth, to preserve areas of "wilderness"and to enjoy the beauty of flowers, plants and animals, birds especially. School gardens and tree planting were encouraged (Fig. 16.17). Arbor days and Gould League Bird Days were to be celebrated by appropriate planting of trees while observation of bird behaviour included the mimicking of their calls.[27] From the middle of the nineteenth century, in Great Britain, America and Australia, books, magazines, lectures and courses of study for teachers, clubs and societies abounded, surveying the world of nature more generally. Many of them advised teachers on schemes of work, object lessons, nature excursions, bird watching, school gardening and tree planting, ways of correlating the painting and drawing of plants and animals with literature and poetry aimed at developing in children a sympathetic understanding of the natural world.

Although restricted access to State Department of Education records prevented identification of the Department teachers who joined the Field Naturalists in Broken Hill in the 1920s (Chaps. 2 and 3), it is likely that at least some of them had been

Fig. 16.17 Pupils gardening at Croydon Park Public School, suburban Sydney c 1910. NSW State Archives Department of Education History Unit; NRS 15051, Photographic collection. NRS-15051-1-10-[530]-1

students at Sydney Teachers' College in the first decades of the twentieth century. They would have studied the obligatory new course in nature study, and, later, have welcomed the opportunity Albert and Margaret Morris and their field naturalist colleagues provided them to consolidate and extend their newly acquired understanding of nature.

In the decades following the First World War - itself such a fearsome assault on the natural world - nature study continued as part of the prescribed or recommended curriculum, but after the Second World War was absorbed by biological sciences and then displaced by physical science when nuclear fission and the technology of space-probing rockets became fashionable themes of school science. However, the emerging field of environmental education became by the 1980s a more substantial and variegated vehicle for the earlier and simpler nature study idea.

We do not know just how effective, or in what ways, the studies and activities by schools early in the twentieth century were in developing a community-wide understanding of the natural world. It is, however, reasonable to suppose that the feelings and insights so encouraged in children by teachers who themselves became active leaders of the new movements did indeed lead to a growth in the whole community of a mentality supportive of nature conservation and preservation, grounded in affection and concern. In schools, environmental education, social studies and parts of the science curriculum have now taken up this responsibility. One of the leading figures in the emergence and spread of environmental education – heir to nature study – is co-editor of the Springer series of which this volume is a part, Annette Gough (Foreword).[28]

16.3.4 *Learning to Value Nature*

Today, there is an undeniable global environmental crisis, fuelled by atmospheric and oceanic pollution, excessive warming of the atmosphere and the oceans, forest and habitat destruction, desertification, species extinction and a huge accumulation of waste from growing levels of production and consumption. We are constantly made aware of these and associated problems of modern life, and there are numerous public and private efforts to address them. However, it seems that the drive for continuous economic growth and societal dynamics of consumer capitalism are, as yet, far more powerful than any countervailing force. Human ingenuity, in combination with a clear understanding of what is required to sustain the natural world in all its glory, beauty and value, may possibly suffice to ensure survival, but it is evident that both study and love in full measure are necessary, not only in schools and through other educational agencies. The contemporary young people's international climate protest movement entails a school strike (the slogan of teenage Greta Thunberg is *Skolstrejk för klimatet*). Appropriately she shares her surname with Carl Peter Thunberg (1743–1828), a pupil of Linnaeus and one of Sweden's most distinguished naturalists. Hers is a universal call requiring a society-wide response, including schooling but reaching far beyond it.

Yet school nature study and the infusion of the whole educational process, at all levels and in all stages, with this loving understanding, enjoyment and appreciation of nature continue to have a vital role to play. What children and young people learn to value and appreciate in their school years, as the early advocates of nature study knew, will be one of the most important pillars of our capacity to achieve not dominion over nature, but an active and productive partnership with her. We are facing a global crisis and in her addresses to world bodies, such as the World Economic Forum, where Greta Thunberg was almost alone in not mincing words: "I want you to act as if our house is on fire. Because it is."[29]

16.4 "They Stole Our Land and (Thereby) Destroyed Our Culture"

In January 1788 Captain Arthur Phillip, sailing from England with a convoy of eleven ships carrying convicts, a military garrison to guard them in their new prison and provisions for a new settlement, landed on the shores of Botany Bay. This was the name given by his predecessor, Captain James Cook, and his botanical shipmate, Joseph Banks, sailing along the east coast of Australia, to acknowledge a shoreline rich with grasses, shrubs and trees new to Western science. Finding no fresh water and inadequate harbourage, Phillip and his party soon headed north to better pickings at a splendid harbour which he named Port Jackson, now familiarly known as Sydney Harbour.

16.4.1 Australian Aboriginals and European Settlers

In some Indigenous communities, it is not possible to mention the names of dead people or look at images of them. People offended by seeing or reading about deceased Indigenous persons should not read this section.

At first, the native inhabitants of New Holland, which was wrongly designated by officialdom as "uninhabited" because "unowned", were curious about the new arrivals, and not overtly hostile. Some even welcomed these strange, pale-skinned creatures, who may have been spiritual beings with their extraordinary white-winged means of transport across the ocean. That early welcome changed, however, as the native people realised that their lands were not being shared, according to custom, but taken over by invaders. The hard-hoofed animals of the settlers were cropping the grasses and herbs to the roots, sending off the kangaroos, emus and other animals which for the Aboriginal inhabitants were not only a food source but part of an elaborate cultural system of totems and ancestral spirits, a system which had perhaps more significance (and value) for them than the panoply of saints and martyrs had for the nominal Christian arrivals.

At the time of Phillip's landing, the Aboriginal population of Australia has been estimated at three quarters of a million, which can only be an approximation. This population was thinly distributed across the entire continent. As a result of disease, whose vectors had travelled with the new arrivals, shooting of native creatures by the invaders, loss of land and, in all probability, the devastation of the spiritual domain from which land occupancy and its ceremonial use were inseparable, that considerable population declined to the point where some believed the Aboriginal population was faced with extinction. That has not happened, although the population of "full-blooded" Aboriginals reached vanishing point during the twentieth century. The population of indigenous communities began to recover, but there was much loss of traditional knowledge through the disruptions along the way.

Biographies of notable Aboriginals living a substantially Western existence are a poignant reminder that notwithstanding the benefits and overall success of that existence, a strong sense of loss and injustice persists: "I've struggled to live up to the whiteman's standard, but here I am, old and good for nothing, and what keeps coming back to me? Dances, singing, stories the old people used to tell. Every night I lie in bed and sing myself to sleep with all my old corroboree songs."[30] Many people of Aboriginal descent through mixed marriages are working hard to maintain ties with ancient life and to alleviate living conditions for fellow Aboriginals whether in the crowded suburbs of the major cities, country towns or remote inland settlements.[31]

Historian, John Ramsland, has recorded the struggle to attain educational provision for Aboriginal people comparable to that for the white community. Some Aboriginals, such as Worimi Elder Les Ridgeway, struggled successfully to overcome the barriers, rising to positions of leadership in their communities and beyond.[32] (Fig. 16.18).

In contemporary Australia, some Aboriginal people live a semi-traditional life in northern, north-eastern and north-western regions, but few if any are either able or find it possible to complete that life by practicing the full repertoire of rituals and ceremonies of their ancient religion, which required the closest possible identification with particular stretches of land. Nor are they able to manage the land as in the past by extensive burning, including the undergrowth of forests and grasslands, and the seasonal harvesting and processing of fruit, roots and other indigenous food sources.

While Aboriginal communities today no longer live traditional ways of life, many are actively seeking new ways of engaging with their heritage. Understanding and valuing traditional approaches to caring for the land has been a significance focus, with relevance for the wider Australian community in the context of contemporary environmental challenges. The Aboriginal culture is now coming to be widely admired - an ancient and continuing history extending over perhaps sixty to seventy thousand years - Aboriginal understanding, care and management of the land, the healthy survival of the indigenous people in the most adverse circumstances, and spiritual richness.[33] Following failure of inept policies aimed at forcibly assimilating Aboriginal people into white, European culture, a national policy of self-determination is in place, within the framework of Australian law and governance.

Recent studies have shown that, despite long-standing characterisations of Aboriginals as nomadic hunter-gatherers whose culture was that of what nineteenth century anthropologists labelled as primitive, at the pre-religious stage of magic, in fact they traditionally farmed the land in a form of proto-agriculture, in places in semi-permanent settlements and had a complex system of religious beliefs and practices comparable to those brought to the country from Europe and Asia.[34]

What did it mean to "farm" the country Aboriginal style? Traditionally, there were two main forms of farming, proto-agriculture and aquaculture. In south-west Victoria, adjacent to Budj Bim (Mt Eccles), are the remains in volcanic stone of an elaborate aquacultural system some 6600 years old. Traps, weirs, channels and ponds extending for more than 30 kilometres around a lake [Lake Condah and Darlot Creek] were constructed by indigenous people to harvest in season and consume eels and fish. The eels were smoked and traded. World Heritage listing of this ancient system occurred in 2019, thanks to research and sustained representations by Aboriginal groups, and local, state and federal governments.

Historian Bill Gammage has amassed a great volume of evidence from the reports and writings of the early explorers, surveyors, travellers and pastoralists which consistently demonstrate that, on arrival, settlers observed vast areas of grassland clear of the shrubbery and trees their European eyes expected to see; the forests, lacking dense, tangled undergrowth, suggested a human hand. While its scale

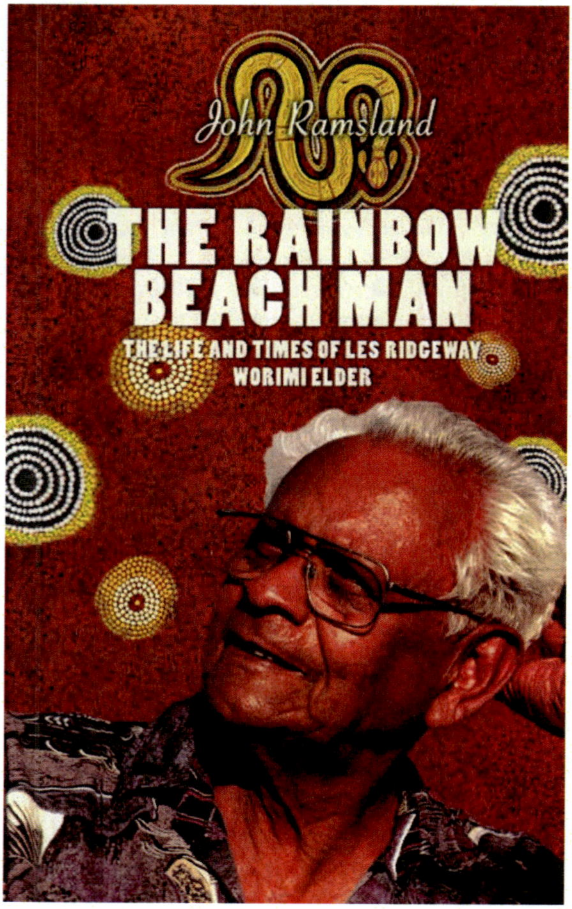

Fig. 16.18 Worimi Elder Les Ridgeway, an Aboriginal leader who achieved success in modern Australia. Cover of John Ramsland (2009) *The Rainbow Beach Man – The life and times of Les Ridgeway, Worimi elder.* Melbourne. Brolga Publishing. Cover design David Khan. Courtesy John Ramsland

is contested,[35] the mosaic pattern of burning provided the indigenous animal hunters both ease of access and cover needed to approach their prey. Also, often commented on, was the widespread Aboriginal practice of what the anthropologist Rhys Jones termed "fire-stick farming," selective cool burning of grassland and shrubs. Appropriately managed, this avoided the huge conflagrations of today's bushfires and, following rain, promoted fresh green growth which in turn attracted kangaroos, emus and other animals that the Aboriginals speared for meat and skins. It has also been suggested that Aboriginals cultivated yams and other roots and bulbous plants, since early colonial observers commented on what may have been regular lines of plants and women using digging sticks in such a way as to leave portions of perennial plants in the ground to re-grow. Bruce Pascoe, in south-eastern Victoria, argues for this and, himself an Aboriginal, practices a form of yam farming in a cool temperate part of the country.[36]

In *The Biggest Estate on Earth*, Gammage provides a telling case study in the Up & Downs block of Bundaleer Station on the Queensland border with NSW east of the Warrego River.[37] This block was depicted in some detail by F.M. Rothery in his remarkable *Atlas of Bundaleer Plains and Tatala* of 1877–78 (Fig. 16.19). The Ups and Downs block had only received an occupation licence in 1876, so its pictured condition would be a close representation of the traditional husbandry of its Aboriginal community. "Thick belts of Gidyah, Pine, Brigalow and small Scrub with broken plains scattered all through with open Coolibah flats. Common and Saltbush more or less all through." Gammage's careful analysis shows how plant distribution within this block was not entirely confined to soil type: "Brigalow and Gidgee are in clumps and belts; grass and scrub alternate." (p.85). He found that within the following decade, which started European management of this area, "the patterns vanished very quickly" with brigalow invading Gidgee plains, mulga and false sandalwood becoming common. Meanwhile, the best grasses, cotton and salt-bush were almost destroyed by drought, heavy stocking and rabbits. Heavy grazing selectively eliminated fodder plants, and with high stocking rates maintained for decades, the land lost significant productivity.

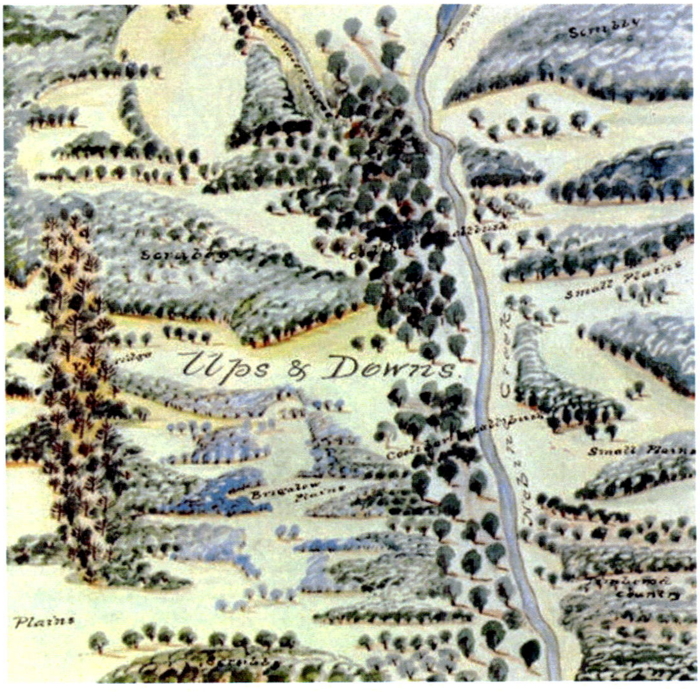

Fig. 16.19 Ups and Downs block, Bundaleer Station, Queensland, 1877–8 in Rothery, F.M. (1970) *Atlas of Bundaleer Plains and Tatala*, Canberra. ANU Press. p.37. Courtesy ANU Library

The overall picture of the experience of Australia's indigenous peoples was one of frequent physical loss and cultural annihilation. Only now is Aboriginal understanding, valuing and caring for the land and its animals and plants as a sacred duty coming to be widely understood and valued by the broader population.

16.4.2 The Dreaming

Philosopher and comparative religionist, Max Charlesworth, working with anthropologists, school teachers and others who had spent decades living among Aborigines in the country's inland, has argued that, so far from being "simple, primitive and magical", traditional Aboriginal beliefs about the cosmos and the good life were highly complex and as sophisticated as any of the world's acknowledged "great religions", surviving not with aid of a book of revelation but as the centrepiece of an ancient oral culture.[38]

The term "Dreaming" is the English word many Aborigines use to conceptualise and express their fundamental beliefs about the universe and long established rituals and customs. It is difficult, probably impossible, for whites nowadays fully to grasp the varied and subtle meanings of Dreaming which must be constantly lived in order to be deeply understood. Referring to Muta, a Murinbata man, anthropologist W.E.H. Stanner wrote: "I can recall one intelligent old man who said to me, with a cadence almost as though he had been speaking verse:

> "White man got no dreaming,
> Him go 'nother way.
> White man, him go different.
> Him got road belong himself."[39]

But anyone seeking to understand Aboriginal attitudes towards nature must make the attempt, since it is in and through the Dreaming that nature is held to be sacred, to have been created by spirit ancestors, and to continue to exist both physically and in and through the minds and actions of Aboriginal people.

Based mainly on anthropological, philosophical and Aboriginal artistic sources as documented in this text, my interpretation, which relates mainly to the theme "love and study of nature", is as follows: metaphysically, the "Dreaming" connotes the cosmic creation process, of a spiritual, moral and natural order of being, originating in a vague mystery beyond time. It seems like an eternal, unchanging backcloth to all existence, except that it suffuses *all* existence, including mundane life, changes in everyday time and gives profound meaning to people's lives. To be is to be enfolded by and absorbed into the Dreaming.

The Dreaming communicates moral and social law to the people, their duties, responsibilities and rights. It has to be sustained by the Aboriginal people, collectively, through conventional rituals and ceremonies. As well, its vitality must be maintained by knowledge, which is not static, and by active care. The land has been rendered sacred by virtue of its creation by the ancestor spirits. Its plants and its creatures are for use but with profound understanding of what *they* need for survival as species. Animals can be hunted for food and use of their skins, bones, feathers and so on, but always in moderation, not to the point of destruction of habitat and species extinction/exhaustion as with ancient megafauna. (Whether Aboriginal hunting was the cause of their extinction is a matter of debate.) Similarly, the vitality of the soil and its organic life must be sustained, so land can and should be burnt to provide clearings for the hunt and fresh grass for the browsers but not to the point of the hugely destructive bushfire conflagrations now common in Australia's dry summers. Small, cool fires in mosaic patterns enable the tiniest creatures to survive.

The Dreaming is one of the world's greatest conservation ethics. It is both understood to be the continuing (or "eternal") ontological process of cosmic creation and a psychologically profound mind-set or constellation of beliefs not unlike those of other great world religious traditions. Like them it has its "fixed" and its "fluid" elements and requires to be re-interpreted as life conditions change. For example, Aborigines lost exclusive control over occupancy and uses of land under colonialism, they moved and were moved out of traditional areas and had to re-imagine their spiritual lives as breaks occurred with long-established patterns of occupancy, ritual and use. It is remarkable that in the course of such disruption and destruction, reinterpretation and renewal have taken place and that, albeit in heavily modified form, major aspects of traditional cultures including languages, symbolic systems, religious beliefs and practices, have been maintained and rebuilt.

Aboriginal people state that, through the Dreaming and their apprehension of it, they are brought into harmonious equilibrium with the universe. The process is continuous, unitary, yet is itself composed of multiple dreamings to embrace the varied beliefs and practices of different groups of people across the whole continent. People are traditionally brought into a sacred tryst thereby to care for and to make productive use of specific areas and regions of the country. These are linked in a vast network of tracks, those of the creative ancestor beings, covering the whole land mass and neighbouring islands. The love and the understanding of nature are, for traditional and many modern Aboriginal people, at the core of their being from birth to death, necessary for the meaning they find in life, their spiritual as much as their material existence. "The particular land or 'country' which has been shaped and impregnated with ancestral spiritual energy or power and given to each Aboriginal people is, so to speak, the conduit through which the ongoing influence of the Ancestor Beings on various groups is effected."[40] Humans, land, its creatures and Ancestral Spirits are parts of one another.

Passionate concern for the well-being of the land is often expressed by modern Aboriginals: "You have to talk to the land. And also, like I said before, the Land's gotta have people talk on its behalf" and "Plants grow with the power of the spirit of country in them."[41]

Continuing relationship to land is deeply spiritual, past, present and future. Thus when Gondiwindi elder Elsie, and grand-daughter August, in *The Yield* fret over impending destruction of their people's land by a mining company, "The land, the earth is the victim now ... She's the one in real trouble",[42] in effect they are bewailing blows to their people's spiritual existence and destruction of their culture.

For traditional Aborigines there was no sacred book, no theological treatises smoothing the paradoxes or contradictions and filling the gaps, no grand congresses of bishops or assemblies of the "saints" who are in authority over the rest of the congregation of true believers. But there are, or were, some equivalents, an order of being grounded in religious knowledge and experience. The "elders" of the clan were, and are, people who have passed through the ceremonial stages that grant them access to occult knowledge, the mysteries, the myths, the understandings and the insights accumulated and sustained over centuries of oral and ceremonial transmission through the generations. Whatever local and regional variations there have been, all Aboriginal communities appear to have believed that creation – of the land, the creatures, the seasons, the sky, the cosmos – was the work of ancestor spirits and supernatural beings, most of them benign but some dangerous to humans. They include the Rainbow Serpent, Lightning Man, the Wagilag Sisters and others about whom an abundance of myths have been woven. Numerous English language versions of these myths have been published, one of them by a leading ethnologist, Charles Mountford, with illustrations by a white artist, and several times reprinted.[43] The ancestor spirits were not worshipped, but their power and presence were and continue to be acknowledged in ceremonies, stories, dance, music and art. Their work of creation, as they moved across the land, yielded sustenance in the form of the animals and plants they created and the distinctive features of the landscape - the hills, rock outcrops, water holes, springs and soaks. In every single thing that they created, the ancestor spirits left a part of themselves when they did not wholly reside in it. In this way, the land and the creatures became not only physical entities, they all were and are parts of the embodied spirit world and thus to be treated with respect and cared for as a sacred duty: "The landscape was not only formed as a byproduct of ancestral action, it was also a result of the transformation of ancient beings' bodies or bodily substances. Sources of ochre throughout Arnhem Land are formed from the blood or fat of ancestral beings."[44]

Thus a sacred bond was established: the spirit ancestor, the land and its creatures, and the Aboriginal people. This is nowhere better demonstrated to modern students of Aboriginal culture than in the highly creative, thriving genre of Aboriginal art through which many of the traditional myths and the identification of clans and people with totemic creatures, plants and sacred places are revealed to public gaze. Their secret meanings are held fast in the minds of the artists and the elders, but thanks to their willingness to offer at least partial public explanations and to the growing literature in English produced by white and Aboriginal art critics, museum curators and anthropologists, an accessible visual landscape of "the Dreaming" has in recent years become available to the wider public.

The designs of the artists can be readily appreciated for their visually striking forms, use of brilliant colour and the intricate working of surfaces but for their spiritual meaning close study and the assistance of explanations by the artists themselves and knowledgeable art critics and anthropologists are required. I have drawn on a variety of publications, as a sample of the scholarly and popular works which thanks to growing public interest are now widely available in museums, galleries and bookshops.[45]

Fig. 16.20 Artist: Tjam Katani Yilkari, Liyagalawumirr clan, Dhuwa moiety, Wagilag Sisters Story, Eastern Arnhem Land, Northern Territory, Australia, 1937. The Donald Thomson Collection. Donated by Mrs. Dorita Thomson to the University of Melbourne. On loan to Museums Victoria. ©Tjam Katani Yilkari/Copyright Agency, 2021

From a large and ever-growing gallery of ancient rock art, bark paintings, modern work in acrylics on board and linen, tools, implements and baskets, let us take just one example with accompanying explanation of Aboriginal art. Unless one has familiarity with the details of myths and the role of images and techniques, an explanation is necessary in order to understand the intended meaning of the art object (Fig. 16.20). Wally Caruana writes of a bark painting *Wagilag* by Tjam Katani Yilkari, painted in 1937:

> Among the most important Dhuwa moiety themes in Central Arnhem Land is that of Wititj the Great Python and the Wagilag Sisters. This chronicles the creative acts of the ancestral sisters as they travel from south-eastern Arnhem Land culminating in an epic encounter with Wititj.
>
> The two Wagilag Sisters have children through incestuous relationships with their clansmen, forcing them to leave their inland country and journey north and then west towards the

sea. As they travel, they come across animals, plants and country on which they confer names, and by so doing, bring them into being. They make camp at Mirarrmina waterhole in Liyagalawumirri country, unaware this is the sacred home of Wititj, the Python. Wititj is angered by their presence. Sucking the waters of the lagoon, he rears erect in the sky and spits to form the rain clouds of the first monsoon season. The Sisters perform songs and dances to avert the deluge and thunderstorm which follow, and take shelter in a bark hut they have made. Wititj suddenly descends and swallows the women, their children and all their belongings. As Wititj raises himself again into the sky, floodwaters cover the earth.

Wititj explains his actions to other ancestral pythons and is made to realise that he has erred in swallowing those who share the same moiety affiliation. The Python sickens and crashes to the ground, leaving the impression of his body in the earth. He vomits out the women and the children, and creates a strong wind which causes the floodwaters to abate. The action of swallowing and vomiting is repeated, but now Wititj regurgitates only the women, who become two large boulders at Mirarrmina.

The songs and dances performed by the women to stop the rain are revealed to the Wagilag men who had followed. The men are instructed to return home and teach these rites to others to guarantee the continuation of the cycles of nature. Although this is just one aspect of the purpose of the ritual, the Wagilag story emphasizes the importance of religious ceremony, a major function of which is to take an individual from one level of spiritual being to a higher one. The acts of swallowing and regurgitation, as seen in the paintings of Nabekeyo and Mawurndjul in West Arnhem Land, once again provide the metaphor for spiritual transformation."[46]

Respect and care for land, plants and animals, loosely "the world of nature," were solemn responsibilities, duties shared by families, kinship groups, clans and by all Aboriginal people right across the country. When indigenous people talk of "country" as they so often do today, they are referring not only to land as physical space owned by someone but of a shared responsibility for the care, management and use of a particular community, of people who do not so much own the land in the sense of title deeds but by birth and history have rights and duties and are responsible for it *ad finitum*, for all eternity.

A modern Aboriginal land rights movement has followed on the determinations of the High Court of Australia that the country was not "terra nullius" as Cook and colonial officialdom had determined but had been, in part at least, continuously occupied according to the beliefs and usage of the indigenous people. The ensuing Native title legislation, the Land Councils and the whole apparatus of dispute resolution, now in place, inevitably mean that "land" has now acquired meaning which draws together Western legalism and Aboriginal spirituality. Some areas of land, especially in northern Australia, are legally owned by Aboriginal groups, some are leased by the Crown to white station owners and certain Aborigines have access and rights of use. Disputes over rights of access, for example by mining companies to Aboriginal reserves, continue. And so on. Our point, here, is not to traverse the legal, financial and political complexities of land rights and issues arising but to reflect on the meaning of land and its care and uses as understood in traditional and elements of contemporary Aboriginal culture. But such understanding is necessarily limited since the traditional culture has long since been disrupted and almost wholly displaced. We are reliant on the aforementioned reports of the early explorers, surveyors, settlers and the interpretations of anthropologists, philosophers, art critics and historians and on what Aboriginals themselves have been prepared to communicate either directly and publicly or to interlocutors.

16.4.3 *Caring for the Land and Its Creatures*

As noted above, Aboriginal people, until the arrival of Europeans and well into the nineteenth century, skilfully and with a deep understanding of conservation and sustainability, cared for and farmed the land, treating it as both a material utility and a spiritual resource. Farming included the controlled harvesting of eels and fish (Fig. 16.21). It is disputed whether burning was beneficial to soil and soil organisms but whether white tillage, monoculture and hard-hoofed animals are better is doubtful. Aborigines did not slash, burn, cultivate then move on but allowed the land and its creatures to recover, perhaps for decades, before returning to farm again. They sustained what was naturally there and used the plant and animal resources in moderation, producing surpluses for their own seasonal uses, not for profit and export around the world. Traditional practices sustained only low population densities, in balance with what the land could support. The alterations to vegetation patterns by controlled burning were indeed designed to facilitate hunting but no less to enhance sustainability in the sense of continued productive use. Aborigines were of course cognisant of the harshness of much of Australia's natural world; their lives depended on practical in-depth knowledge of seasonal variability, wind patterns, water scarcity inland, fierce sun and dry atmosphere, high combustibility of the ubiquitous eucalyptus species and the role of undergrowth in fuelling fire.

The spiritual significance attributed to land reflects a profound, pre-scientific understanding of nature. Aborigines hunted, killed, cooked and ate a wide variety of animals, birds and fish, eels, turtles, crabs, insects and so on, in addition to their basically vegetable and fruit diet. But, apart from the loss of ancient huge megafauna – marsupials, as noted above, and that is debatable - they are not believed to have hunted any species to extinction. Nor are they known to have so intensively "farmed" the food plants they apparently cultivated and harvested – yams, bulbs, water plants, fruits, berries, nuts, seed plants – as to endanger species survival. They used no artificial fertilisers, poisonous sprays or cultivation methods that destroyed the precious ecology of the land and its intimate web of life. Developing a modern understanding by both whites and contemporary Aboriginals of the traditional uses and products of native plants is one of the aims not only of botanists such as Philip Clarke at the South Australian Museum but of local Aboriginal people such as Bruce Pascoe in south eastern Victoria, and a group at Byron Bay in New South Wales.[47] There is much from these sources to build into the Western stock of knowledge.

There may be in this account some sense that traditional Aboriginal attitudes towards and behaviour in the natural world corresponded to or anticipated enlightened modern concepts of sustainability and conservation, and contrasted positively in every respect with how we Europeans behaved over more than two centuries of

Fig. 16.21 A striking image of ancient, traditional Aboriginal stone fish traps. Aboriginal fisheries, Darling River, New South Wales. Glass plate negative, full plate, 85/1286–721. Unattributed studio, Sydney, NSW, 1880–1923. Collection: Museum of Applied Arts and Sciences. Gift of Australian Consolidated Press under the Taxation Incentives for the Arts Scheme, 1985. Photographer unknown

occupation. That would be to suppose that everything in traditional Aboriginal life worked according to the Dreamtime order, or paralleled modern notions concerning species survival, ecology, conservation, sustainability and so forth. It would also, in reifying traditional knowledge, derogate the enormous developments in biology, zoology, agriculture and other scientific domains without which modern life, for Aboriginals as well as Europeans, could not be sustained.

Nevertheless, there is much more of value a Western style of life can learn from a fuller understanding and appreciation of traditional Aboriginal practices than European settlers generally afforded themselves, at least until relatively recently. Whereas it was once thought unimportant for children and young people at school to learn much about Aboriginal culture and that from the standpoint of the superiority of white culture, in recent decades positive steps have been taken both in individual state systems of education and at the level of national government to encourage a knowledgeable understanding and appreciation of both traditional and modern Aboriginal ways of life. Syllabuses such as those recommended to, or required of, teachers of science in the Northern Territory and a substantial course of study produced by the national Curriculum Development Centre for use by schools across Australia are examples of efforts to educate students in the values, ways of living, beliefs and customs of living Aboriginal communities.[48] Similarly, as mentioned above, Marcia Langton, academic and Aboriginal activist, is encouraging Australians of all ages to learn more about Aboriginal culture, both traditional and modern.[49]

As we have seen, for the colonial era settlers, Aboriginal knowledge and skills were not completely ignored even though not finding a place in the school curriculum, such as it was. The understanding of the land shown by black trackers was admired and used by the settlers, their knowledge of inland sources of fresh water and food saved white lives and they guided land-takers to the best pasturelands and the safest routes through trackless marshland. Had the cross-continent explorers, Robert O'Hara Burke and William John Wills in 1861, known more about the ways in which Aboriginals processed indigenous food plants, they may not have starved to death.

Aboriginal people have played an indispensable role in the early development of botany and zoology in Australia, when acting as guides and collectors for expeditions by whites intent on adding to the store of scientific knowledge (Fig. 16.22). Sometimes this was acknowledged in their reports and scientific papers, but whereas we have the names of the European members and leaders of the expeditions, it is rare to find adequate acknowledgement, let alone naming, of the Aboriginals whose observations, and gathering, collecting and retrieval of specimens made possible the growth of natural science. An exception is the expedition of William Blandowski, German zoologist, who in the 1850s employed Aboriginals to collect zoological specimens along the Murray River and freely acknowledged their expertise.[50] Another is that of Ludwig Leichhardt whose notebook records of his expedition in northern Australia, inspired by his idol, Alexander von Humboldt, pay tribute to the expertise and knowledge of the land, its plants and creatures, and of the Aboriginal members of the party.[51]

However, indigenous knowledge and experience were not sought by European settlers as a foundation for colonial Australian land management, any more than in other European colonies around the world. Apart from a few far-sighted critics, settlers did not pause to ask themselves whether introducing northern hemisphere species of plants and animals, including rabbits and foxes, pasture grasses, exotic fish, birds and toads, was likely to harm or benefit what was already here. We did not, for a very long time, inquire into Aboriginal ways of "farming" the land, harvesting its crops and creatures and sustaining its resources. Presumably this is because over the centuries a deep sense of superiority and comparative advantage closed the European mind to the alternative Aboriginal modes of existence, except as subjects of curiosity and intellectual investigation, or to be commercially and politically exploited while the people were being "saved" by missionaries from their heathen deities. Also, much settler agriculture was based on sheep and cattle, with which Aboriginals had no familiarity; and assumed technologies which were very different with regard to cultivating crops. As we have become more conscious of the destruction we have wrought, along with our innumerable achievements, the mentality of "we are the greatest" or "will be great again" is being moderated, at any rate in those parts of

Fig. 16.22 Indigenous hunters of the Herbert River area, Far North Queensland were critical to the success of Norwegian scientist Carl Lumholtz's 14 month collecting trip to the rainforests in the early 1880s. Here, they start out on an expedition to the mountains. "Start for an Expedition", plate opposite p.188 in Lumholtz, C. (1889) *Among Cannibals: An Account of Four Years' Travels in Australia and of Camp Life with the Aborigines of Queensland* . National Library of Australia. nla.obj-2,409,812,159

Western society and in our personal being, where limits and the risks of excess are well recognised and other ways of achieving happiness than the constant quest for more and yet more are understood and valued.

A return to completely traditional ways of life, except perhaps for a handful of people, is not now possible; Aboriginals, and Europeans, are long since past the stage where such a life could ever become either desirable or a practical possibility for a whole society. Many Aboriginal people have, however, kept in touch with their Dreamtime stories and take part in ceremonies connecting them with traditional life. Due in large part to the activities of missionaries into late in the twentieth century, Christianity has been blended with Aboriginal spirituality in the lives of some communities. We can all learn from and embrace the ancient nature ethic, its close engagement with and knowledge of nature and its deep understanding that we are part of nature and have a responsibility for the land, its creatures, the atmosphere. Our continuing health, mental as well as physical, and our very existence are challenged by our reckless ways. It is an ethic that we all can join together in taking to our hearts and minds.

From this reflection on the life traditionally lived by the Aboriginal people, principles for sustainable living for all Australians today, not only those of Aboriginal descent, can be drawn. We need to acknowledge and seek to better understand the mutual dependence of the different elements and processes of the natural world.

That ecological understanding depends on the growth of scientific knowledge but is equally a matter of the activation of ethical, aesthetic, emotional and, in a broad sense, religious values and insights and specific kinds of practical and local action informed by them. We cannot continue to take more from the earth than we give in return, so we must conserve and re-create. In order to achieve the necessary balance, we need to adjust our wants and felt needs in accordance with our understanding of what the earth is capable of sustaining. It is ironical that in the first two centuries of European presence in Australia, so much was done to destroy or diminish the existing indigenous culture and its way of living in accordance with principles and values that are now slowly gaining acceptance and whose issue in everyday practice for all of us is vital to our very survival.

Notes

[1]Quoted in *Hortus* No. 130 Summer 2019 p. 108.

[2]Allingham, H. and Dick, S. (1909) *The Cottage Homes of England.* London: Edward Arnold.

[3]"An Enquiry Into the Changes of Taste in Landscape Gardens" quoted in Hadfield, M. (1969) *A History of British Gardening.* London: Spring Books. p. 245.

[4]Computer images of Williamsburg gardens.

[5]Dudgeon, P. (ed) (1991) *The English Vicarage Garden.* London: Mermaid Books. Introduction by Miss Read p. 7

[6]Dudgeon, *op.cit.* p. 20.

[7]Dudgeon, *op.cit.* p. 101.

[8]Fox, R. L. (1982) *Better Gardening.* Oxford: R&L.

[9]Hadfield, *op.cit.*p. 373.

[10]Dudgeon, *op.cit.* pp. 50–51.

[11]Gilbert, K. L. (n.d.) *The Orchid Man. The Life, Work and Memoirs of the Revd H.M.R. Rupp 1872–1956.* Kenthurst, NSW.: Kangaroo Press.

[12]Rupert Brooke *The Old Vicarage, Grantchester.* Café des Westens, Berlin, May 1912.

[13]Hareuveni, N. (1980) *Nature in Our Biblical Heritage* trans. Helen Frenkley. Kiryat Ono, Israel: Neot Kedumim Ltd.

[14]A notable example is political radical William Godwin: *An Enquiry Concerning Political Justice* (1796) ch IV The characters of men originate in their external circumstances. Godwin's feminist daughter, Mary Shelley (nee Mary Wollstonecraft Godwin) was author of the Gothic horror novel Frankenstein, a demonstration of circumstantial creation .

[15]Barzun, J. (1962) *Classic, Romantic and Modern.* London: Secker and Warburg; Lovejoy, A.O. "On the Discrimination of Romantics" in Thorlby, A.K. (1966) *The Romantic Movement.* London: Longmans.

[16]Adamson, J. (1919) *A Short History of Education.* Cambridge: At the University Press. ch XVI The Establishment of a National Elementary System.

[17]Monroe, P. (1915) *A Textbook in the History of Education.* New York: The Macmillan Company. chs X-XIV presents an early twentieth century appraisal of preceding "tendencies" in educational thought and practice. On the naturalistic, psychological, scientific sociological, eclectic ideas which culminated in "the new education"; see Cremin, L.A. (1961) *The Transformation of the*

School. Progressivism in American Education 1876–1957. New York: Alfred K. Knopf; Selleck, R.W. (1968) *The New Education, the English Background 1870–1914.* Melbourne: Pitman & Sons; Connell, W.F. (1980) *A History of Education in the Twentieth Century World.* Canberra: Curriculum Development Centre. Part 1 Educational Awakening c. 1900–1916.

[18]Cole, M. (1945) *Beatrice Webb 1848–1943.* London: Longman Green p. 46 fn. Webb was addressing the English Association of Technical Institutes in 1909.

[19]For an introduction to Dewey's thought together with extracts from his writing: Skilbeck, M. (1970) *John Dewey.* London: Collier-MacMillan Limited. Dewey is the foremost theorist of the New Education. Dewey, J. (1900) *The School and* Society. Chicago: Chicago University Press; Dewey, J. (1916) *Democracy and Education.* New York: The Macmillan Company; Holmes, E. (1911) *What Is and What Might Be.* London: Constable.

[20]Laurie, S.S. (1902; first ed. 1901) *The Training of Teachers and Methods of Instruction – Selected Papers.* Cambridge at the University Press. second edn.

[21]It was said of Parker's impact on American elementary schooling that "The external machinery was there, but it needed to be taken possession of by the spirit of life" (Parker, Francis W. (1885) Notes of *talks on teaching given by Francis W. Parker, at the Martha's vineyard summer institute, July 17 to August 19, 1882/ reported by Leiia E. Patridge.* New York. E.L. Kellogg & Co.).

[22]Kass, D. (2018) *Educational Reform and Environmental Concern. A History of School Nature Study in Australia.* London: Routledge.

[23]Turney, C. (1962) *The History of Education in New South Wales 1788–1900* Vol.3 Ph.D University of Sydney p. 1180.

[24]Ramsland, J. (1979) "William Catton Grasby: pioneer of progressive and comparative education in South Australia" *South Australiana* 18.1 March pp. 45–68.

[25]Wulf, A. (2016) *The Invention of Nature: the adventures of Alexander von Humboldt, the lost hero of science.* London: John Murray; von Humboldt, A. (2014) *Views of Nature* edited by Jackson, S.T. and Walls, L.D. trans by Person, M.W. Chicago: University of Chicago Press.

[26]Sand, C., & Clement, G. (1995) *Le jardin Romantique de George Sand.* Paris: Albin Michel.

[27]A typical example of this literature is Stubbin, J.C. (1914) *The School Garden. Hints on Tree Planting, Gardening – Fruit Growing, and – Elementary Agriculture.* Brisbane. By Authority Anthony James Cumming, Government Printer: "Every school should possess its own garden, and in it children should learn to love trees and flowers and follow their life stories. There is no form of work which permits the child so extensively to make his own observations, draw his own conclusions, and discover things for himself." p. 8.

[28]Gough, A, (1997) *Education and the Environment: Policy, Trends and Problems of Marginalisation.* Melbourne: Australian Council for Educational Research.; Greenall, A. (1981) *Environmental Education in Australia: Phenomenon of the Seventies.* Occasional Paper No.7. Canberra: Curriculum Development Centre.

[29]Address to the World Economic Forum, Davos, 25th January 2019 in Thunberg, G. (2019) *No One is Too Small to Make A Difference.* UK Penguin. Random House.

[30]Morgan, S. (1989) *Wanamurraganya. The Story of Jack McPhee.* South Fremantle, W.A.: Fremantle Arts Centre Press.

[31]One is the teacher, Native Title Holder and Elder in Central Australia, Margaret Kemarra Turner; Turner, M. K. OAM (2010) *Iwele Tyerrtye – what it means to be an Aboriginal person* as told to B. M.Terrarle with translations by V. P. Dobson. Alice Springs, N.T.: I.A.D. Press; another is Les Ridgeway, Elder and former manager of Aboriginal settlements on Australia's East coast. Ramsland, J. (2009) *The Rainbow Beach Man. The Life and Times of Les Ridgeway, Worimi Elder.* Melbourne: Brolga Publishing.

[32]Ramsland, J. (2009) *The Rainbow Beach Man – the life and times of Les Ridgeway, Worimi Elder.* Melbourne: Brolga Publishing.

[33]For adults and children alike, an excellent introductory overview, with a focus on Aboriginal uses of and attitudes toward the land is Australian Geographic Geography (2019) *Custodians of the Land. Connections to Country. Use of Resources, Sustainable Land Practices.* Sydney: Australian Geographic. See also footnote 48 below.

[34]Gammage, Bill (2011) *The Biggest Estate on Earth.* Sydney Allen and Unwin; Pascoe, Bruce (2014) *Dark Emu.* Broome W.A. Magabala Books Aboriginal Corporation; Charlesworth, Max, Kimber, Richard, Wallace, Noel (1990) *Ancestor Spirits. Aspects of Australian Aboriginal Life and Spirituality.* Geelong. Deakin University Press.

[35]Fensham, R. (2012) "Review of 'The biggest estate on earth: how Aborigines made Australia'" *Australian Geographer* Vol. 43, No.3, pp. 325–329. September.

[36]Gammage, B. *op.cit.*; Pascoe, B. *op.cit.*

[37]Gammage, B. *op.cit.* pp. 85–86; Rothery, F.M. (1970) *Atlas of Bundaleer Plains and Tatala.* Canberra: ANU Press.

[38]"The Invention of Australian Aboriginal Religion" in Charlesworth, M. (1997) *Religious Inventions. Four Essays.* Cambridge: Cambridge University Press; Charlesworth, M., Kinder, R. and Wallace, N. (1990) *op,cit.*

[39]Stanner, W.E.H. (1979) *White Man Got No Dreaming – Essays 1938–1973.* Canberra: ANU Press. p. 24. This book reprints the essay "The Dreaming" (pp. 23–40) which Stanner first published in 1953.

[40]Charlesworth, M. Introduction to Charlesworth, M, Dussart, F., and Morphy, H. (eds) (n.d.) *Aboriginal Religions in Australia. An Anthology of Recent Writings.* Burlington, VT: Ashgate.

[41]Turner, M. K. *op.cit.*, pp. 116–117 and p. 133.

[42]Winch, T.J. (2019) *The Yield.* Australia: Hamish Hamilton. p. 93.

[43]Roberts, A. and Mountford, C. (1971) *The Dreamtime Book.* Adelaide: Rigby.

[44]Morphy, H. (1998) *Aboriginal Art.* London: Phaidon pp. 69–71.

[45]Caruana, W. (2003) *Aboriginal Art.* London: Thames and Hudson; Chaloupka, G. (1982) *Burrunguy-Nourlangie* Rock (Arnhem Land) Kakadu National Park: Northart; McCarthy F. D. (1952) *Australian Aboriginal Decorative Art.* Sydney: The Australian Museum; Morphy, H. (1991) *Ancestral Connections. Art and an Aboriginal System of Knowledge.* Chicago & London: University of Chicago Press; Morphy, H. (1998) *Aboriginal Art.* London: Phaidon; Mountford, C. P. (1964) *Aboriginal Paintings from Australia.* New York: The New American Library of World Literature by arrangement with Unesco; Mundine, D., OAM (n.d.) *The Native Born. Objects and representations from Ramingining, Arnhem Land.* Sydney: Museum of Contemporary Art in association with Bula'bula Arts, Ramingining; Thomson, D. (2005) *Donald Thomson in Arnhem Land.* Compiled and introduced by Nicolas Peterson. Carlton, Vic.: The Miegunyah Press;The Donald Thomson Collection (2010) *Ancestral Powers and the Aesthetic: Arnhem Land paintings and objects from the Donald Thomson Collection.* Melbourne: Museum Victoria.

[46]Caruana, W. (2003) *Aboriginal Art.* London: Thames and Hudson. pp. 48–49.

[47]Clarke, P. (2007) *Aboriginal People and Their Plants.* Dural NSW: Rosenberg; Pascoe, B., *op. cit.*; Low, T. (2003) *Place of Plenty. Culturally Useful Plants Around Byron Bay.* Sydney: Department of Environment and Conservation (NSW).

[48]Michie, M., Aniezark, J. and Uibe, D. (n.d.) "Beyond bush tucker: implementing indigenous perspectives through the science curriculum." Darwin: Northern Territory Department of Education. Mimeo; Fidock, A. (ed) (1982) *Introducing Aboriginal Australians.* Canberra: Curriculum Development Centre.

[49]Langton, M. (2018) *Welcome to Country. A Travel Guide to Indigenous Australia.* Melbourne: Hardie Grant Travel. (and teacher's notes for years 7–10 students on-line); Langton, M. (2019) *Welcome to Country: An Introduction to our First Peoples for Young Australians.*

[50]Allen, H. (ed) (2010) *Australia. William Blandowski's Illustrated Encyclopaedia of Aboriginal Australia.* Canberra: Aboriginal Studies Press; See also Olsen, P. and Russell, L. (2019) *Australia's*

First Naturalists. Indigenous People's Contribution to Early Zoology. Canberra: National Library of Australia Publishing. Ch.3 Naturalists 1838–1888.

[51] Durragh, T.A. and Fensham, R.J. (eds) (2013) "The Leichhardt Diaries. Early Travels in Australia during 1842–44." *Memoirs of the Queensland Museum/* Culture. Vol.7 (1) Brisbane: Queensland Government; Fensham, R.J. ed. (2013) "The Leichhardt Papers. Reflections on his life and legacy." *Memoirs of the Queensland Museum/ Culture*. Vol. 7 (2). Brisbane: Queensland Government.

Complete Listing for Wikimedia Commons Images

16.1 A new and accurate plan of Blenheim Palace – L'Art de Créer les Jardins (1835), pl. 1 – BL. https://commons.wikimedia.org/wiki/File:A_new_and_accurate_plan_of_Blenheim_Palace_-_L%27Art_de_Cr%C3%A9er_les_Jardins_(1835),_pl._1_-_BL.jpg Attribution: Nicolas Vergnaud, CC0, via Wikimedia Commons.

16.2 Blenheim Palace Park & Lake (6093445394). Author: Simon Q. https://commons.wikimedia.org/wiki/File:Blenheim_Palace_Park_%26_Lake_(6093445394).jpg Attribution: Simon Q from United Kingdom, CC BY 2.0 <https://creativecommons.org/licenses/by/2.0>, via Wikimedia Commons.

16.3 Humphry Repton00. https://commons.wikimedia.org/wiki/File:Humphry_Repton00.jpg Attribution: Humphry Repton (1752–1818), Public domain, via Wikimedia Commons.

16.4 Humphry Repton01. https://commons.wikimedia.org/wiki/File:Humphry_Repton01.jpg Attribution: Humphry Repton (1752–1818), Public domain, via Wikimedia Commons.

16.5 Portrait of Humphry Repton. https://commons.wikimedia.org/wiki/File:Portrait_of_Humphry_Repton.jpg Attribution: Northmetpit at the English Wikipedia, CC BY-SA 3.0 <http://creativecommons.org/licenses/by-sa/3.0/>, via Wikimedia Commons.

16.10 Jean-Jacques Rousseau (painted portrait). By Maurice Quentin de La Tour. Photographer unknown. https://commons.wikimedia.org/wiki/File:Jean-Jacques_Rousseau_(painted_portrait).jpg Attribution: Maurice Quentin de La Tour, Public domain, via Wikimedia Commons.

16.11 Frederick-Froebel-Bardeen. https://commons.wikimedia.org/wiki/File:Frederick-Froebel-Bardeen.jpeg Attribution: C.W. Bardeen, publisher, Public domain, via Wikimedia Commons. Source: Library of Congress Prints and Photographs division.

16.16 Alexandre humboldt. Photo: Karin März. https://commons.wikimedia.org/wiki/File:Alexandre_humboldt.jpg Attribution: Friedrich Georg Weitsch, Public domain, via Wikimedia Commons.

Chapter 17
Reaching an Understanding and Love of Nature

... l'éléphant est d'abord une véritable cathédrale vivante

(... the elephant is first of all a true living cathedral)[1]

When the cathedral of Notre Dame in Paris caught fire in May 2019, the President of France immediately announced its rebuilding, money no object. This was not primarily for reasons of religion, it was because Notre Dame symbolised the French nation, its glory and power, and indeed its tourism income. For the world of nature, its glory and power, the elephant is an equally fitting symbol. But do the threats to living nature that now abound command comparable action at the highest levels of state? Presumably they do not because in practice, we value present and future natural life less than we do the stones and rituals of a mediaeval Gothic structure. In an ethic of nature and culture alike, they do, or should, have equal place. To achieve that inseparability, human feelings and imagination must be brought into play together with scientific understanding; indeed, that understanding must be infused with all the resources of the human head and its heart. In this, the concluding chapter, we reflect on the themes developed throughout the book, with a plea for a reconstructed ethic that recognises our place in the natural world while celebrating the Earth, our ultimate home.

17.1 A Continuing Quest

Throughout this book evidence has been sought in the Western world – Europe and its outreach around the globe – of efforts to study, understand, value and love nature. Moreover, reaching into ancient Persia, the Mogul Empire and Australian Aboriginal culture, the inquiry has not been confined to Europe. Notice has been taken of attitudes and frames of mind that tend, consciously or not, to privilege humanity over all other forms of life and the Earth itself. This privilege has been at times rooted in religious beliefs about a special relationship with an all powerful deity, to draw from that deity freedom to exercise dominion over all other creatures and the whole natural world, fellow humans as well, in slavery and serfdom. That belief has at times, and in the minds of notable Christians such as Francis of Assisi, been ameliorated by a profound sense of community, brotherhood and a recognition that we best see

M. Skilbeck, *Loving and Studying Nature*, International Explorations in Outdoor and Environmental Education 10, https://doi.org/10.1007/978-3-030-80751-1_17

ourselves not as masters but stewards, charged to care, not exploit. In the immediately preceding chapters especially, but also throughout, we have found evidence of attitudes, values and practices that demonstrate increased understanding and valuing of the land and its creatures. Inevitably this is but a small selection and its compass has not been extended far beyond the Western world.

Little enough has been said about the threats; for example, our wasteful, ever-growing material consumption, spurred by a global economic system now appropriately designated "consumer capitalism" as it was in the nineteenth and early twentieth centuries, "industrial capitalism". This economic system has conferred great benefits but, no less, has visited great damage and harm on the natural world and its swarming human population.[2] The command and control systems such as that of the former Soviet Union, are of course similarly open to criticism and there is no question that China's current Belt and Road drive is a global force accelerating consumption and increasing pressure on natural resources.

The great intellectual leaps of science from the time of the Renaissance to the present have fostered a new sense, at times worryingly prideful, of mastery over nature as its mysteries have been unveiled, its secrets disclosed. How, then, are we to come to know and understand the profundity of the natural world while avoiding hubris? What is it in our nature that may lead us to care, to respectfully admire and enjoy the natural world, enhancing its bounty and enabling other living creatures to do the same while ensuring that their enjoyment is life enhancing, not life destroying?

We can be optimistic, as when reviewing the care and concern for the Earth shown by those earnest Anglican vicars tending their gardens, the Israeli botanists creating a park to celebrate ancient history and modern botanical science, the Mogul emperors and the land-loving, spiritually rich traditional Aboriginals from whom we have much to learn. It is in childhood that awareness of the natural world starts and must be nourished. Small bands of teachers and enlightened educational leaders late in the nineteenth century not only recognised this; they set to work, as we saw in the preceding chapter, to make it a reality by teaching nature study in the new elementary schools. As one English writer remarked of a later, wider transformation of schooling, a silent social revolution occurred. It is certainly one to sustain since there are now indications that, if we do not, the great saga of life on earth may fall into an irreversible decline.

Western capitalism, liberalism, colonialism and consumerism are regularly denounced by critics ranging from political and cultural theorists through environmentalists to anarchists. They are debated in schools and universities, especially by young people impatient with the mistakes and fumbling reform efforts of their elders. Over the centuries, European practices and ideas have been refreshed at times by revolutions and whole landscapes ruined by warfare and mismanagement, but there has been a continuing thread of renewal and reform, most often inspired by the quest for knowledge, the refinement of moral sensibility and the perennial quest for beauty. We can all too easily succumb to gloom and despair when contemplating the numerous failures. Or, more optimistically, we can continue to search for ever deeper understanding and moral purpose. Our natural heritage, the very source of our existence, calls for nothing less. The slave trade, an abominable blot on civilisation, was abolished, at least in the West. Prolonged and patient reform efforts were required, notably those of active Christians in the United Kingdom and America who toiled against economic and political headwinds.

17.2 The Gathering Tide of Secularism and Religious Surges; Dominion Reinforced

Until at least the eighteenth century, nature and the natural world were commonly perceived and understood by Europeans within the pervasive, supernatural world view of the Judaeo-Christian religion, mediated by early theologians, the humanists and scientists of post Mediaeval Europe, the Renaissance prelates of Catholic Rome or that nursery of the Church of England, the clergy of eighteenth century Anglican Oxford. Reaching far beyond their metropolitan heartlands, the European voyagers and the crowds of missionary and settler followers spread Christian beliefs and rituals around the planet in the centuries following the Portuguese ventures along the west coast of Africa. The human world, in Europe itself and in its furthest reaches in the Americas, Africa or Australasia, was not always conceived in terms of the first chapter of *Genesis* as fallen, and inherently depraved, although indigenous peoples far from the European heartland often were so seen. That was so until the intellectuals and romantics of the eighteenth century discovered the "noble savage" and humanitarian voyagers like James Cook and overlords in the British Admiralty and their opposite numbers across Europe called for respect and humane treatment of indigenous people. The progressive movements led by such as Lord Shaftesbury, William Wilberforce, the London-based, pious Clapham sect, Thomas Paine, Continental reformers and American emancipists eventually resulted in the abolition of the slave trade. Gradually, a world of people, of common humanity, yet also of diverse beliefs, customs and practices, came to be recognised, if not always valued. Human nature by degrees, and even now incompletely, has come to be morally accepted as universal and common to all, as in the UN Universal Declaration of Human Rights, adopted in 1948, in the aftermath of the Second World War.

The claim of Christianity to a pre-eminent and unique place among the world's proclaimed grand truths has fallen away, through common human experience of everyday life in, and awareness of, an expanding natural world. From the eighteenth century onwards, historical and biblical scholarship and philosophical and scientific scrutiny led to a crumbling of foundations that had been carefully constructed over millennia of religious thought and practice. Not least, the growth of newspapers and magazines reporting on everyday life, of novels dealing with individual life and social affairs and popular literature generally, sustained and strengthened an expanding secular life.

From Old Testament times onwards even in its own texts, first those of Judaism and then those of Christianity, belief was warmly contested, criticised and doubted. Religious doctrines and beliefs were not just debated, they were blasphemed and ridiculed. Faithful followers were conspicuous as much for their worldly failings as for their devotional achievements, if not more so. Ridicule, as Voltaire and Scottish poet Robbie Burns realised, was a most effective weapon of attack. *Holy Willie's Prayer*, a devastating attack on a hypocritical church elder, is not itself anti-religious but left many wondering about leading Christian practitioners,

... O Lord, Thou kens what zeal I bear,
When drinkers drink, an' swearers swear,
An' singing here, an' dancing there,
Wi' great and sma';
For I am keepit by Thy fear
Free frae them a'...

Alternative ways of thinking, believing and living have progressively forced or edged religion either into defensive by-ways or accommodation with secular existence and "worldly" ways of life. Christian theologians now commonly concern themselves with contemporary issues: social justice, poverty, liberation theology, threats to nature, feminism and eco-feminism.[3]

During the Renaissance it was not the pious, leading a modest, virtuous Christian life who were the cultural heroes. Even the proud popes of the Catholic church, from the aristocratic families of the Medicis, the Borgias and the della Roveres lived in princely luxury, fathered children and patronised the arts by no means solely to honour the Gospels. The exalted, educated person of the Renaissance was not a priest or theologian but an aristocrat or well-bred man-of-the-world (and man not woman was the Renaissance cultural hero). His education in the classics of ancient Greece and Rome was to be borne lightly, urbanely. He was politically, socially and culturally engaged, a city-dweller with a country estate maintained by servants and slaves. The popes and cardinals had their country villas and palaces. Happiness and cheerfulness, pleasures of the flesh, not mediaeval ecclesiastical sobriety and gloom, or knightly valour on the field of combat, were cultivated, as in the sumptuous garden house-casino (*Casa Gioiosa* was the "house of joy" provided for Vittorina da Feltre's schooling in Mantua of children of the Gonzaga family, other noble families as well as poor boys chosen for their talent).

Whereas the Reformation leader, Martin Luther, thought that ancient learning might harm Christian devotion, many of his humanist contemporaries and forebears sought a union of the revived culture of Antiquity and Christianity purged of supernatural fantasies. For Erasmus, leading Renaissance thinker and educator, it was the individual free of prejudice and dogmatic beliefs, pursuing his earthly opportunities and behaving as an active, responsible citizen, who set the standard. Neither scholarship nor piety – although each contributed – was the determining purpose. Rather, the human quest was for self-understanding and expression, worldly success but not vanity, action, glory and a refined character. Against this ideal, the older Christian world views shrank into those of a pious minority which even the upsurge of energetic devotion in the Reformation and Counter-Reformation struggled in vain to revive.

Even so, the deeply rooted Christian narrative of a quest for perfection inspired by the vision of eternal life in heaven persisted, surviving attacks and promoted in its various post-Reformation branches and sects by a large and powerful priesthood, widely respected and accepted even when not actively supported by the ever more powerful secular authorities in Europe and its colonies. Humans, whether the faithful or the heathen, were constantly enjoined to live according to God's will, in humility and obedience, seeking pardon for sins and errors. Life in the earthly city remained, for all its mundanity, but a prelude to eternal heavenly life for the saved and, for the unredeemed, eternal damnation. This was the destiny of souls which, however, only humans possessed in full measure. Nature, as a whole, was lacking in "soul".

Although the birth of a holy saviour, Jesus Christ, heralded hope, peace, goodwill on earth and potential happiness for all people, the achievement of comfortable, pleasurable earthly existence while reverencing all life on earth was not at all the goal of true believers. It did, however, become a goal, if always elusive, of scientific and social reformers both secular and religious, working in the natural world. Important steps were taken to understand, even to cherish, the Earth, as for example by the English statesman and philosopher, Francis Bacon, and by the founders of the Royal Society. Bacon, in his *New Atlantis* and other philosophical not to say utopian works, founded a movement of secular ideas designed to improve human society through knowledge grounded not in biblical and other ancient texts but in observation and experiments with natural phenomena and forces. Bacon's goal had a religious ring, but his means were empirical and earthly; his grand project was restoration of humanity's dominion over nature, so damaged in Adam's Fall, and the reign of God's image in the heart of men, through the study of Nature's Book. That this could be a thoroughly pleasurable endeavour was shown by Bacon in his celebrated essay *The Garden*.

Neither Bacon's philosophy, nor the establishment of the Royal Society in 1660, nor the natural philosophy of an early president, Isaac Newton, and the savants who joined him, were intentionally anti-theistic. But the covert expression of scepticism, disbelief and even atheism multiplied in the Christian nations as the ancient pagan cultures were more widely taught, read and discussed in the revival of Classical arts and learning.[4]

In the undermining and displacement of theological foundations of human understanding and of theological authority, the *Bible* was seen increasingly as a human artefact, and religious institutions as "relics of man's ancient bondage". Philosophers, from Baruch Spinoza (1622–77), who equated God with the force of nature or the state of the universe, to Pierre Bayle, Voltaire and the French *philosophes,* rejected the truth claims of Christianity and its subservience of humankind to God. But there have been religious surges around the globe and a reinforcement of fundamentalism within Judaism, Christianity and Islam.

Ironically, it was the intellectual deconstruction of the biblical doctrine of man's superior place in nature, at the behest of a supernatural god which had dominated European thought and much of its daily life over more than a millennium that in turn elevated humanity's power and dominance. The superseding of the supernatural by the natural in European thought was at the same time the fall of God the creator and the rise of his creation, man, the new creator, of science, mathematics, philosophy, history and a new economic and social order. This was to result in a more powerful and pervasive idea of dominion over nature, on the one hand scientific, on the other economic. Increasingly, and from the perspectives and works of military strategists, navigators, explorers, herders, farmers and artificers in towns and villages, scientists, historians and philosophers, the elements of the natural world thereby gained in significance for everyday life. Their value was of a practical nature to which religious interpretations were generally quite irrelevant.

17.3 Experiencing, Knowing and Using the Natural World

If wild nature was dangerous, urban settlement and built structures provided a bulwark against its threats. If labour on the land was arduous, its burden could be alleviated by the harnessing of animals, the invention of tools and machinery and the systematic pursuit of technical knowledge. If the frontiers of the known world reached ever further into distant and dangerous places, fresh experience and new knowledge could be incorporated into an enriched metropolitan understanding. The West, its ways of knowing and living, confirmed itself (and not God and his Heaven) as the heartland of being and value.

From the time herdspeople became agriculturists, minerals were systematically extracted from the earth and smelted into metal, and forests were felled for firewood and timber, arable fields and pasture, the Earth was seen as an exploitable resource. The fertility of farmland might decline with use, deserts expand, forests be cleared but more remained to be cleared, and the fruits of global voyages included vast, new territories from whose plundering the Europeans could benefit indefinitely. Opportunities were seized and limits ignored and this mentality was facilitated by, and in turn hastened, the growth of modern capitalism.

Wealth was amassed not only for present pleasure but invested for the sake of even greater future pleasure. The Dutch invention of the stock exchange helped to regularise commercial transactions while it boosted wealth accumulation. If there were crises in the investment cycle as in the collapse in 1637 of the market in tulip bulbs in Amsterdam and in the notorious South Sea bubble in London in 1720, there were compensating booms such as the East India Company profits from Indian tea, and profits from West Indian sugar plantations whose slave labourers were to be nourished by Captain Bligh's transport of Pacific breadfruit. Moreover, a succession of political economists from Physiocrats in France, and Adam Smith and Ricardo in England were constructing ingenious theories whose application would ensure an ever increasing (if unequally distributed) stream of national wealth.

From the time of the Renaissance, the world expanded beyond measure, through growing awareness of its physical scale and complexity, its place in the cosmos, its cultural diversity, and the uses to which its elements could be put, through human intervention, ingenuity and imagination. Religion maintained some of its ancient grip on human imagination and thought but as one complex of elements among many, which even its adherents increasingly treated as largely incidental to mundane secular life.

Early in the seventeenth century, French mathematician and philosopher, René Descartes, launched a new movement in philosophy, prelude to critical modern philosophy as distinct from the settled survivals of ancient and mediaeval ideas. His mathematical and scientific achievements strongly influenced subsequent thought as did his argument that humans and animals have machine bodies. Humans additionally had soul; animals were mere automata. In the great debates – from John Locke's rejection of innate ideas and sinfulness to David Hume's scepticism, the attacks on religion by the *philosophes* of eighteenth century France, to the great divide in the nineteenth century between spiritually-minded Idealists and the socially and politically-minded Utilitarians, Positivists and Pragmatists, notwithstanding – religious belief was a continuing undercurrent, sometimes a well ordered torrent as in Leibniz, at other times rejected as by the atheist, Schopenhauer. God was either a postulate, a ghostly figure or, occasionally, part of an elaborate metaphysical structure. Theology came to a turning point, shifting away from conjectures about the nature of the human soul and its destiny and the obscure metaphysics of the Trinity of father God, son Jesus and the mysterious holy ghost or spirit, towards inquiry into measures to combat the deprivations and injustices of contemporary world-wide life, the good life for all on earth. A personal Christian ethic, based on the *New Testament* Sermon on the Mount did not wholly supplant the ecclesiastical rituals and disciplines enjoined by the established churches but was part of a progressive shift from the kingdom of a supernatural heaven and the adoption of ways of life suitable to the republic of a thoroughly earthly character.

For the few, this meant an enhanced spirituality and rejection of materialism; for the many, an acceptance of ordinary, material life as experienced by the vast majority of people but elevated as may be through the practice of a Christian ethic of fairness, compassion, charity and peaceful co-existence.

In the course of the nineteenth century, a wide range of critics focussed their reforming zeal on investing those conditions of industrialisation and congested urban life that could not be materially overcome (as Marx and Engels proclaimed inevitably they would be) with spiritual and aesthetic values. John Ruskin (1819–1900), architect, artist, art historian and social critic, was an immensely influential moral force across the English-speaking world. In a stream of popular lectures and scholarly publications, he urged a new respect for the dignity of labour, destroyed by factory employment and urban conditions, and the adoption of architecture and decoration based on natural forms. One of his numerous followers was socialist, artist and designer, William Morris (1834–1896), who read to fellow undergraduates at Oxford passages from Ruskin's *Stones of Venice* and *Seven Lamps of Architecture*, part of his own formation as a lover of natural form and expression. Morris, projecting an ideal life close to nature in *News From Nowhere* (an idyllic rural community of the highest moral order), happily combined spiritualised art and social philosophy with great skill in crafting materials, church windows and tapestries among them. Morris' craftsmanship extended to gardens and he is credited with laying the foundation of the arts and crafts garden, forerunner of the much loved, carefully crafted modern English garden style. The *Red House*, that he had designed by architect, Philip Webb, as his own abode in South London, "sparkles with a bright determination to forge a new harmony between man and nature … The house and its garden were inseparable ideologically, as a fusion of art and nature ... the house and its garden were to be used as a single entity."[5]

17.4 A Return to Nature – A Physical Quest and a Moral Ideal

The work of scientist, geographer, traveller, and polymath, Alexander von Humboldt, has been designated "the invention of nature". This is understandable given the efforts to produce a general theory, embracing many branches of developing early nineteenth century science, and the enormous popularity of his writings.[6]

This attitude towards the natural world was expressed in a broader "return to nature" movement, in philosophy, literature and the arts generally. Its foundations, often attributed to the Swiss, Jean-Jacques Rousseau, and the Romantic movement in Germany and England, lie deep in human experience and times long past; for example, in the worship of nature gods, in holy places such as mountains, forest groves and springs, central to the ancient pagan nature religions. Mad monks and mystics sought a simple, unrefined existence free of the travails of society. What later writers and artists did was to articulate human sensibility, trappings and dream-like feeling into definite states of keen awareness and consciousness which seemed essentially spiritual and anti-materialist. As indicated above, the arts and crafts garden of the mid to late nineteenth century embraced nature as both a physical source of plant life and a moral force in the organisation of social life and idealized personal living. Followers of Rousseau and devotees of the poetry of Wordsworth and Coleridge in England, of Emerson and Thoreau in America, and the music and art of German musicians and French painters turned to the concrete natural world, rambles in the English and German countryside, visits to famous beauty spots in the Lake Country, the Italian lakes, the Alps or the Black Forest. Writers like George Sand were also enthusiastic amateur naturalists, botanising, collecting and drying wildflowers, or unearthing fossils along coastlines or in ancient valleys. This was not primarily a dispassionate scientific analytic movement but one that drew on the personal and subjective expression of people's feelings for nature which became elevated not only into the love of plants, animals and natural scenery but into a desire to conserve and protect the natural world against the inroads of industrialisation and urbanisation: the spread of artifice, the threat to spontaneity, the dominance of the analytic over the imaginative mind.

As we observed in Chap. 3, the field naturalist movement, thriving in England and Australia in the second half of the nineteenth and into the twentieth century, drew ideas and practices alike from science, the literary and artistic well springs of European Romanticism, and the fruits of collectors who scoured the new world(s) for plants, animals and mineral specimens.

In the German-speaking countries and particularly (at first) in Switzerland, the romantic sentiment and the quest for a more "natural" (i.e. less mannered, citified way of life) led to waves of educational reform across continental Europe and the English-speaking countries. As we saw in Chap. 16, Rousseau, in his celebrated treatise *Emile*, on the education of a boy and young man, conjured up an educational regime of diverse experience including practical training for a trade. He urged that educators should not impose worldly knowledge and adult wisdom but instead seek to instil virtue and to study and seek to understand vital forces in the growing child. They should gear their teaching, he argued, to the developing talents and interests of the individual learner with happy, not tearful, results. Despite its appeal, this is an ideal which mass systems of highly regimented schooling have ever since struggled to attain, even when they manage to accept and understand its ideals and principles. Friedrich Froebel carried Rousseau's spirit of individualism into the classroom for infants and young children – the kindergarten (or child's garden), replete with structured games and activities ("gifts and occupations") and outdoor play and excursions, all to engage and respect vital energies and emerging interests of childhood. Pestalozzi, among later reformers, such as medical practitioner, Maria Montessori in Italy; theosophist, Rudolf Steiner in Austria and philosopher, John Dewey in the United States, further developed these ideas in efforts to turn schooling from a type of factory producing dull, docile citizens through drill and memorisation into free, active and socially engaged young women and men.

Twentieth century psychologists, following the Frenchman, Jean Piaget, in the spirit of Rousseau, recognised that learning is a developmental process and that in childhood and youth there are distinct phases, emerging capabilities that should set directions for schooling, even more than the social and economic goals of the state and its authorities. For his numerous readers around the globe, Charles Dickens in *Hard Times* had demonstrated the iniquities and failures of Mr. Gradgrind's cruel regime.

As outlined in Chap. 16, the nature study movement in public schools in England, continental Europe, America and Australia, late in the nineteenth and early twentieth centuries, was a further elaboration of the ideas and ideals of the earlier Romantic movement. Perhaps because of the challenge these ideas represented to the conventions and artifices of sophisticated society, as much as to the exigencies of mass schooling systems, they are today mostly submerged where social and political requirements lead to a focus on pupil attainments in science, mathematics and literacy as attested in national and international testing regimes.

It was Immanuel Kant, the great German philosopher who, as he said, bestirred himself from dogmatic slumber to confront David Hume's scepticism about the foundations of human knowledge, in order to demonstrate its nature and limits. Kant was also an admirer of Rousseau's achievement in showing the fundamental role of feelings in ethical judgements. Kant did not reject religion, as did many philosophers of his time and thereafter, but it did not play an important role in his thinking. He applied his formidable intellect to demolishing purely intellectual arguments for the existence of God but felt obliged to invoke Providence, one of the "ideas of reason", and the reality of an after life to ensure the justice of the moral law. This was hardly a line of argument that would interest the ordinary run of believers or unbelievers, whereas his ethical claim that all humans should be treated as ends in themselves has had a profound impact on theories and manifestos of human rights.

17.5 Darwinism, the Agnostic Challenge and Ecology

The strongest, most damaging challenge to the fundamental religious doctrines concerning nature and creation came from the *philosophes* in eighteenth century France and the sceptical, empirical strain in British philosophy, rather than from science, at least until Darwin. Religion and science have learnt to co-habit in modern intellectual space but this settlement follows the turmoil engendered by early geological and archaeological discoveries and, above all, the publication in 1859 of Charles Darwin's *The Origin of Species*. Darwin's great treatise was a powerful destabilising force in nineteenth century thought and social life since it offered a rich body of evidence in support of the thesis that humans are part of an inclusive chain of life and in a line of descent from the great apes. Neither the idea of the connectedness of life nor that of evolution originated with Darwin. "Connectedness", for example, was an idea that persisted in the form of "the great chain of being" and the search for a mechanism to explain how evolution occurred was well under way even before the nineteenth century. Darwin's great achievement was the convincing hypothesis of a mechanism, evolution through natural selection in the competitive conditions of nature.

Published in 1836, Charles Lyell's epochal work, *The Principles of Geology*, was among the scientific challenges to the narrative of "6 to 7 days of creation" which still prevailed in the Judaeo-Christian world. Lyell's work was a prized possession of the young Charles Darwin, accompanying him in the great cabin of *HMS Beagle* that he shared with the captain, in the circumnavigation of the globe, together with his equipment, books and specimens. Lyell, and Joseph Hooker as well as Huxley, read a draft of Darwin's text of *The Origin of Species* and encouraged immediate publication of a summary in the dispute over whether he or Alfred Wallace could claim precedence in the discovery.

By that time, natural history (renamed "science" in 1833 by Oxford historian, scientist, Anglican priest and polymath, William Whewell) was making advances on a wide front (leading to re-appraisals of religious claims concerning origins of life and growing uncertainty among the more thoughtful clergy). On the title page of Darwin's *The Origin of Species* there was an ambiguous quotation from Whewell. It gives a clue to Darwin's ambivalence over the relationship of old religion to new science, "But with regard to the material world, we can at least go so far as this – we can perceive that events are brought about, not by insulated interpositions of Divine power, exerted in each particular case, but by the establishment of general laws." Thereby, room could be found, according to preference, for both an ultimate creator, God, and the natural, secular evolutionary process of species diversity and environmental adaptation. A shifting, at times uneasy, alliance of science and religion has continued to the present day, including the Darwinian interpretations of nature's working, except for those (mainly north American) fundamentalists and creationists for whom the theory of evolution in any form is anathema. But, for many thinkers of the time, faith was wavering, as in Mathew Arnold's evocative poem, *Dover Beach*,

... The Sea of Faith
Was once, too, at the full, and round earth's shore
Lay like the folds of a bright girdle furled.
But now I only hear
Its melancholy, long, withdrawing roar, ...

The chief and most effective protagonist of the Darwinian theory of evolution by natural selection was Thomas Henry Huxley, "Darwin's bulldog". He coined the term "agnostic" to underline his view that the theory, no matter how plausible, was yet a testable hypothesis. As such it could be supported. Another neologism of the time was "ecology" coined in 1866 by Darwin's German disciple, Ernst Haeckel, a biologist who found in Darwin's work confirmation of the interrelatedness and mutual dependence of the vast array of forms of life and the multiplicity of environments and habitats which support life. Modern ecologists have taken our understanding of the extent and complexity of these relationships far beyond what Darwin, Haeckel and their colleagues and early followers achieved. In doing so they have become acutely aware of the risks and threats of human intervention in natural processes, broadly speaking of the multiple and varied uses we make of the natural world. Some have become outspoken critics of the continuing human destruction of life-supporting habitats for other species and leaders in international movements to conserve and restore whole environments threatened by human population growth and spread, deforestation, mining, urbanisation and pollution of rivers, oceans and the atmosphere beyond. Modern ecology is a good example of how carefully considered advances in science provide the intellectual foundations and motor for an enhanced capacity for social and political action; what humans are presently destroying, they can, through the advancement and diffusion of knowledge and forethought, learn to sustain and restore. Climate science is another, if more contested example. As a society we must actively and vigorously counter ignorance, prejudice and narrow self interest which stand in the way of the widely beneficent changes it is within our power and general interest to make. Hence the contemporary "environmental crisis" and the popularity of the shibboleths "sustainability" and "sustainable growth", which risk obscuring the most complex and difficult issues that have yet to be resolved as we continue in practice to plunder the earth and poison our environment.

17.6 Scientific Inquiry Flourishes while Religious Faith Vacillates

Darwin's theory of evolution, Lyell's geology and Alexander von Humboldt's multi-volume *Kosmos* might have suggested the possibility of a global unifying theory of nature both organic and inorganic, but such notions gave way to the fracturing of natural history and nineteenth century European and American science into the distinct disciplines of physics, chemistry, biology, geology, astronomy and so forth. They all surged ahead especially in universities in the latter part of the nineteenth century, but it fell to the emerging, multi-disciplinary ecology to find bridges between biological, physical, chemical and social sciences, blending into them ethical concepts, aesthetic values and the understanding that emotional force, and not only abstract intellectual analysis, must be brought into play in reconciling humanity with nature.

While diverse religious feeling and expression have continued to revitalise belief and practice, in many countries active church attendance has steadily declined. The distinguished church historian, Diarmuid McCullough, subtitled his monumental study *Christianity, the first three thousand years* (my underlining), anticipating at least another millennium. Whether or not that third millennium is one of continued vitality of Christianity, or of its further decline, just what most people continue to believe and how they map out ways of living, religious life will change profoundly as our understanding (and destruction) of the natural world continues to grow.

As science becomes more esoteric, the need for a 'general public understanding' increases. The need is partly met by magazines, popular books by scientists and science journalists, television, film, radio and the internet, as it was in the nineteenth century by lectures given by Faraday and others at the Royal Institution, Mechanics Institutes, summer camps and such vehicles of popular adult education. As a result, it is no longer possible to speak in public with the confidence of early theologians about a literal heaven, whose features echo in exalted ways earth and physical human life. God, as some mystical being, beyond time, and creator and lord of the universe, has been decisively and irredeemably toppled even if in mysterious ways we find that, for some theologians, Nietzsche's pronouncement that *God is dead* does not quite mean the end of godhood. But as a dominant, if never quite exclusive world-view, religion has given way to secularism at least in the intellectual Western world view. Traditional religious beliefs about the origins and nature of life, the nature of the cosmos, the destiny of humanity and the forces that govern behaviour have thereby lost credibility even if still clung to by people to whom they give psychological comfort and cognitive assurance. The situation is different for followers of Islam, but that also is shifting at least for those Muslims in the West as the Islamic community grows and spreads. Indeed, as we saw in the chapter on the Mogul Empire, Islamic culture has a history of rich diversity in belief and action that is belied by contemporary Islamic fundamentalism.

What has not lost credibility and remains one of the most secure foundations of western civilisation, is a faith, often philosophically argued for, in the value and significance of life itself. While this view is most often declared on behalf of human life, a great modern achievement, led by a small band of philosophers and eco scientists is to show why life in all of its forms is to be valued, for human benefit, yes, but no less for itself.[7]

This concern for life is no mere intellectual abstraction or sentimental avowal. It is manifested in many different ways which taken together constitute what may be termed a culture, a civilisation, a way of life and being, all created and sustained by people living together in reasonable amity. If this seems like a dream and a delusion in a world of chaos, no better for practical purposes than the succession of utopias from Plato's *Republic* to William Morris' *News from Nowhere,* to Karl Marx and Friedrich Engels' *Communist Manifesto*, it sets a direction for generous and capable human action. Heirs to one of the world's greatest civilisations, intellectually, ethically, culturally and technically vital, Westerners of today can see its continued renewal and advance as a meaningful purpose worthy of our best effort. This cannot be on the battleground of contemporary political extremism which must be countered but across the vast and expanding field of civil and cultural life that we have created for ourselves.

17.7 The West and its Values

Understanding, conserving, protecting, developing the idea of Western Civilisation, as already noted, has become highly contentious, with extreme forms of nationalism and political extremism spreading across Europe, the United Kingdom, the Americas, Australasia, Africa and Asia. All regions and countries are faced by critical challenges arising from our treatment of the common world of nature which is our fundamental habitat. The upsurge of virulent forms of nationalistic belief and behaviour threatens to shift the needed focus on strengthening, not destroying, the natural world and on redressing the failings in our socio-political affairs. Perhaps these tendencies are more closely linked than we imagine. Political national boundaries of themselves pay no heed to natural boundaries. France prides itself on its politically and militarily delineated hexagonal shape, a figure that brings emotional satisfaction, but the north shores of the Mediterranean stretch uninterruptedly from Spain to Lebanon and the ocean from Normandy and Aquitaine stretches around the globe. So the French response to climate change, how they manage their waste, how they treat the air above and the fields below are all globally significant. Their minority of flag-waving nationalists are an irritating, potentially significant (while socially obnoxious), interruption in a quest to find ways of life apt for the whole globe, continuous in nature.

Whatever may be our views about its achievements and its failures, we are all heirs to one of the world's greatest civilisations and cultural achievements, variously known as "Western" or "European". These are loose terms, nowadays scarcely applicable since many of their elements and aspects have become global, not only because of their spread through worldwide settlement by Europeans and their colonial outreach but also because of their success in meeting common human needs and preferences. We are increasingly aware of the injustices and destruction wrought by Western beliefs and practices, including colonialism and its capitalist and militarist partners. But we must also be ready to acknowledge the benefits to everyone of the spread of democratic institutions, reasoned ideas of justice, liberty and equality, healthful living, great artistic, literary and musical achievements, highly organised social and cultural institutions and procedures, organised scientific inquiry, and the technical harnessing of stupendous natural forces. The popular and often deserved denunciations of the cultural costs and dangers in pursuit of these immense developments need to be tempered by acknowledgement of the achievements and benefits. The Western idea has yet to achieve fulfilment. It can only be realised through a universal vision drawing together in peaceful co-existence the cultures of the world. Such was the aspiration in the establishment of the UN following the evils and devastation of the Second World War. Do the ideas of "the West" need defence? No, but a just appraisal is of value as we seek to shape a fair and equitable future, not only for humans but the living planet in all its diversity and beauty. These achievements and their attendant virtues (and shortcomings) are of course not unique to Western civilisation but they are among its more notable features.

The major purpose underlying/guiding this book has been to trace some of the ways in which Western peoples have perceived and responded to their understanding of the two poles of St Augustine's great work: the earthly (nature) and the heavenly (supernature). At the same time there has been acknowledgement that other civilisations, specifically that of Islam and Australian Aboriginal religion, have shared that love and study of nature and its supernatural counterpointing in major world religions. There is of course room for quite diverse naturalistic interpretation of one of the West's greatest texts, the Judaeo-Christian *Bible*, but there can be no doubt that one of the most influential from earliest times is the supernaturalistic, that the human species is special, created by God to act on his behalf and at his behest in ruling over nature and disposing of its resources according to unrestrained choice and felt need. This, the thesis of dominion, is a contentious interpretation among biblical scholars, but the continuing use and exploitation by humans of earth's huge – but finite – resources, together with the role played by religion in the global European expansion story, testify to its continuing impact.

Combined with the gathering force of post Renaissance capitalism, its drive for unrestrained growth and its relentless urge towards ever greater consumerism, this misguided religious assertion of human dominion in service to God has only in recent decades been revealed as potentially fatal, to itself and the earth itself. But there can be no doubting that the history of Western religion is one of humanity's most searching, imaginative and sustained attempts to understand the natural world and the place of humanity in it. Natural history in the past and science in modern times came to rival and then very largely to supplant religious interpretations of the world, not only in intellectual conviction but also in imaginative force. The mysteries and quirkiness revealed by atomic and quantum mechanics, particle physics, astronomy, cosmology from late niineteenth century onwards equal or frequently exceed the mysteries evoked by *Old Testament* prophets. The explanatory power of the scientists equals when it does not exceed that of the theologians. The Western inheritance contains and acknowledges the force of both as continuing sources of imaginative speculation and moral and intellectual creativity.

But in our uses of science and religion, have we not also learnt to despoil even to despise raw, uncultivated nature? Until environmentalists taught a better lesson, "wilderness" was a pejorative term. Is it possible, in the modern world, to live a just and sustainable life in harmony instead of at odds with nature? Need humans continue to gain a sense of value and worth by seeing themselves as a unique and superior species? Do we not gain more by finding ways to conserve and sustain rather than recklessly and unthinkingly exploit bountiful nature? Are we able and willing to moderate our consumption and to care for rather than disdain what we think of as the lowliest of creatures and their behaviour in the struggle for a place in the natural world?

For the religiously minded, care, compassion and intelligence, and a belief system which accepts and appreciates the value of scientific inquiry and of the continuing, reflective appraisal of belief, need not depend on an outmoded if not preposterous mental archaeology of heaven above and perdition below. Can all religiously-minded people not accept this conversion of their faith, as many are now doing from a supernatural to a natural foundation? Attack and condemnation, in the manner of Oxford professor of biology, Richard Dawkins, and writer, Christopher Hitchens, may be intellectually gratifying but it is not the way to bring about the naturalistic turn in fundamentalist religious life that is needed if old style creationism is to give way to well-grounded scientific attitudes. That is a longer term, patient process of awareness and education in which all forms of modern communication must play a part. In this regard, popular natural science TV programs like those of David Attenborough (Fig. 17.1) and Brian Cox could be having more impact on public attitudes and beliefs than either teaching or denouncing the theory of evolution in high school and university.

We have a concrete reality to address now, the living Earth and threats to it, without resorting to ancient myths and fables, except as means of stirring the imagination by pleasurable reading. The paradises constructed by the Muslim Moguls in North India and Kashmir and those in the imaginary world of secular socialist poet, William Morris, and the long tradition of Utopian communities are reminders that the idea of an earthly paradise expresses human hope whether in religious or secular terms. Respecting and creating rather than destroying or demeaning nature and natural ways of living is a never-ending challenge.

As the ancient Christian doctrine of man's dominion is being progressively displaced by a growing acceptance that human life is but one form and of no greater interest or fundamental value than all others, so did the sense of humans' importance in the scheme of things and the superiority of the human species over all others expand in other spheres. Science, a grand project in de-mystifying the natural world and providing material instead of spiritual mechanics for its operations, demonstrates intellectual mastery by humans far beyond that of any other species. Production, commerce and world trade are decisive demonstrations of human capability. The

Fig. 17.1 David Attenborough, leading conservationist, at Great Barrier Reef c.2015 during the making of his Great Barrier Reef series. Australian Department of Foreign Affairs and Trade website – www.dfat.gov.au, CC BY 3.0 AU, via Wikimedia Commons

economic and political system of modern industrial capitalism and its material consumption have further strengthened a sense of human mastery, with a disregard of its consequences for all other inhabitants of the natural world. Is there an alternate way?

In one sense, man or, rather, humanity, does of course have dominion. This is by virtue of mind, of intellect, and possession of a moral sense. We have consciousness, and an ability to understand, to reflect on our actions and their likely consequences. We can and do empathise, and enter both emotionally and intellectually into the lives of others. We can and do love and reason. The challenge we humans face is to make the fullest, most conscious and collective use of these distinctive qualities.

With gathering pace since late in the twentieth century, awareness and concern over dangerous features of modern life have grown, in a virtual cascade of scientific reports, books, pamphlets, television and radio shows, protest movements, conservation and preservation legislation and regulation and organised lobby and activist groups ranging from Greenpeace protecting whales from the Japanese harpoonists in the Southern Ocean to Friends of the Earth campaigning on the steps of parliament against coalmining and fracking for underground gas. These are but two among thousands of movements that are aiming to better protect and use the earth's resources. So, it seems that awareness of damage and destruction is no longer the point, whereas changing political attitudes and behaviour have become the crucial issue. We know on the best evidence that the global climate is changing and have well-founded belief that human activity is accelerating it. We have good and adequate (scientifically informed) reasons for believing that human activities including the extraction and burning of fossil fuels, if substantially reduced, would significantly reduce the adverse impact of climate change on human life. Bodies like the

UN's Intergovernmental Panel on Climate Change (IPCC) are amassing the necessary scientific evidence; public figures ranging from Swedish school student to heir to Great Britain's throne have campaigned for greater political action to protect the earth. Greta Thunberg's strident calls at international meetings have been rejected by world leaders, including former US President Trump (Fig. 17.2). Yet her list of rules and recommendations for followers in her global campaign of school strikes are a model of moderation and good sense:

> A list of rules and recommendations for those on schoolstrike for climate: no violence, no damage, no littering, no profit, no hate, minimise your carbon footprint, always refer to science.
>
> Our demand: Follow the Paris Agreement [2015] and IPCC Reports. Stay below 1.5 degrees. Focus on the aspect of equity and climate justice, clearly stated throughout the Paris Agreement, because no manifesto can be more radical than that. Unite behind the science. Greta Thunberg, writing to those on school strike for climate.[8]

Greta Thunberg has encouraged direct action including civil disobedience by students around the world. Others are doing likewise, frustrated by the inadequacy of government action. The passionate nature lover, broadcaster, writer and activist, Dara McAnulty in Northern Ireland is another youth leader calling for immediate steps to save the natural world.[9] Extinction Rebellion is one of many expressions of public anger over the impending crisis, not only climate but massive species loss.

In December 2019, climate champions, Greta Thunberg, then aged 16 and David Attenborough, then aged 93, joined forces on UK radio calling for governments to step up action on climate change. Their voices, intelligent, knowledgeable, passionate and clear are heard around the world, even when brushed aside by insensitive politicians and business leaders.

Fig. 17.2 Climate champion Greta Thunberg speaking at Climate March of 27 September 2019 in Montreal. Lëa-Kim Châteauneuf, CC BY-SA 4.0, via Wikimedia Commons

Through co-operative international action, as in the work of the IPCC and the 2015 Paris Accord, under the auspices of the UN, some of the necessary action is being undertaken. But, due to the complexities and pressures of everyday social and economic life and the power of national policy making compared with our international instruments, the action we are taking to maintain the resources of the earth and the atmosphere increasingly appears to be inadequate to the tasks that need to be accomplished.[10] Are we then, as a species, unwilling or unable to sustain the conditions that we know to be necessary for our continued survival and that of innumerable forms of life? Are we so utterly, recklessly stupid? Human intelligence and the aspiration to perpetuate the species may in the decades ahead find ways to avoid the disaster predicted by the sharpest critics. The enduring love of life in the natural world and the huge growth of knowledge from its study inspire some hope and provide grounds for confidence. But we must rapidly expand and enrich our fund of knowledge through research and, in no less measure, community-wide learning. The potential for disaster has increased since H.G. Wells penned his famous race between education and catastrophe. We have since then, with difficulty, survived two global wars and decades and more of the possibility of nuclear winter and the human destruction of all forms of life on earth. If civilisation is to live up to its highest ideals and most exalted standards, we now have no choice but to save and cherish what we have for too long been unthinkingly destroying.

Notes

[1] Barbault, R. (2014) *Au Nom du Vivant. Plaidoyer pour réconcilier l'homme et la nature.* Paris. Buchet/Chastel. p. 96.

[2] Klein, N. (2014) *This Changes Everything. Capitalism vs Climate.* London: Allen Lane. Ch 2 Hot Money. How Free Market Fundamentalism Helped Overheat the Planet.

[3] Kennedy, P. (2010) *Twentieth-Century Theologians. A New Introduction to Modern Christian Thought.* London: I.B. Tauris.

[4] MacCulloch, D. (2009) *Christianity. The First Three Thousand Years.* New York: Viking. Ch21 Enlightenment: Ally or Enemy? (1492–1815) pp769–816.

[5] Hitchmough, W. (1997) *Arts and Crafts Gardens.* London: Pavillion. p. 32.

[6] Wulf, A. (2016) *The Invention of Nature: the adventures of Alexander von Humboldt, the lost hero of science.* London: John Murray; von Humboldt, A. (2014) *Views of Nature* edited by Jackson, S.T. and Walls, L.D., trans by Person, M.W. Chicago: University of Chicago Press.

[7] Birch, C. and Cobb, J. B. (1982) *Liberation of Life: from the cell to the community.* Cambridge: University Press.

[8] https://climatesafety.info/thesustainablehour257/ Accessed March 23, 2021.

[9] McAnulty, D. (2020) *Diary of a Young Naturalist.* Dorset, U.K.: Little Toller Books.

[10] Flannery, T. (2020) *The Climate Cure. Solving the Climate Emergency in the Era of Covid-19.* Melbourne: Text Publishing. Part 1 The Great Australian Tragedy.

Complete Listing for Wikimedia Commons Images

17.1 David Attenborough at Great Barrier Reef. c.2015. https://commons.wikimedia.org/wiki/File:David_Attenborough_at_Great_Barrier_Reef.jpg Attribution: Department of Foreign Affairs and Trade website – www.dfat.gov.au, CC BY 3.0 AU <https://creativecommons.org/licenses/by/3.0/au/deed.en>, via Wikimedia Commons.

17.2 Marche pour le climat 27-09-2019 (Montréal) 13. Author: Lëa-Kim Châteauneuf Source: own work. https://commons.wikimedia.org/wiki/File:Marche_pour_le_climat_27-09-2019_(Montr%C3%A9al)_13.jpg Attribution: Lëa-Kim Châteauneuf, CC BY-SA 4.0 <https://creativecommons.org/licenses/by-sa/4.0>, via Wikimedia Commons.

Bibliography

Abu-L-Fazl. (1973). *The Akbar Nama of Abu-L-Fazl. History of the reign of Akbar including an account of his predecessors.* Translated from the Persian by H. Beveridge. Vols 1–3. Ess Publications (1973 reprint).

Adamson, J. (1919). *A short history of education*. Cambridge University Press.

Aldo, L. (1949). *A Sand County Almanac: And sketches here and there*. Oxford University Press.

Allen, H. (Ed.). (2010). *Australia: William Blandowski's illustrated encyclopaedia of Aboriginal Australia*. Aboriginal Studies Press.

Anderson, S. (2009). *Pelletier. The forgotten castaway of Cape York.* With ethnographic commentary by Athol Chase. Melbourne Books.

Angell, S. W., & Dandelion, B. P. (Eds.). (2013). *The Oxford handbook of Quaker studies*. Oxford University Press.

Angus, M. (1975). *The world of Olegas Truchanas*. O.B.M. Pty. Ltd.

Aristotle. (1959). *Politics* (H. Rackham, Trans). William Heinemann Ltd./Harvard University Press.

Aristotle. (1968). *Parts of animals, movement of animals, progression of animals*. Peck, A.L., & Forster, E.S. (transl.) The Loeb Classical Library. William Heinemann Ltd./Harvard University Press.

Attenborough, D. (1987). *The first Eden: The Mediterranean world and man*. Collins.

Attenborough, D. (2020). *A life on our planet. My witness statement and a vision for the future*. Witness Books.

Augustine. (1957–1972). *The city of God against the pagans* (7 vols). William Heinemann Ltd./ Harvard University Press.

Augustine. (1965). *St Augustine select letters*. William Heinemann Ltd./Harvard University Press.

Augustine. (1977). *St Augustine's confessions* (Vol. I). Harvard University Press/William Heinemann Ltd.

Barbault, R. (2014). *Au Nom du Vivant. Plaidoyer pour réconcilier l'homme et la nature*. Buchet/ Chastel.

Barzun, J. (1962). *Classic, romantic and modern*. Secker and Warburg.

Bataille, G. (2005). *The cradle of humanity: Prehistoric art and culture*. Zone Books.

Bates, D. (1944). *The passing of the Aborigines*. John Murray.

Bean, C. E. W. (1963). *On the wool track*. Angus and Robertson Ltd/Sirius Books.

Berger, J. (1972). *Ways of seeing*. BBC/Penguin.

Berger, J. (2009). *Why look at animals*. Penguin Books.

M. Skilbeck, *Loving and Studying Nature*, International Explorations in Outdoor and Environmental Education 10, https://doi.org/10.1007/978-3-030-80751-1

Beveridge, A. S. (1922/2019). *The Babur-Nama in English (Memoirs of Babur)* 2 Vols. Translated from the original Turki text of Zahiru'd-din Muhammad Babur Padshah Ghazi. Alpha Editions. Also available on Internet Archive: https://archive.org/stream/in.ernet.dli.2015.85352/2015.85352.The-Babur-Nama-In-English-Vol-I_djvu.txt. Accessed 4 Mar 2021.

Bible., (King James Bible, English Revised Version) *The Holy Bible* Containing the Old and New Testaments and the Apocrypha. Translated out of the Original Tongues: and with the former Translations diligently compared and revised by His Majesty's special command. Appointed to be read in Churches. Oxford University Press.

Birch, C., & Cobb, J. B. (1982). *Liberation of life: From the cell to the community*. Cambridge University Press.

Blainey, G. (1963). *The rush that never ended*. Melbourne University Press.

Blainey, G. (1966). *The tyranny of distance: How distance shaped Australia's history*. Sun Books.

Blainey, G. (1968). *The rise of Broken Hill*. Macmillan.

Blainey, G. (2001). *2001 Boyer lectures: This land is all horizons – Australian fears and visions*. ABC Books for Australian Broadcasting Corporation.

Bolton, G. (1981). *Spoils and spoilers. Australians make their environment 1788–1980*. Allen and Unwin.

Boucharlat, R. (2009). The "Paradise" of Cyrus at Pasargadae, the core of the Royal ostentation. In J. Ganzert & J. Wolschke-Bulmahn (Eds.), *Bau und Gartenkultur zwischen "Orient" und "Okzident"* (pp. 47–64). Martin Meidenbauer Verlag.

Boucharlat, R. (2016). *Qanat* and *Falaj*: polycentric and multi-period innovations (chapter 17). In A. N. Angelakis et al. (Eds.), *Underground aqueducts handbook* (pp. 279–301). CRC Press Taylor & Francis Group.

Boucharlat, R. (2018). Cyrus and Pasargadae: Forging an empire – Fashioning "paradise". In M. R. Shayegan (Ed.), *Cyrus the great – Life and lore* (pp. 131–149). Ilex Foundation.

Brown, J. (1982). *Gardens of a golden afternoon. The story of a partnership: Edwin Lutyens and Gertrude Jekyll*. Penguin Books.

Brown, J. (1999). *The pursuit of paradise – A social history of gardens and gardening*. Harper Collins.

Brown, R. (1958a). *Fifty animal stories of Saint Francis as told by his companions*. Franciscan Herald Press.

Brown, R. (1958b). *The little flowers of St Francis*. Doubleday.

Brunel, F. (n.d.). *The splendour of Indian miniatures*. Clarion Book/Editions Delroisse.

Burckhardt, T. (1976). *Art of Islam: Language and meaning*. World of Islam Festival Publishing Company.

Caruana, W. (2003). *Aboriginal art*. Thames and Hudson.

Chambers, D. C. C. (1993). *The planters of the English landscape garden. Botany, trees and the georgics*. Yale University Press.

Charlesworth, M. (1997). *Religious inventions. Four essays*. Cambridge University Press.

Charlesworth, M., Kimber, R., & Wallace, N. (1990). *Ancestor spirits. Aspects of Australian Aboriginal life and spirituality*. Deakin University Press.

Charlesworth, M., Dussart, F., & Morphy, H. (Eds.). (2005). *Aboriginal religions in Australia. An Anthology of recent writings*. Ashgate.

Chaucer, G. (1986). *The Canterbury Tales* (F. N. Robinson, Ed., D. Wright, Trans.). The Folio Society.

Clarke, P. (2007). *Aboriginal people and their plants*. Rosenberg.

Clarke, P. A. (2008). *Aboriginal plant collectors. Botanists and Australian Aboriginal people in the nineteenth century*. Rosenberg.

Cole, M. (1945). *Beatrice Webb 1848–1943*. Longman Green.

Connell, W. F. (1980). *A history of education in the twentieth century world*. Curriculum Development Centre.

Constable, A. (1891/1968). 2nd ed. *Travels in the Mogul Empire by Francois Bernier AD 1656–1668*. A revised and improved edition based upon Irving Brock's translation. S. Chard and Company.
Coulton, G. G. (1959). *Ten medieval studies*. Beacon Press.
Coulton, G.G. (1961). *Medieval Panorama* (2 vols). Collins.
Cremin, L. A. (1961). *The transformation of the school. Progressivism in American education 1876–1957*. Alfred K. Knopf.
Cresson, O. (2005). *Quakers and the environment: Three options*. Posted June 18, 2005 in Blog Posts. Accessed 17 Feb 2020. http://www.nontheistfriends.org/article/2005/06
Crone, G. R. (1958). *Maps and their makers*. Hutchinson's University Library.
Crowe, S., & Haywood, S. (1973). *The gardens of Mughul India – A history and a guide*. Vikas Publishing House.
Cunningham, P., (1827). *Two Years in New South Wales. Comprising sketches of the actual state of society in that colony; of its peculiar advantages to emigrants; of its topography, natural history, & c &c.* In Two Volumes 2nd edition. Vol. 1, Henry Colburn, New Burlington St.
Curtis, G. (2006). *The cave painters. Probing the mysteries of the world's first artists*. Knopf.
Darwin, C. (1859). *The origin of species*. John Murray.
Davis, R. C. (Ed.). (2002). *The central Australian expedition 1844–46. The Journals of Charles Sturt*. The Hakluyt Society.
de Clavijo, R. G. (1859). *Narrative of the Embassy of Ruy Gonzalez de Clavijo to the Court of Timour at Samarcand, AD 1403–06* (C. R. Markham, Trans.). Hakluyt Society.
Dewey, J. (1900). *The school and society*. Chicago University Press.
Dewey, J. (1916). *Democracy and education*. The Macmillan Company.
Diamond, J. (1999). *Guns, germs and steel – The fates of human societies*. W.W. Norton & Co.
Fensham, R. J. (Ed.). (2013). The Leichhardt Papers. Reflections on his life and legacy. *Memoirs of the Queensland Museum/Culture, 7*(2). Queensland Government.
Fidock, A. (Ed.). (1982). *Introducing Aboriginal Australians*. Curriculum Development Centre.
Flannery, T. (1994). *The future eaters – An ecological history of the Australasian lands and people*. Reed Books.
Flannery, T. (2020). *The climate cure. Solving the climate emergency in the era of Covid-19*. Text Publishing.
Fox, R. L. (1982). *Better gardening*. R&L.
Gammage, W. L. (2011). *The biggest estate on earth: How Aborigines made Australia*. Allen and Unwin.
Gibbon, E. (1985). *The history of the decline and fall of the Roman empire* (Vol. 2 and 3). The Folio Society.
Gilbert, K. L. (n.d.). *The orchid man. The life, work and memoirs of the Revd H.M.R. Rupp 1872–1956*. Kangaroo Press.
Gombrich, E. H. (1972). *The story of art*. Phaidon.
Gooding, M., Mabberley, D., & Studholme, J. (2017). *Joseph Banks' Florilegium. Botanical treasures from Cook's first voyage*. Thames and Hudson.
Gough, A. (1997). *Education and the environment: Policy, trends and problems of marginalisation*. Australian Council for Educational Research.
Greenall, A. (1981). *Environmental education in Australia: Phenomenon of the seventies* (Occasional paper no.7). Curriculum Development Centre.
Grey, E. B. (Ed.) (1892). *The travels of Pietro della Valle in India*. From the Old English Translation of 1644, by G. Havers. In two volumes. Printed for the Hakluyt Society.
Gribbin, M., & Gribbin, J. (2008). *Flower Hunters*. Oxford University Press.
Griffiths, T. (1996). *Hunters and gatherers. The antiquarian imagination in Australia*. Cambridge University Press.
Grigson, G. (1957). *The painted caves*. Phoenix House.
Gul-Baden, Begam (Princess Rose-Body) *The History of Humayun (Humayun-nama)* translated with Introduction by Annette S. Beveridge (1902). Royal Asiatic Society.

Hadfield, M. (1969). *A history of British gardening*. Spring Books.
Harari, Y. N. (2011). *Sapiens. A brief history of humankind*. Vintage Books.
Hardy, B. (1969/1981). *West of the Darling*. Jacaranda Press.
Hardy, B. (1976). *Lament for the Barkindji. The vanished tribes of the Darling River Region*. Rigby.
Hareuveni, N. (1980). *Nature in our Biblical heritage* (H. Frenkley, Trans.). Neot Kedumim Ltd.
Harris, T. Y. (1966). *Wild flowers of Australia*. Angus and Robertson.
Hasluck, A. (2002). *Georgiana Molloy. Portrait with background*. Fremantle Arts Press.
Hayden, R. (1980). *Mrs Delany. Her life and her flowers*. British Museum Press.
Hazzard, M., & Hewson, H. (1982). *Flower paintings of Ellis Rowan*. National Library of Australia.
Heidegger, M. (1959). *An introduction to metaphysics*. Yale University Press.
Hermerén, G. (1969). *Interpretation and meaning in the visual arts*. Scandinavian University Books.
Heywood, W. (1906). *The little flowers of the glorious Messer St Francis and of his friars*. Methuen.
Hitchmough, W. (1997). *Arts and Crafts Gardens*. Pavillion.
Hobhouse, P. (2003). *Gardens of Persia*. Florilegium.
Holmes, E. (1911). *What is and what might be*. Constable.
Hoyles, M. (1991). *The story of gardening*. Journeyman Press.
Hunt, J. D., & Willis, P. (Eds.). (1975). *The genius of the place. The English landscape garden 1620–1820*. Harper and Row.
Hurd, D. L., & Kipling, J. J. (1964). *The origins and growth of physical science* (Vol. 2). Penguin Books.
Hutton, D., & Connors, L. (1999). *A history of the Australian environmental movement*. Cambridge University Press.
Huyghe, R. (Ed.). (1966). *Larousse encyclopedia of prehistoric and ancient art*. Hamlyn.
Idriess, I. (1941). *The great boomerang*. Angus & Robertson Ltd..
Ipsiroglu, M. S. (1969). *Painting and culture of the Mongols*. Trans. With Introduction by D.S. Phillips. Thames and Hudson.
Jaeger, W. (1944). *Paideia: The ideals of Greek culture. Vol. II. In search of the divine centre*. Basil Blackwell.
James, W. (2002). *Varieties of religious experience. A study of human nature* (Centenary Edition). Routledge.
Jellicoe, G., Jellicoe, S., Good, P., & Lancaster, M. (Eds.). (1986). *The Oxford companion to gardens*. Oxford University Press.
Jervis, J. (1948). The West Darling country: Its exploration and development. Part III. *Journal of the Royal Australian Historical Society, XXXIV*.
Jones, D. (2011). Re-greening 'The Hill'. Albert Morris and the transformation of the Broken Hill landscape. *History of Gardens and Designed Landscapes, 31*(3), 181–195.
Kass, D. (2018). *Educational reform and environmental concern. A history of school nature study in Australia*. Routledge.
Kennedy, P. (2010). *Twentieth-century theologians. A new introduction to modern Christian thought*. I.B. Tauris.
Keynes, J. M. (1933). "Newton the Man", an addendum to *Essays in Biography* (G. Keynes, Ed., 1963). W.W. Norton and Co.
King, R. (1979). *The quest for paradise. A history of the world's gardens*. Mayflower Books.
Kingsley, C. (1885). *The water babies*. Macmillan & Co.
Klein, N. (2014). *This changes everything. Capitalism vs the climate*. Allen Lane. Penguin.
Kloss, W. (1985). *Treasures from the National Museum of American Art*. Smithsonian Institution Press.
Krishna, R. A. (1981). *Early Mughal painting*. Lalit Kala Akademi.
Kynaston, E. (1981). *A man on the edge. A life of Sir Ferdinand von Mueller*. Allen Lane/Penguin.
Lacour-Gayet, R. (1976). *A concise history of Australia*. Penguin Books Australia.
Laird, M. (2015). *A natural history of English gardening*. Yale University Press.
Langer, S. (1957). *The problems of art*. Routledge and Kegan Paul.

Langer, S. (1976/1941). *Philosophy in a new key. A study in the symbolism of reason, rite and art* (3rd ed.). Harvard University Press.
Langton, M. (2018). *Welcome to country. A travel guide to indigenous Australia*. Hardie Grant Travel. (and teacher's notes for years 7–10 students on-line).
Langton, M. (2019). *Welcome to country: An introduction to our first peoples for young Australians*. Hardie Grant Travel.
Le Goff, J. (1985). *Les intellectuels au Moyen Age*. Editions du Seuil.
Leclercq, J. (1957). *L'Amour des Lettres et le Désir de Dieu. Initiation aux auteurs monastiques du Moyen Age*. Ed. du Cerf.
Leighton, A. (1986). *American gardens in the eighteenth century – 'for use or for delight'*. The University of Massachusetts Press.
Lewis-Williams, D. (2002). *The mind in the cave. Consciousness and the origins of art*. Thames and Hudson.
Lin, Y. (1937). *The importance of living*. Reynal and Hitchcock.
Losly, J. P. (1986). *Indian paintings in the British library*. Lalit Kala Akademi.
Lovejoy, A. O. (1964/1936). *The great chain of being: A study of the history of an idea*. Harvard University Press.
Lukacs, G. (1962). *The historical novel*. Merlin Press.
Mabberley, D. J. (2019). *Botanical revelation. European encounters with Australian plants before Darwin*. New South Publishing.
Mabey, R. (1986). *Gilbert White. A biography of the author of The Natural History of Selborne*. Century Hutchinson.
Mabey, R. (Ed.). (1995). *The Oxford book of nature writing*. Oxford University Press.
MacCulloch, D. (2009). *Christianity, the first three thousand years*. Viking Penguin.
Majumdar, R. C. (Ed.). (1984). *The Mughul empire*. Bharatiya Vidya Bhavan.
Martin, S. (1993). *A new land. European perceptions of Australia 1788–1850*. Allen and Unwin/ State Library of New South Wales.
Mattingley, C. (2010). *For the love of nature: E.E. Gostelow's birds and flowers*. The National Library of Australia.
McCalman, I. (2010/2009). *Darwin's Armada. Four voyagers to the southern oceans and their battle for the theory of evolution*. Pocket Books.
McCarthy, F. D. (1952). *Australia Aboriginal decorative art*. The Australian Museum.
McCurdy, E. (Ed.). (1938). *The notebooks of Leonardo da Vinci* (Vol. 1). The Reprint Society.
McNeill, W. H. (1965). *The rise of the west*. University of Chicago Press.
Minne-Sève, V. (1991). *La France Romane*. Nathan.
Monroe, P. (1915). *A textbook in the history of education*. The Macmillan Company.
Morphy, H. (1991). *Ancestral connections. Art and an Aboriginal system of knowledge*. University of Chicago Press.
Morphy, H. (1998). *Aboriginal art*. Phaidon.
Morris, A. (1966). *Plant life of the West Darling. Compiled by the Barrier Field Naturalists Club*. Barrier Field Naturalists Club.
Mountford, C. P. (1964). *Aboriginal paintings from Australia*. The New American Library of World Literature by arrangement with UNESCO.
Moynihan, E. (1982). *Paradise as garden in Persia and Mughal India*. Scholar Press.
Muir, J. (1982). *The Wilderness world of John Muir* (E. W. Teale, Ed.). Houghton Mifflin.
Muir, J. (2009). *Journeys in the Wilderness – A John Muir Reader*. Introdn White, G. Birlinn Ltd.
Musée du Louvre. (1989). *Arabesques et jardins de paradis: collection francais d'art Islamique*. Ministere de la Culture, de la Communication, des Grands Travaux et des Bicentennaire.
Myers, B., & Copplestone, T. (Eds.). (1985). *The history of art*. Hamlyn.
Nath, A. (2006). *Dome over India – Rashtrapati Bhavan*. India Book House.
Newton, E. (1960). *The arts of man*. Thames and Hudson.
North, M. (1980). *A vision of Eden – The life and work of Marianne North*. Webb & Bower.

O'Donovan, P. (1984). *Saint Patrick of Nursia*. Collins. English rendition of Nigg, W. *Benedikt von Nursia*. Verlag Herder.

Olsen, P. (2013). *Collecting ladies – Ferdinand von Mueller and women botanical artists*. National Library of Australia Publishing.

Olsen, P., & Russell, L. (2019). *Australia's first naturalists. Indigenous people's contribution to early zoology*. National Library of Australia Publishing.

Parker, F. W. (1885). *Notes of talks on teaching given by Francis W. Parker, at the Martha's vineyard summer institute, July 17 to August 19, 1882/reported by Leila E Patridge*. E.L. Kellogg & Co.

Pascoe, B. (2014). *Dark Emu*. Magabala Books Aboriginal Corporation.

Passmore, J. (1970). *The perfectibility of man*. Duckworth.

Passmore, J. (1974). *Man's responsibility for nature*. Duckworth.

Plato. (1961). *The collected dialogues of Plato* (Hamilton, E., & Cairns, H. Eds.). Princeton University Press.

Ponsonby, L. (2000). *Marianne North at Kew Gardens*. Royal Botanic Gardens, Kew.

Pope, A. *An essay on man: Epistle II*. https://www.poetryfoundation.org/poems/44900/an-essay-on-man-epistle-ii. Accessed 3 Mar 2021.

Powell, T. G. E. (1966). *Prehistoric art*. Thames and Hudson.

Praeger, R. L. (1937/1980). *The way that I went – An Irishman in Ireland*. Allen Figgis.

Prest, J. (1981). *The garden of Eden*. Yale University Press.

Q'ran translated into English from the original Arabic. *The Koran: commonly called The Alcoran of Mohammed* (G. Sale, Trans.). William Tegg, 1865.

Quakers in Britain. (2009a). *The Quaker Testimonies*. "The earth and environment." Produced by Testimonies Committee of Quaker Peace and Social Witness, published by Quaker Books.

Quakers in Britain. (2009b). *The Quaker Testimonies*. "Our beliefs" Produced by Testimonies Committee of Quaker Peace and Social Witness, this edition published by Quaker Books 2009.

Quest-Ritson, C. (1996). *The English garden abroad*. Penguin Books.

Ramsland, J. (2009). *The Rainbow Beach man. The life and times of Les Ridgeway, Worimi Elder*. Brolga Publishing.

Rappaport, A. (Ed.). (1972). *Australia as a human setting*. Angus and Robertson.

Ratcliffe, F. (1938). *Flying Fox and drifting sand*. Chatto and Windus.

Read, H. (1945). *Art and society*. Faber and Faber.

Regan, T. (1983). *The case for animal rights*. University of California Press.

Roberts, A., & Mountford, C. (1971). *The dreamtime book*. Rigby.

Rowan, E. (1987). *Flower Paintings of Ellis Rowan*. Introduction by M. Hazzard. Notes on the flowers by H. Hewson. National Library of Australia Publishing.

Russell, B. (1968). *The autobiography of Bertrand Russell* (Vol. II). George Allen & Unwin Ltd..

Russell, B. (1971). *The autobiography of Bertrand Russell* (Vol. III). George Allen & Unwin Ltd.

Schama, S. (1995). *Landscape and memory*. Harper Collins.

Scott, E. (Ed.). (1929). *Australian discovery by land*. J.M. Dent and Sons Ltd..

Sedding, J. D. (1891). *Garden craft old and new*. K. Paul, Trench, Trubner and Co.

Seddon, G. (2005). *The old country. Australian landscapes, plants and people*. Cambridge University Press.

Selleck, R. W. (1968). *The new education, the English background 1870–1914*. Pitman & Sons.

Short, P. (2012). *In pursuit of plants*. University of Western Australia Press.

Singer, P. (1975). *Animal liberation. A new ethics for our treatment of animals*. New York Review/ Random House.

Skilbeck, M. (1970). *John Dewey*. Collier-MacMillan Limited.

Snell, T. (Ed.). (2019). *Nikulinsky naturally – An Artist's life*. Lawrence Wilson Art Gallery, University of Western Australia.

Solomon, R. J. (1988/2008). *The Richest Lode. Broken Hill 1883–1988*. Austral Books. (reprint of 1988 edition Sydney: Hale & Iremonger).

Spate, O. H. K. (1979). *The Spanish Lake*. Australian National University Press.

Stanner, W. E. H. (1979). *White man got no dreaming – Essays 1938–1973*. ANU Press.

Strabo, W. (1966). *Hortulus* Trans. By Raef Payne, commentary by Wilfrid Blunt. The Hunt Botanical Library.
Stronach, D. (1994). Parterres and stone watercourses in Pasargadae. Notes on the Iranian contributions to the evolution of the garden designs. *Journal of Garden History, 14*, 1994.
Sturt, C. (1834). *Two expeditions into the interior of Australia during the years 1828, 1829, 1830 and 1831. With observations on the soil, climate and general seasons of the colony of New South Wales.* 2 volumes. Smith, Elder & Co. 2nd ed. Vol. 1.
Sturt, C. (1848). *Narrative of an expedition to Central Australia 1844–46*. T and V. Booth.
Symonds, J. A. (1907). *Renaissance in Italy. The revival of learning*. London.
Tawney, R. H. (1926). *Religion and the rise of capitalism*. John Murray.
Teilhard de Chardin, P. (2002-1955). *The phenomenon of man*. Harper Perennial.
Thacker, C. (1979). *The history of gardens*. A.H. and A.W. Reed.
Thomas, K. (1984). *Man and the natural world. Changing attitudes in England 1500–1800*. Penguin Books.
Thomson, D. (2005). *Donald Thomson in Arnhem Land.* Compiled and introduced by Nicolas Peterson. The Miegunyah Press.
Thomson, D [The Donald Thomson Collection] (2010). Ancestral powers and the aesthetic: Arnhem land paintings and objects from the Donald Thomson collection. : Museum Victoria.
Thoreau, H. (1965). *The Portable Thoreau*. Ed. and Introdn by Bode, C. Viking Press.
Thorlby, A. K. (1966). *The romantic movement*. Longmans.
Thunberg, G. (2019). *No one is too small to make a difference*. Penguin Random House UK.
Turney, C. (1962). *The history of education in New South Wales 1788–1900* (Vol. 3). Ph.D University of Sydney.
Vellacott, H. (Ed.). (1986). *Some recollections of a happy life – Marianne North in Australia and New Zealand*. Edward Arnold Australia.
Villiers Stuart, C. M. (1913). *The gardens of the great Mughals*. A & C Black.
Vines, E. (2010). *Broken Hill. A guide to the Silver City*. Broken Hill City Council.
von Humboldt, A. (2014). *Views of nature* (S. T. Jackson, & L. D. Walls, Eds., Mark W. Person, Trans.). University of Chicago Press.
Webb, J. (1998). *Thistle Harris. A biography of Thistle Yolette Stead*. Beatty and Sons.
Webber, H. (1992). *The greening of the hill. Revegetation around Broken Hill in the 1930s*. Hyland House.
Weber, Max (1920/2002). *The Protestant ethic and the spirit of capitalism*. Introduction and Translation by Stephen Kalberg. Roxbury Publishing Company.
Welch, S. C. (1975). *Imperial Mughal Painting*. George Brazziler.
Western Division of New South Wales Royal Commission to Inquire into the Condition of Crown Tenants, appointed by the Legislative Assembly 11 August 1900. The Report and Summary of Evidence were published in Sydney by William Applegate Gullick, Government Printer, in 1901.
White, G. (1802). The works in natural history by the Late Rev. Gilbert White, A. M. Fellow of Oriel College Oxford in 2 vols London. Printed for J. White, Fleet Street, By T. Bernsley Bolt Court 1802.
White, G. (1938). *The Writings of Gilbert White of Selborne* (H. J. Massingham, Ed.). The Nonesuch Press.
White, G. (1975). Garden Kalendar, 1751–1771. Reproduced in facsimile from the manuscript in the British Library. Introdn and notes by John Clegg. The Scolar Press.
White, L. (1967). The historical roots of our ecological crises. *Science, 155*(3767).
Whitehead, A. N. (1926). *Religion in the making* (Lowell Institute lectures 1926). The Macmillan Company.
Whitehead, A. N. (1948). *Adventures of ideas*. Penguin Books.
Whitehead, A. N. (1950). *The aims of education and other essays*. Ernest Benn Ltd..
Whitehead, A. N. (1953). *Science and the modern world*. Cambridge University Press.

Whitehead, A.N. (1978). *Process and reality* (Gifford lectures 1927–28*)*. Griffin, D., & Sherburne, D. (Eds.). The Free Press.

Wilbers, D. C. (1962). *Persian gardens and garden pavilions*. Charles E. Tuttle Company.

Wilson, E. O. (2003). *The future of life*. Random House.

Winchester, S. (2021). *Land*. William Collins.

Woodward, W. H. (1906). *Studies in education during the age of the renaissance*. Cambridge University Press.

Wordsworth, W. (1850). *The Prelude or, Growth of a Poet's Mind; An Autobiographical Poem*. London: Edward Moxon. https://archive.org/details/prelude00unkngoog/page/n9/mode/2up?view=theater Internet Archive. Retrieved March 28, 2021.

Wulf, A. (2009). *The brother gardeners. Botany, empire and the birth of an obsession*. Windmill Books.

Wulf, A. (2016). *The invention of nature: The adventures of Alexander von Humboldt, the lost hero of science*. John Murray.

Xenophon. (1968). *Memorabilia and Oeconomicus* (The Loeb classical library). William Heinemann Ltd./Harvard University Press.

Young, A. (2000). *Environmental change in Australia since 1778*. Oxford University Press.

Young, G. M. (1963). *Victorian England. Portrait of an age* (Oxford paperbacks no.12). Oxford University Press.

Zwart, H. (2008). *Understanding nature. Case studies in comparative epistemology*. Springer.

Index

M. Skilbeck, *Loving and Studying Nature*, International Explorations in Outdoor and Environmental Education 10, https://doi.org/10.1007/978-3-030-80751-1

N

R

S